Praise for *The Thief at the End of the World*

"Fascinating . . . Mr. Jackson relates the tale of Wickham's Amazon miseries and the scramble for the seeds with an admirable mix of botanical detail and storyteller's verve." —*The Wall Street Journal*

"A fabulous story, well told, with excerpts from Wickham's own writings and Violet's memoirs. Jackson is especially good on the magical and dangerous nature of jungle and forest. . . . The book works best when we're with Wickham, chasing his El Dorados. . . . This story moves and haunts. Perhaps man can't, in the end, control nature; he can't even control himself." —*Los Angeles Times*

"Jackson has made a first-rate book out of Wickham's story. . . . He has done a heroic amount of research, made a coherent story out of a huge mass of material, and identified the larger themes that give the story its resonance. His writing is lucid, occasionally vivid, and he brings to the enterprise a welcome sense of humor, as well as, when it is useful, a sense of the ridiculous. *The Thief at the End of the World* not merely is informative and instructive, it also is immensely entertaining, an attribute always to be welcomed." —*The Washington Post*

"*The Thief at the End of the World* is clearly the product of remarkable research and a journalist's feel for what to keep and what to leave out. Wickham's heroic failures and his neglected love are narrated with poignancy and humor. There are touches of Joseph Conrad's *Heart of Darkness*. . . . But more than Conrad, Jackson obviously has a good time telling a terrific story." —*The Seattle Times*

"Jackson vividly portrays the rigors of life in the tropics." —*BookPage*

Joe Jackson is the author of four works of nonfiction and a novel. He was an investigative reporter for the *Norfolk Virginian-Pilot* for twelve years, covering criminal justice and the state's death row. He lives in Virginia Beach, Virginia. He can be reached through his Web site, www.joejacksonbooks.com.

THE THIEF AT THE
END OF THE WORLD

*Rubber, Power, and the
Seeds of Empire*

JOE JACKSON

PENGUIN BOOKS

PENGUIN BOOKS
Published by the Penguin Group
Penguin Group (USA) Inc., 375 Hudson Street, New York, New York 10014, U.S.A. • Penguin
Group (Canada), 90 Eglinton Avenue East, Suite 700, Toronto, Ontario, Canada M4P 2Y3 (a division
of Pearson Penguin Canada Inc.) • Penguin Books Ltd, 80 Strand, London WC2R 0RL,
England • Penguin Ireland, 25 St Stephen's Green, Dublin 2, Ireland (a division of Penguin Books
Ltd) • Penguin Group (Australia), 250 Camberwell Road, Camberwell, Victoria 3124, Australia (a
division of Pearson Australia Group Pty Ltd) • Penguin Books India Pvt Ltd, 11 Community
Centre, Panchsheel Park, New Delhi – 110 017, India • Penguin Group (NZ), 67 Apollo Drive,
Rosedale, North Shore 0632, New Zealand (a division of Pearson New Zealand Ltd) • Penguin
Books (South Africa) (Pty) Ltd, 24 Sturdee Avenue, Rosebank, Johannesburg 2196, South Africa

Penguin Books Ltd, Registered Offices:
80 Strand, London WC2R 0RL, England

First published in the United States of America by Viking Penguin,
a member of Penguin Group (USA) Inc. 2008
Published in Penguin Books 2009

10 9 8 7 6 5 4 3 2 1

Image credits appear on page 395

THE LIBRARY OF CONGRESS HAS CATALOGED THE HARDCOVER EDITION AS FOLLOWS:
Jackson, Joe.
    The thief at the end of the world : rubber, power, and the seeds of empire / by Joe Jackson.
        p. cm.
    Includes bibliographical references and index.
    ISBN 978-0-670-01853-6 (hc.)
    ISBN 978-0-14-311461-1 (pbk.)
    1. Rubber industry and trade—Amazon River Region—History. I. Title.
HD9161.A562J33 2008
338.4'767820981I—dc22   2007019872

Printed in the United States of America
Designed by Carla Bolte • Set in Scala

AS ALWAYS,

TO KATHY AND NICK

God is great, but the forest is greater.

—Amazonian proverb

CONTENTS

HENRY'S DREAM

Deep in the forest grew a ruinous tree. From its veins flowed a milky sap that magically shielded man from the elements. From its branches drooped enough fruit to feed entire villages. Birds perched in its crown, filling the forest with song. Those who lived beneath it knew they were blessed: While others toiled and suffered, this tree of life filled their every need. They were the Chosen Ones.

Yet divine gifts come with a warning: This treasure must be hidden from the world. Each tribe who told the tale—the Tupabaya and Mundurucú of the Lower Amazon, the Motoco of Gran Chaco, the Carib and Arawak of Guiana—considered itself the tree's sacred guardian. Their holy men warned them to stay vigilant: If strangers ever found the tree, they prophesied, they'd cut deep into its heart to reveal the source of riches, but instead of gold would release a deluge. The Father of Waters would never stop, destroying the entire forest.

Drowning all mankind.

In December 1928, a vision of white bobbed on the water off Villa Santarém.

The tiny port was centrally located on the Lower Amazon, hunched between the moiling river and forested highlands, and hadn't changed much in the last century. The wide sandy beach where *caboclo* women washed their laundry, the white adobe church with two square turrets, the one-story houses with their blue fronts and red roofs—all were comfortably stuck in the past, the town's wireless station one of the few concessions to modern times. Plenty of fortune hunters had passed Santarém over the decades, each searching out the newest El Dorado hidden deep in the forest, but what set the smart white steamer *Lake Ormoc* and the huge barge *Lake LaFarge,* towed behind it, apart from the others was the promise of industry stored in their holds. The two had left Dearborn, Michigan, four months earlier and arrived without mishap, besting the mouth of the Amazon with its treacherous currents, where a seagoing freighter had run her bow on the bank and turned turtle, where another wrecked and the captain blew out his brains. They passed ghost towns in the wilderness: Pará (now called Belém), with its cavernous warehouses and empty "Parisian" boulevards; small towns at the edge of the jungle like Santana and Monte Alegre, where paved roads were swallowed by rain forest and lavish city centers were now home for bird-eating spiders.

The marvels of the *Lake Ormoc* included a diesel tugboat, a narrow-gauge locomotive and tracks, a steam shovel, a pile driver, tractors, motorboats, stump pullers, components of a powerhouse, a disassembled sawmill, prefabricated stores and houses, and ice-making machines. Her crew dripped sweat in their wool sailor suits as they stood at the rail; the Danish captain puffed on his pipe, an act he'd heard would repel mosquitoes. Fragile Portuguese fishing boats surrounded the ship, darting past with their red-and-blue sails and raked prows.

Here at Santarém, where the Amazon met the blue Tapajós, its fifth-largest tributary, nearly 173,000 cubic yards of water passed by each second; when the Amazon slammed into the Atlantic, that increased to 216,000 cubic yards, accounting for nearly one fifth of all fresh water received annually into the earth's oceans. Millions of tons of debris and sediment streamed past Santarém. If one stood on the beach at night and stared up at the Southern Cross, that unfamiliar sound at the edge of thought was the hiss of the river as it coursed east to the ocean, five hundred miles away. A little of everything washed up on this beach. There were still tales of the day thousands of alligators died on the Tapajós and floated down to Santarém: "So great was the stench of their decomposing carcasses," wrote British naturalist Richard Spruce, that the town's main merchants "had all their boats and men employed for some weeks in towing them down the river to a safe distance below the town."

From 1850 to 1913, the Amazon Valley had been the world's single source of high-quality rubber, and the ambitions of the Great Powers had transformed the jungle. Great Britain was the first to realize rubber's geopolitical promise. Rubber was essential for the production of gaskets for steam engines, and by 1870, London saw the need for giant turbines to drive ironclads across the seas. This obscure material, whose chemical composition no one yet understood, accompanied iron and steel wherever railroads, factory machinery, and mining pumps found footing; it was essential for machine belting and tubing, as buffers between railroad cars, and soon for "pneumatic tyres." Progress meant mobility, and world power depended on uninterrupted access to the three strategic resources necessary for that freedom: the oil, steel, and rubber essential for the production and maintenance of trains, ships, autos, and planes.

Long before the rise of Standard Oil and the Seven Sisters, before the trusts, trustbusters, and the "great carve-up" of the Middle East into oil-rich puppets, the economy and politics of rubber created for a very few the riches of Midas while deciding the fate of nations and enslaving or exterminating legions of natives. For sixty-three years, the Amazon Valley owned the world market in rubber, but then the bubble broke, as so many do. The Amazon Rubber Boom turned into a bust in one short year. In 1913, the rubber from seventy thousand seeds smuggled from Brazil and planted in Britain's Asian plantations flooded the market, outselling the more expensive "wild" rubber and tossing it from the stage. The bust dealt the Amazon Valley a blow from which it has never recovered: In 1900, the region produced 95 percent of the world's rubber. By 1928, when the *Lake Ormoc* floated off Santarém, the Amazon produced barely 2.3 percent of its needs.

Now Great Britain pulled the strings. In 1922, Britain's rubber plantations in Indochina were still reeling from the economic shock of World War I. Since four out of five plantations did not pay dividends, a government panel called for price restrictions, initiating the twentieth century's first global experiment in resource control. As expected, rubber prices rose. Hardest hit was the United States, which consumed 70–80 percent of the world's rubber, most of it going to the gigantic automobile industry. The industrial boom of the Jazz Age rested on the fortunes of Detroit, and the autos rolling from her factories rested on rubber tires. U.S. Secretary of Commerce Herbert Hoover ordered that, in the interest of national security, the nation must wean itself from foreign rubber and develop her own supply.

The *Lake Ormoc's* mission was to swing the world rubber market in America's direction and shift the balance of power. It seemed a tall order for any ship, much less a humble freighter, but such was the faith in her owner that anything seemed possible. That owner was Henry Ford.

No man more fully epitomized American ingenuity and industry in 1928; no one more truly believed in the gospel of work and salvation by capitalism. He was portrayed in the press as a modern Prometheus, bringing the light of cheap goods and high wages to the masses. His 1913 implementation of the moving assembly line made "Fordism" the most imitated business concept in the world. With his fair skin, pale blue eyes, and sharp profile, he projected a jaunty self-confidence. He was small and wiry, built like a sprinter, famous for racing employees across his factory floor. In 1928, at age sixty-three, he was still at the top of his game and eager, he said, to conquer the Amazon. It seemed inevitable that he would come to this jungle, the original source of the rubber used for his tires. Detroit's consumption of the rubber known as "Pará fine" was the reason for the growth of the Federated Malay States, capital of Britain's rubber-plantation industry and center of its world monopoly. In 1914 alone, Detroit consumed 1.8 million tires, and of these, Ford used 1.25 million. He envisioned a southern extension of his flagship River Rouge plant carved from the jungle. The Amazon was a chance to build from scratch his vision of an "agro-industrial utopia" modeled after small-town America. It was an opportunity to create, as he said in his Ford Sunday Evening Hour Sermonettes, a world in which workers had "one foot in industry and one foot on the land." To do this, he purchased 2.5 million acres of rainforest along the Tapajós, an area 82 percent the size of Connecticut. He planned to build an entire American city smack in the middle of the jungle, and like other empire builders, he would name it for himself.

Fordlandia was the logical extension of Ford's career. It is hard today to remember the hold that the "romance of rubber" had on the imaginations and pocketbooks of the Western world. Ford's life was bracketed by what brokers in the commodities markets of London and New York had dubbed the "Rubber Age." He was born in 1863, the height of the American Civil War, the first modern war to make extensive military use of India rubber and its close cousin, gutta-percha, for everything from coats, capes, caps, buttons, haversacks, and canteens to pontoon boats, cartridge belts, and horse syringes. He grew into manhood as the bicycle craze exploded across Europe and America. In 1885, when Ford was twenty-two, Carl Benz made the first differential gear tricycle, an oversized three-wheeler with a small engine turning the rear axle. By 1891, a vibrant automobile industry was emerging in Europe, made up primarily of former bicycle makers—Opel in Germany; Clément, Darracq, and Peugeot in France; and Humbert, Morris, and Rover in Great Britain. By the twentieth century, auto production had shifted to the United States: From 1908 to 1927, Ford alone had sold 15 million black Model Ts, the linchpin to his empire, which he'd christened the "people's car."

To those living in Santarém, neither Fordlandia nor the arrival of the *Lake Ormoc* came as any great surprise. For the past six years, a half dozen "secret" U.S. expeditions had searched the Basin for likely spots to establish vast rubber plantations as a way to undercut the British price controls. There were few secrets out here. Land speculators fanned out ahead of the Americans. The most successful was Jorge Villares, a Brazilian coffee producer and elevator manufacturer who heard rumors of Ford's quest and secretly planted half a million rubber seedlings along a seventy-five-mile stretch of forest on the east bank of the Tapajós River. He left behind armed guards as protection against rival speculators and contacted the leader of the most important American mission, the University of Michigan agronomist Carl LaRue.

The well-connected LaRue had been appointed by Herbert Hoover to explore the junction of the Tapajós and Amazon, reputed source of the stolen rubber that gave birth to the British monopoly. If a lone Briton could succeed in this malarial backwater, think what could be accomplished by American money and know-how. All that was needed was a reliable source of labor: "A million Chinese in the rubber section of Brazil would be a godsend to that country," LaRue wrote in 1924. Three years later, he convinced Ford to buy Villares's concession for $125,000; the next year, Ford learned that the state of Pará had been willing to give him the land for free.

The Villares payout had all the hallmarks of a swindle, and according to several accounts, Ford never trusted LaRue again. Nevertheless, the terms were favorable, and Ford forged ahead. Pará was happy to forgive Ford's taxes for fifty years; in return, Ford Inc. promised the government a 7-percent return on all profits after twelve years. Ford vowed to begin the industrial conquest of the Amazon and expected his sons and grandsons to complete the job in future decades. As the *Lake Ormoc* steamed from Dearborn, the last piece fell into place for Ford.

Fordlandia would replace Boa Vista, a small jungle village whose Portuguese name meant "beautiful view." One final tract remained to complete the package, a spit of land at the point where Ford planned to build a giant pier capable of off-loading oceangoing steamers. But this spot was still owned by a family named Franco, and Old Man Franco held out for more. After Ford closed the deal with the state, his representatives met Franco in his kitchen and began stacking money on the table. The old man's eyes grew wider as he watched: He'd never seen so much money in one place at one time. When the stacks reached over his head, Ford's men stepped back. Old Man Franco looked at the stacks, looked at the men, and shoveled the money into a canvas sack. He didn't say a word but stuck out his hand. The deal was done.

North Americans greeted the news with a fervor bordering on religious ecstasy: "While there may be a difference of opinion as to the prospects of rubber growing in the Philippines and in Africa, there is none as to the Amazon Valley," crowed the *Outlook,* a New York business magazine. "Brazil is the native home of the rubber plant, and the possibilities of extending production are almost unlimited." Nothing could go wrong.

Yet from the outset, the venture seemed plagued by troubling missteps and unchallenged assumptions. Those in Santarém familiar with Boa Vista knew the surrounding jungle as some of the most undesirable terrain in Pará, a landscape of deep ravines, sandy soil, and steep, rocky slopes, geography inimical to growing rubber trees. A seasoned agronomist like LaRue should have seen the place for what it was, but instead he praised Boa Vista, and rumors of a kickback began to taint the venture. More baffling than this was the fact that, to oversee the work, Ford had hired not an agronomist but a company man versed in the assembly line and factory floor. As the *Lake Ormoc* began its sixty-mile trip up the Tapajós to Boa Vista, the mercury edged into the nineties, making life miserable for the cool-weather Michigan managers on board. It was quickly discovered that the captain had miscalculated their passage. Although a vessel of four hundred tons like the *Lake Ormoc* could travel a thousand miles up the Amazon at any time of year, passage on the smaller Tapajós was seasonal: Its water level could rise and fall as much as forty feet between the wet and dry season. In December, Fordlandia was outside the dry-season limit for oceangoing freighters.

But that problem was easily fixed: It was simply a matter recalculating draft, child's play for engineers, and so they returned to Santarém to reload everything on smaller river steamers. During the delay, Ford's managers discussed the planting, and those they'd hired as guides and translators began to grow concerned. These Americans planned to clear-cut the forest, burn out the underbrush and bulldoze everything flat, as if they were building a factory instead of preparing the land for a vast grove of trees. By clear-cutting the landscape, Ford would deplete that thin layer of soil and nutrients upon which the tropical forest depended, and every Amazonian knew through generations of experience that the forest itself was the weather engine creating rain. By cutting down the trees in hopes of introducing a machinelike predictability, Ford might very well create a desert instead.

Finally there was a problem that no one truly understood. The storied source of the valuable rubber known as "Pará fine" was *Hevea brasiliensis* (hereafter simply called hevea, unless the name of the genus or species is meant). It is a sixty- to one-hundred-foot giant of the spurge family (Euphorbiaceae), which grew naturally in only one place on earth—a wide swath of forest running in an east–west ellipse along the Amazon. Most Amazonians knew that hevea was a jungle tree that could not survive on open land exposed to long dry spells and pelting rain. A greater threat was microscopic, a blight that hit the young trees once their branches reared up and reached to one another in an unbroken canopy. Those who'd seen traces of this "rust" on their small plantations tried warning the Ford managers: Perhaps the young trees should be planted far apart; perhaps the primary forest surrounding Boa Vista should not be leveled, nor the thin layer of humus and fungus forming the topsoil scraped away. But the company men did not listen. Conditions out here were so blessed, they said, that even if they stood hevea's branches in the ground, they would "take root almost without fail." What were the fears of some backwater planters against the genius of Henry Ford?

The Brazilians shrugged their shoulders and said nothing more, but in private they laughed among themselves. It soon became obvious that Ford intended to run Fordlandia by remote control from Dearborn. His managers, who knew nothing of tropical plants or the rain forest, would not listen to those who did. What could one do with such people?

And so Ford's minions reloaded their equipment and started up the river again. They steamed past the high sandstone bluffs lining the Tapajós. A beaked freshwater dolphin occasionally leaped beside the curling wake of the bow. About twenty miles upriver they passed the tiny village of Aveiro, a long, narrow town with the jungle on its doorstep; before that Boim, on the opposite shore, its white church to St. Ignatius greeting visitors as they climbed the riverbank. Nearby stood the abandoned trading houses of Jewish merchants who'd come from Morocco at the beginning of the Rubber Boom. So many people and nationalities came together in this lonely stretch of jungle: Britons, Indians, former Barbadoan slaves, displaced Confederates from the American Civil War, and *caboclos*, Spanish priests and killers—and now one of the richest men in the world. There were ironies in such meetings. Even here, Ford's hatred of Jews was infamous, but many of the seeds that formed the British monopoly were purchased from these Jewish merchants, who'd bought them in turn from the Indians deep in the jungle. Now Ford was bringing the fruit of those smuggled seeds back. The world was a very small place, especially in the tropics. Everything came full circle, if you lived long enough. On this river, though, that was never assured.

One can never be too careful on the Amazon. Lives can be forgotten, swallowed by the forest. Pride is a calamitous misstep. Visions of riches shed their luster in an instant, transforming suddenly into fatal dreams.

I came to the Amazon in October 2005, tracking the remnants of two such dreams. One was Henry Ford's. Fordlandia still stands along the river, a sparsely inhabited ghost town. I chugged up the Tapajós in a boat little bigger than the *African Queen*. The dusty streets are empty save for a few farmers and their children; the old water tower overlooks the river, but the FORD, INC. logo was painted over long ago. To many, such ruins seemed incomprehensible, so pervasive was Ford's myth—the man who never failed. Approaching the huge, deserted pier that still thrusts into the Tapajós, one cannot help but remember the signatory lines from Percy Shelley's "Ozymandias": "Look on my works, ye Mighty, and despair!"

Ford's metropolis in the jungle never materialized as he saw it. The marching rows of hevea have long rotted away, replaced by secondary forest. The scrubbed white hospital devoted to combating tropical disease sits empty on the hillside, patient records strewn across the concrete floor. Ford dreamed of ten thousand residents. Today there are 150 at best, clinging to a red-dust hillside facing the wide Tapajós. Behind them, the jungle creeps closer each day.

The second dream I chased was Henry Wickham's, and his is the stranger tale. It is the story of how one man—likeable, if prone to bombast; determined, if not particularly talented or wise—pulled off one of the most successful and far-reaching acts of biological piracy in world history, and of the personal costs it entailed. Where experts failed at the task, this amateur, who'd stumbled into the jungle after gaudy feathers for the "ladies' hat trade," succeeded in smuggling a species of plant thought incapable of growth anywhere but in the Amazon to the British plantations of Malaya and Ceylon. He did so, he boasted, in sight of a Brazilian gunboat, loading the seeds at night aboard a commandeered Liverpool-to-Manaus freighter stripped of its cargo by unscrupulous officers. There was always an air of the fantastic to Wickham's exploits, an extravagant blend of Edgar Rice Burroughs and Lord Dunsany, and even today, historians seem uncertain what to make of him. Was he a patriot or an opportunist, a visionary or an extraordinarily lucky buffoon? A 1976 article in the *Times of London* linked his presence in Santarém to deliberate Victorian espionage. A British cruise-boat tourist recently claimed that Wickham had smuggled rubber seeds in the false soles of his shoes. For all the fanciful portrayals, the fact remains that in May 1876 he secretly loaded seventy thousand rubber seeds into the hold of the ocean-bound *Amazonas* and accompanied them to the Royal Botanic Gardens at Kew. Though only two thousand would germinate, these saplings changed the world.

The story of the exploitation of rubber is the story of the Amazon itself. No other river on Earth compares to this continent-bisecting waterway and its countless tributaries. No other has such a history of fortune hunting and attempted conquest. The Amazon drains an area of 2.72 million square miles; although the Mississippi-Missouri river is longer, it drains little more than half the same area. All the rivers of Europe constitute barely a trickle to this flood. It is a congress of rivers; during the rainy season a vast reservoir forms that reaches four hundred to five hundred miles in width and drains more than a third of South America. The river and its branches include ten thousand miles of navigable waterway. Land travel is nearly impossible, though plenty have tried.

It is extraordinarily easy to get lost forever in such a land, yet a fascination with the rain forest cannot be denied. Perhaps it is the realization that, in a heartbeat, life can switch from sensational beauty to the sweet breath of death and decay. A whole host of biblical nasties lie in wait for the unwary: amoebic and bacillary dysentery, yellow and dengue fever, malaria, cholera, typhoid, hepatitis, and tuberculosis. There are places in the Amazon experiencing a plague of vampire bats, where entire families catch rabies. There is river blindness, transmitted by blackflies and caused by worms that migrate to the eyes. The giant catfish is said to chomp off swimmers' feet, while a miniature catfish, the *candiru,* or "toothpick fish," swims up the stream of uric acid released by bigger fish, then lodges in the gills or cloaca with the help of some sturdy, sharp spines. It does the same to humans unwise enough to use the river as a urinal. Drawn by the nitrogen in urine, the *candiru* swims up the urethra like an eel, then sets its spines.

Still, I told myself, if I kept my eyes open, watched where I stepped, took my malaria pills, and didn't pee in the pool, the chances were better than even that I'd return home unscathed. Such are the dangers of hubris: Just when I thought I'd tamed the jungle, the ordinary taught me otherwise.

Monuments are not well marked in the Amazon: They are quickly covered by underbrush or succumb to mold, termites, and decay. In 1849, three of the era's preeminent naturalists converged on Santarém, drawn by tales of hevea and other forest wonders. Alfred Russel Wallace, Henry Walter Bates, and Richard Spruce were invited to a private estate named *Sitio de Taperhina* on the nearby Rio Ayaya to pursue their investigations. Their journal descriptions of hevea set off a rush that would be as frenzied as the one then occurring at Nevada's Comstock Lode. Wickham may have lingered here too; his wife's memoirs suggest that he may have been nursed back to health in the sugarcane plantation when he nearly died. The old house was still said to stand, and so my guide, some friends, and I took a Land Rover as far as we could into the primary forest, then split up when the path forked. I took my walking stick for snakes, but forgot two other essentials—my hat and water bottle.

One does not mess with the sun along the equator: It beats down from straight overhead instead of the more forgiving oblique angle in the temperate zones. Heat exhaustion comes quickly, especially when coupled with dehydration and exertion; dizziness and nausea, if unheeded, can lead to heatstroke and seizure. I knew this from experience, but under the green shade of the canopy the hat did not seem necessary. As for the water—hey, I would not be long.

But the path went on, steepening from the jungle plateau to the river plain below, and when I found Taperhina I was overjoyed. Until that moment, there'd been no problem: It was on the return trip *up* the escarpment to get the others that my troubles began. There is a good reason why Amazonians dive indoors for midday siestas. Halfway up the slope, I felt dry and woozy; three quarters up, I saw spots, so I found a patch of shade. When everyone appeared, we headed back downhill.

I took a rest, found some shade, and drank some water, then decided to photograph the famous ruin, maybe look inside. But one's judgment doesn't always return as fast as one's sense of physical well-being. As I stepped close to the house, I noticed what seemed to be huge spots of mold clinging to the adobe walls. As I checked the angle through my camera's viewfinder, a strange thing happened. The mold spots began to move.

I must still be dizzy, I thought, lowering the camera. But the mold crawled and contracted, and as I placed my foot on the first step, a drone rose around me. In the field before the house, a swarm of red-and-black hornets rose like mist from the grass and hovered.

The possibility of an aerial assault by hornets does wonders to focus one's attention. I realized in that second that the old house was a gigantic nest, that the crawling mold was hundreds of hornets tracing arabesques on the walls. Inside, there were probably thousands, if not millions, cling-ing to each other like glistening fruit, and that vision drove all further thoughts of exploration straight from my mind. Luckily, the hornets seemed as doped by the sun as I had been earlier: They hung in the air as I retraced my steps, then the drone diminished, and they sank to the grass again. I was more than happy to snap pictures from afar with a tele-photo lens.

Incursions into the Amazon region have a habit of repeating this pat-tern, stopped dead in their tracks by the sudden, unexpected, and mundane. Wickham and Ford both considered themselves instruments for the spread of empire, but the practice was far more bitter than the theory. Slack dry-season currents meant stagnant air and fetid water; the main difference between regions was judged by the prevalence of death and fever. Hookworm enfeebled the populace so much that in the 1940s Pará recorded an 80-percent prevalence rate, while in neighboring Amazonas, 96.4 percent of the population was infected. During the same period in Pará, leprosy claimed two thousand victims. With death-dealing insects and disease came new details of daily life that drove Europeans to the limits of sanity: a steam-bath atmosphere in which biscuits sprouted a whiskerlike mold, salt dissolved into sludge, and sugar turned into syrup. Guns that were loaded overnight could not be fired the next day, as gunpowder loaded into canisters liquefied.

Long before OPEC, Wickham's biopiracy handed Britain the first worldwide monopoly of a strategic resource in human history, yet the government he served—from his spymaster, Joseph Hooker, Director of Kew Gardens, to Queen Victoria herself—found him distasteful. He never partook of the riches he brought to his masters. The Wickhams called themselves pioneer planters, but in reality they lived as exiles from empire like that other, more famous couple whose taste of forbidden fruit doomed them to wander the earth's far places for the rest of their lives.

Few modern fables demonstrate so lucidly the means by which an individual affects the course of history—and then the consequences that lurch out of control. Rubber became an end in itself, and the horrors committed in its name were the unintended byproduct of the quest for a greater good. Henry Wickham saw his theft as an act of both patriotic and personal salvation. Henry Ford tried to tip the global scales back in favor of the United States and hoped to do so by building a mini-America in the jungle. Both became parables of the use and misuse of nature in the quest for power.

Today, the ruins prevail. Fordlandia rots among the trees. Wickham got his wish, but not in the way he planned. Rubber brought him honors and fame, but his celebrated seed snatch rained economic apocalypse upon the Amazon. Rubber destroyed his friends and family and killed everything he loved.

She wanted me to tell her what I did at first when I came out here; what other men found to do when they came out—where they went, what was likely to happen to them—as if I could guess and foretell from my experience the fates of men who come out here with a hundred different projects, for hundreds of different reasons—no reason but restlessness—who come, and go, and disappear!

—Joseph Conrad, "The Planter of Malata"

PART I

THE NEED

Men are estranged from what is most familiar, and they must seek out what is in itself evident.

—Heracleitus

CHAPTER 1

THE FORTUNATE SON

Later, when his schemes lay in ruin, all the lives lost and loves departed, he would sit in his club in London among the other old imperialists, embellish his sole victory, and call it justified. By then, the legend of Henry Wickham had become iconic, his deception for queen and country part of imperial lore. His lined and sunburned face stared from the newspapers and magazines toward the vague distance, his white shock of hair floating in a nimbus around his oversize head. Detractors claimed he was coarse and self-serving, nothing more than an opportunist who'd been in the right place at the right time. To others, he was an embarrassing reminder of the empire's rapacity. But these were minority reports, out of touch with popular opinion. By living as long as he had at the ends of the earth, he wore the prestige of the unknown like a medal. He was a force, and forces scoff at analysis. They simply *are*.

Joseph Conrad knew men like him, and the cost of their ambition. "I will tell you what I believe," he wrote in his 1913 novella "The Planter of Malata," a tale some said was modeled after Wickham and his last failed venture. "I believe that when your heart is set on some object, you are a man that doesn't count the cost to yourself and others." But one suspects that Henry knew the cost too well. Photos snapped of him in his triumph show a man who looks distant and somehow unsatisfied. The old cravings still reign uppermost, whether in London tweed or white drill suit as he poses beside a sixty-foot rubber tree. He never seems at peace. There was one portrait given to his niece shortly after he was knighted, a private photo never meant for dissemination. For once the expression is soft. He lowers his prominent chin and doesn't glower. He is relaxed, and descendants insist they can detect the faint shadow of a smile.

If so, it is touched with rue.

Two moments of peace stand recorded in the long life of Henry Alexander Wickham. One was in the jungle. The earlier was in a tamer place—the hills and spacious meadows of Hampstead Heath in North London, his first home.

Henry was born on Friday, May 29, 1846, in Grove Cottage, Haverstock Hill, four miles northwest of St. Paul's dome in London. It was a good time to be alive—if you were English, one of the "middling" classes, and part of the club. Today's Britain has been called "a crowded island where towns and cities rub up against one another like rocks in an old stone wall." Although the island had not yet reached that point, London was growing fast, spreading like an oil slick to absorb the ancient surrounding villages, a foretaste of things to come. For midcentury writers, London had no beginning or end—it was a "province covered with houses," "a state"; it was Gargantua, absorbing and excreting vast quantities of people and goods. All the world was here, or at least her glories: raw sugar from the West Indies, tea and silk from China, hides and skins from Patagonia, rubber from Brazil.

Victoria had been on her throne for thirteen years. Historians call the 1850s a time of relative peace and prosperity, but the mid-Victorian *Pax Britannica* was a relative term. During the "long peace" of Victoria's reign, not a single year passed in which British soldiers were not fighting somewhere in the world for the empire's greater glory. These were the "savage wars of peace," as Rudyard Kipling called them, and the year of Henry's birth saw headlines of the First Sikh War, the War of the Ax, and the Siege of Aden. Such far-flung conflict could not be helped: It was the price that had to be paid to save and civilize the world.

"No one will ever understand Victorian England," wrote historian Robert C. K. Ensor, "who does not appreciate that among highly civilized . . . countries it was one of the most religious that the world has known." Victorians believed that they were "God's elect," a belief infusing them with the fervor to spread their brand of civilization through the world. This was their right and duty, their "white man's burden." The British Empire was engaged in a righteous mission; though it might stumble, its intentions were ultimately honorable. If the individual faltered or lost heart while engaged in this quest, his nation and empire would come to his aid.

The empire's world-shaping creed rested on two broad pedestals. First, a call to spread Christianity and save men's souls. Second, the spread of free trade. Broadly defined as a belief in the free play of the market without government interference or restriction, free trade was considered an instrument of "world betterment" and peace: The spread of British trade and investment overseas was intrinsically good, since it brought enterprise and progress to the world. If enterprise and the work ethic were civilizing values, capitalism was a moral force: Although greed and self-interest entered the equation, they were secondary, at least in theory. The majority of Victorians may not have understood the dynamics of British investment or quite grasped the actual geography of their empire, but they responded with zeal and ardor to those grand celebrations announcing their empire's standing in the world. "For them," said historian John Gardiner, "the empire was hazily exotic but no less a matter of real pride."

The instruments of empire are many and diverse; it sometimes takes generations for their importance to be recognized. Such was the case with rubber. Its exploitation in the nineteenth century seemed to explode overnight and involved a cast of characters ranging from ragged adventurers like Henry to investors, inventors, imperialists, and hucksters. All owed allegiance to Columbus, since by most accounts the wayward mariner was the first European to take note of the strange elastic material. He commented on the white milk oozing from felled trees and bushes. Those following him described a legendary ball game in which two teams of Indians pursued a dark, round sphere that leapt wildly, bounced higher than seemed possible, and did so without there "being need for any inflation." It was played in many places under many names, from *batey* to

Vok-a-tok, and Cortés found the Aztecs playing their own version in the court of King Montezuma II, which they called *tlatchl.* Others dipped their feet in the milk and held them over a smoking fire to create an instant waterproof shoe. Some called the tree *cao o'chu,* or "weeping tree." This was transformed by the Spanish into *cauchu,* which the French eventually turned into *caoutchouc,* the term they use today.

The French were rubber's first press agents, beginning with French geographer Charles Marie de la Condamine's 1735 journey to the New World to determine the true shape of the earth. When he returned, he brought with him samples of rubber and details of its botanical characteristics. He coined the term "latex" from the Spanish for "milk," and found that rubber was an excellent protection for his delicate scientific instruments during the long sea voyage back home. In 1775, French botanist Jean Baptiste Fusée Aublet described the genus and its first species, *Hevea guianensis,* a variety found in French Guiana. Although the Spanish and Portuguese had the most opportunity for commercial gain due to their long control of the New World, rubber did not smack of instant wealth, and they remained uninterested. Until the end of the Enlightenment, rubber was seen in Europe as a novelty, restricted to toys and Indian-made artifacts.

It awaited modern chemistry to tease out its secrets. The English discoverer of oxygen, Joseph Priestley, named the strange stuff "India rubber" in 1770 after observing that a sample from India was "excellently adapted to the purpose of wiping from paper the marks of a black lead pencil." It was also cheap: A "cubical piece of about half an inch [sold] for three shillings" and lasted several years. In 1790, Antoine François de Fourcroy, one of the founders of modern chemistry, tested ways to dissolve rubber, and in 1791, the Englishman Samuel Peal was granted the first English patent for a process that infused rubber into "all kinds of leather, cotton, linen and woolen cloths, silk stuffs, paper, wood," making them "perfectly waterproof," he said. Peal's process seemed simple: He dissolved solid chunks of rubber in a bath of turpentine, then spread the cloth with the tacky sludge. Once dry, it was waterproof—but it never dried completely, creating a garment that, while warm and waterproof, was also smelly and sticky.

Despite such drawbacks, entrepreneurs were drawn to the odd material. At Pará, near the mouth of the Amazon, Portuguese colonial authorities promoted *seringa*, or syringe rubber, named after its earliest application. By the 1750s, army boots, knapsacks, and other military items flowed to Pará from Lisbon for waterproofing. By 1800, New England merchants were placing orders for shoes made of *seringa*. In 1825, the Scotsman Charles Macintosh discovered that rubber would dissolve in the flammable solvent naphtha; he sandwiched his rubber sludge between multiple layers of cloth to create a waterproof garment that was more durable than Peal's. Macintosh's success began a rush of small competitors in England, France, and the United States, but rubberized garments still grew tacky in heat and brittle in cold, and they were often returned in a half-melted state by angry customers. By the 1830s, most of these early companies had failed.

Brazil was the world's principal supplier of raw rubber, but demand was not great—by 1827, total exports only tallied eight tons a year. Rubber was an enigma, a natural product that chemists and industrialists sensed had wide use, but its chemical instability doomed every investor. There was, however, an alternative to the dissolved sludge of Macintosh and Peal. In 1820, Thomas Hancock opened England's first rubber factory; his process was based upon the effects of maceration and heat rather than a liquid solution. Hancock recycled rubber in a masticating machine, which, for security's sake, he called a "pickler," a hollow wooden drum studded with teeth. One day while turning the pickler, Hancock discovered that the heat generated by the process melted the waste rubber into a ball that was uniform, hot, and almost as good as new. Masticated rubber dissolved more readily in naphtha than the "crude" shipped from the jungle. Industrialists awoke to the fact that this heated, pliable rubber could be molded into any shape—the first cheap plastic. England was soon awash in rubber rollers, printer's blankets, drive belts, billiard table cushions, and surgical instruments. In 1827, the first rubber fire hose was used to put out a fire at Fresh Wharf, London. That same year, brewers added rubber to their beer (although *why* someone would ever imagine doing this is not chronicled). Rubber imparted a "vile taste," but after soaking it in waste liquid from the brewing process, the brewmasters discovered that it subtly sweetened the flavor. In 1830, Hancock and Macintosh merged their companies; in 1835, Hancock's rubber factory was the world's largest, and he was using three to four tons of caoutchouc each year.

Yet the problem of stiffening in cold and melting in heat remained, a problem so serious as to seem insurmountable until the 1839 breakthrough of former hardware dealer Charles Goodyear. Like Wickham, there was something driven about Goodyear. Like Wickham, his history would consist of one great triumph and a multitude of failures. Goodyear considered himself an "instrument in the hands of his Maker": his decision to turn his attention to rubber was an act of Providence, he'd later claim. So single-minded was his quest that his life became a litany of imprisonment, beggary, and lawsuits, all in the name of rubber. The black polymer was more than a natural resource: It assumed the characteristics of a religious icon. "While yet a schoolboy," he wrote in his *Gum-Elastic*, "the wonderful and mysterious properties of this substance attracted my attention and made a strong impression on my mind." That first impression never left him. It was a craving so deep that it approached the divine:

> The most remarkable quality of this gum, is its wonderful elasticity. In this consists the great difference between it and other substances. It can be extended to eight times its ordinary length, without breaking, when it will again assume its original form. There is probably no other inert substance, the properties of which excite in the human mind . . . an equal amount of curiosity, surprise, and admiration. Who can examine, and reflect upon this property of gum-elastic, without admiring the wisdom of the Creator?

It would take the famous accident of 1839 to reveal to Goodyear the "cure" for rubber's instability. According to his account, he was trying the effect of heat on a mixture of rubber, sulfur, and white lead when he spilled some of the concoction on a hot stove. To his surprise, the mixture charred but did not melt. Goodyear tried it again, this time before an open fire. There was charring in the center, but along the edges he found an uncharred section that seemed perfectly cured. Tests showed that the new substance did not harden in cold or melt in heat, and it withstood every solvent that had previously dissolved the native gum. He'd found the object of his long search. In time, the process would be dubbed *vulcanization* after the Roman god Vulcan, master of the forge.

Although Goodyear's discovery was an accident, he was ready to understand it after years of preparation. As he later wrote, it was "one of those cases where the leading of the Creator providentially aids his creatures by what we termed accident, to attain those things which are not attainable by the powers of reasoning he has conferred on them."

The Creator may have led Goodyear to his discovery, but He did not reveal the chemical secrets behind the miracle. No one really knew what occurred during vulcanization until the 1960s and 1970s. Rubber is a hydrocarbon, a polymer of isoprene, and is elastic because of its atomic organization into long, crumpled, repeating chains. These are interlinked at a few distant points, and between each pair of links the hydrogen and carbon building blocks rotate freely about their neighbors. This results in a wide range of shapes, like a very loose rope attached to a pair of fixed points on a rock wall. With vulcanization, however, this elasticity is compromised. The polymer chains are joined together by sulfur bridges that create a three-dimensional network: Now there are more bridges between the chains than in the "uncured" state, making each free section of chain shorter and subject to a quicker tightening under strain. This results in a rubber that is harder, less pliable, and far less likely to deteriorate in extremes of temperature.

With Goodyear's discovery, rubber turned into a new kind of gold. It was soon the preferred material for the plethora of gaskets essential to steam engines. What had been an obscure and slightly exotic raw material now began to form a triumvirate with iron and steel in factories, railroads, and mines. The railroads used it for air bumpers and coach interiors, and in engines for gaskets, hoses, and belts. Factories used it for machine belting and tubing, in assembly lines, and on floors, where it made a safe, nonskid, and electrically insulated surface. Rubber hoses pumped air, gas, and water out of mines. For the average consumer, rubber softened the ride and protected the wheels of coaches and buggies. It shielded people from wind and rain with rubber boots and slickers, and it provided the balls for their baseball, football, soccer, tennis, golf, and other sports. Office work became a little easier, thanks to rubber bands, erasers, and gloves. Latex condoms became available midcentury, taking some of the guesswork out of family planning.

In spring 1850, when Henry Wickham was four, his life changed forever. Neither he nor his parents could foresee it as they stood on the crest of Haverstock Hill, enjoying the breeze. Henry was one of the few Londoners privileged to live in what today is considered the suburbs. What now is associated with sameness and sprawl then conjured images of ease and health, a life merely dreamed of by most city dwellers. Haverstock Hill and Hampstead Heath were idyllic spots, the destination for harried Londoners who took the train from the city to walk the country paths or ride the donkeys penned nearby. On their rambles, Henry and his parents could not help but laugh at the otherwise respectable gentlemen who could not control their steeds. Charles Dickens enjoyed the donkeys; Karl Marx showed more enthusiasm in the saddle than skill. Though Haverstock Hill and the Heath stood on the edge of development, herds of cattle and sheep still grazed in the meadows, and thirty ponds dotted the landscape, including the six large ones created to serve as London's water supply. Henry would gaze at London sprawling beneath him, where the dome of St. Paul's stuck up like a giant gold thumb. The contrast between the two worlds was mysterious and not a little exciting for a boy of four. Here, the open heath was a mid-Victorian Paradise where nature was tame and beneficent, a cultivated reflection of God's ordered plan. Down there, the chaotic city tumbled and scattered at his feet like the piles of wooden blocks strewn across his nursery floor.

Although their home life seemed tidy and quiet, the world around the Wickhams roared forward in full gear. That year, Britain's Railway Mania reached its peak with 272 acts of Parliament setting up new railroad companies. The Electric Telegraph Company was founded. The potato crop failed in Ireland, and Irish vagrants filled London's streets. Henry's father, also named Henry, was a London solicitor, which meant that the small family was comfortably middle class. His mother, Harriette Johnson, was young and attractive, with flashing black eyes and a steady gaze. The

Wickhams believed their family was descended from William of Wykeham, the first Bishop of Winchester, who founded New College, Oxford. By the 1700s, the Wickhams had exchanged religious vestments for military red and blue. Ancestors fought and governed in the American Revolution and in the West Indies; Henry's grandfather, Joseph, nearly lost his life during an amphibious landing against the French during the 1801 Battle of Abukir. Though his leg was torn off by a cannonball, he stopped the bleeding with a tourniquet made from his own sash and the bayonet of a dead soldier beside him. He was fitted with a wooden leg and discharged. Ten years later, he married Sophie Phillips, whose father had been ruined by King George IV in a shady racing wager. According to *Gentleman's Magazine,* the Prince of Wales weighted the jockey's pockets before the race; although the deception was unmasked, Sophie's father still lost the bet—and his estate. The young couple immediately moved to London, where Henry's father was born on July 14, 1814, in St. Marylebone Parish in Westminster, known as the "richest and most populous metropolitan parish" in the growing City of London.

Unlike his predecessors, Henry's father was not of martial blood. He joined the bar and became a solicitor at age twenty-three. Seven years later, in 1845, he married Harriette, a dark-haired milliner from Wales. Henry was born a year later; a sister, Harriette Jane, was born in 1848. By summer 1850, Henry's mother expected a third.

By then, the Wickhams were firmly ensconced in the country. Haverstock Hill had its bohemian side, and that was part of the appeal. The painter William Charles Thomas Dobson, who would influence the preRaphaelites, lived nearby at No. 5 Chalcot Villa, as did the eminent Egyptologist Samuel Birch, who lived at No. 17. Several rich courtesans built retirement homes in Haverstock Hill, including Moll King, the model for Daniel Defoe's *Moll Flanders.* But there was a better reason to move from the city, and that was to escape disease. The epidemic diseases of the nineteenth century—tuberculosis, typhoid fever, smallpox, and cholera—were associated with crowded conditions. According to *The Lancet,* the death rate in 1880 from smallpox in the "open and airy slope of Hampstead" was 12.6 per 1,000, compared to 22.2 per 1,000 "in the metropolis generally."

The worst by far was cholera. The speed with which it killed was spectacular: a healthy person could die within two or three hours, making it the most rapidly fatal disease known to man. By 1850, the "Asiatic cholera" had swept through London twice, first in 1832, when it was called the scourge of the impious and dissolute, then in 1848–49, when it was evident that this was a worldwide plague. Since the cholera bacteria thrived in the water supply, everyone was vulnerable. Isolating oneself from the crowded city, where the most violent outbreaks occurred, seemed the best course, and Harriette Wickham and her two young children did just that in the healthy spaces of Haverstock Hill.

It is harder to track the movements of Henry's father. As a solicitor, he ranged across the city and surrounding suburbs, catching the thirty-minute omnibus ride into London, visiting his one-legged father in Marylebone, conducting business near Bloomsbury and the British Museum, a stone's throw from St. Giles parish, London's most famous slum. A popular shilling guidebook, *A Week in London*, described St. Giles as a place "occupied by the very lowest class of society and through which it is hardly safe to pass alone in the day-time." In *Sketches by Boz*, the young Charles Dickens called it a land of "wretched houses with broken windows [and] starvation in the alleys." The poor were a dangerous "tribe," the bestial Other lurking at the edge of civilization, slaves to violence, unnatural desire, and disease.

But Henry's father did not have to go as far as the city to encounter the poor. Hampstead was an urban center in itself, paved and lit by gas, a polling place for county elections and site of "court leets and baron," a medieval system that settled disputes and maintained common pastures and ditches. And it had its own slums. The poor lived "in alleys and courts without drainage or water supply," said an 1848 report by the Metropolitan Association for Improving Dwellings of the Industrial Class. Overcrowding was common, and a workhouse and soup kitchen were built near Haverstock Hill. All the conditions for cholera were present; they lay in the path of many London solicitors.

Thus, it is no surprise that Henry's father contracted the disease. The microscopic viper acted fast, with a timing that was particularly cruel. The Wickhams survived the winnowing of 1848–49; there would not be another full-blown epidemic until 1854. They must have sighed a breath of relief, for Harriette was due to give birth in the summer. They'd taken every precaution imaginable, but that was not enough. Henry's father ingested the deadly germ in the prime of his life at age thirty-five.

The disease left victims and their loved ones little time in which to prepare. Nothing in Henry's memoirs ever touched upon his father's death, but it would have been one of his first memories. He was four, the age of basic impressions, and his father was dying in the bed-sitting room. The psychological shock was far-reaching. Mortality always lurked. One might seek safety in the assurance of sanitation or class, but they were never enough. Something of fundamental importance was always out of reach that would somehow make things right. Death lurked in the shadows and waited around the corner.

You were always on the run.

The death of Henry's father changed everything. When John Joseph, the third child, was born a few months after her husband's funeral, Harriette had three mouths to feed. What had seemed an exciting future, with husband, security, a home in the country, and bohemian friends, now looked like a future of endless worry and toil. Since she could no longer afford the pretty villa at Haverstock Hill, she soon moved. According to the 1871 census, all four listed their address as 25 Fitzroy Road in Marylebone, close to Joseph Wickham, the warhorse grandfather. Marylebone was one of the city's five wealthiest districts, where the average annual rental rate went far beyond her newly modest means. In all likelihood, Grandfather Joseph became Harriette's benefactor; they became the "poor relations," an embarrassing drop in status keenly felt in class-conscious London.

Harriette returned to millinery, the only skill she knew from before her marriage and "set up a not very successful millinery business in Sackville Street," according to Wickham's only biographer, historian Edward Valentine Lane. Millinery was one of the only trades at the time in which a woman might be a sole proprietress and hire employees. It was the kind of work a middle-class woman down on her luck might take up, yet the hours were brutal and the pay pitifully low. As late as 1903, Jack London reported the case of a destitute, seventy-two-year-old milliner who applied for relief, since the amount she charged per straw hat—two and a quarter pence—was simply too low to survive.

Yet it was also believed that a well-dressed woman never ventured out without a hat or bonnet, so there was always trade. Small touches made the difference—a smartly-placed rose or bird-of-paradise feather could make or break a sale. But feathers from exotic birds cut into the milliner's profits. Unless Harriette had a tropical source, she could only gaze in envy at the trendy shops that could afford to stock such finery.

As the oldest child, Henry would normally have been saddled with the most responsibility, but he seemed to evade this burden. "Henry Alexander, deprived of a father's guidance, grew up something of a spoilt, harum-scarum boy," Lane reported, apparently quoting family tradition. Henry was not a great scholar and showed little interest in any subject save art. He liked to wander more than anything, and a school-age portrait shows a slightly chubby boy, dressed in black school uniform, with dark, wavy hair and intense, blue-gray eyes. He has a distant look on his face and stands angled forward, as if ready to bolt out the door.

He longed to be outside. Regent Street, one block west of Harriette's shop, was the heart of the shopping district, the spring of a new kind of leisure. Shops were a medley of brass, gas, and glass—retailing had become an art, borrowing techniques from the theater to catch and hold the eye. Shopping and window-shopping had become the Victorian woman's entry into the routine of the city; they passed like gaudy moths, and the fine goods or art prints displayed in a window were thought to entrap them like flypaper. The material reverie was said to leave them vulnerable, a hypnotic state charged with sexuality. Predatory males stalked at the edges of sight, ambushing young women with what the press called "The Rape of the Glances." Pickpockets worked the crowds. One side of Regent Street had become the midday haunt of well-dressed prostitutes who ogled and were ogled. New forms of the old predations were taking shape. London's "inconvenient populousness" made living and working a daily adventure.

On July 30, 1850, as Henry struggled with the new world of his father's death, ground was broken in Hyde Park for the wonder of the age: the

Great Exhibition and its centerpiece, the Crystal Palace. Two months later, the first column was raised; the glass edifice rose up until the building covered eighteen acres and enclosed the tallest elms beneath its arches. Within seventeen weeks, nearly a million feet of glass were fastened to a lattice of 3,300 columns and 2,300 girders. It rose so fast because each prefabricated column, girder, gutter, and steel bar was identical. The voices of critics rose with it: The Exhibition would serve as a rallying point for riffraff; the Palace's sparkling roof would be porous, allowing the droppings of fifty million sparrows to rain down. The voices were ignored. The Crystal Palace would be the showplace for the globe's raw materials, machines, and plastic arts, a temple devoted to "the working bees of the world's hive." World peace would radiate from the Crystal Palace like power from a dynamo, for in this assemblage of "industry and skill, countries would find a new brotherhood."

But beneath the steel and glass lay a deeper theme—the industrialization of the natural world. The Palace's inspiration was a plant, the giant water lily *Victoria amazonica,* discovered in British Guiana in 1837, thought to be the largest flowering plant in the world. The water lily's flower was bigger than a head of cabbage and smelled like a pineapple; its leaves measured six feet across and sported a multitude of ribs that could support impossible weights. On a lark, Joseph Paxton, the Duke of Devonshire's head gardener, dressed his seven-year-old daughter as a fairy and posed her standing on the water lily. The photo became the national rage, and piqued Paxton's curiosity about the strength of those ribs. He found he could load a leaf with five children, the equivalent of three hundred pounds. In 1850, Paxton modeled the Crystal Palace after *Victoria amazonica,* its cantilevered trussing of iron girders supporting nearly three hundred thousand glass panes.

This homage to what one botanist called "a vegetable wonder" bespoke the horticultural mania that gripped Britain. The last vestiges of Holland's "tulipomania"—the botanical madness of the 1500s when entire fortunes were spent on forty rare tulips from Constantinople—could still be found among the English royalty. A bulb of the "Miss Fanny Kemble" sold for seventy-five pounds, more than the lifetime wage of a shopgirl. Flowers were so adored that "the language of flowers" grew to cult status; wax flower arrangement became a new art; flowers made of human hair commemorated the dead.

On May 1, 1851, the queen opened the Great Exhibition. A six hundred-voice choir burst into the "Hallelujah Chorus" after a short opening prayer. The event was "the *greatest* day in our history, the most *beautiful* and *imposing* and *touching* spectacle ever seen," Victoria wrote in her diary. Thousands visited the Exhibition: Commentators noted in particular the pairings of the very old and young, the generation that built the empire to its present state and the one that would take it to further glory. Many headed for the cacophonous Machinery Court. Farmers in their smocks crowded around a giant reaping machine from the United States. The queen herself admired a medal-stamping machine that could produce in a week fifty million medals embossed with her image.

But it was the exhibition of rubber that drew the greatest crowds. Charles Goodyear spent thirty thousand dollars of his own money to create Goodyear's Vulcanite Court, a three-room display in which everything was made of rubber: furniture, draperies, rugs, and fixtures—even the walls and a large "Elizabethan" sideboard. There were rubber inkstands, knife handles, stockings, bandages, hot water bottles, syringes, dolls, and air cushions. Goodyear spoke of air-inflated saddles, inflatable boxing jackets that could absorb a punch, and a rubber dress for ministers performing full-body baptisms. He envisioned the day that rubber life preservers would eliminate drowning as one of the evils of the world. Rubber could be prefabricated into any shape; it was a protean substance, hailing from nature but perfected by science. It was the material of the future, a true gift from God.

The high spirits of the exhibition were but a brief respite for Harriette and her family. In 1855, Grandfather Joseph stumbled as he left his club and fell, splintering his wooden leg. The old war wound caught up to him four decades late. He died from the injuries on June 30, at age seventy-four.

His death was a doubly cruel blow. For Harriette, there was no longer the financial backup upon which she depended, no champion in the savage wars of class, where "poor relations" were an embarrassment and a financial drain. For the children, and for nine-year-old Henry in particular, there was no longer a father figure on whom to depend. While his sister Harriette Jane helped as best she could in the millinery shop, Henry wandered the London streets and dreamed. His world was a narrow-laned warren of coal soot, sweatshops, and criminal gangs. His mother feared that he would be easily tempted, pushed by the slightest breeze. One slip, one stupid prank, and he'd be destined for the Borstal, another youth sentenced to the workhouse that guaranteed an education in crime.

Parents always worry about their children: after 1857, Harriette's worries took a new form. That year, the Indian Mutiny burst upon the psychic landscape of the English, and the world that looked bright and promising in the Great Exhibition now seemed sinister. Suddenly, there were many comparisons made between London's poor and bloodthirsty heathens in foreign lands. Both were described in phrases such as "wild tribes" and "savage races," often invoked by reform-minded writers like Henry Mayhew and James Greenwood to portray the impoverished. Travelogues treated the city's rookeries as a new kind of jungle. Shop windows featured a steady stream of books like *Legends of the Savage Life, Wild Tribes of London, Low Life Deeps,* and *The Wild Man at Home.* Such "savage" districts as Whitechapel and New Holborn were inhabited by "strange and neglected races"—the feral child, the screaming harridan, the murderer, the Jew. Every district was an unchristian wilderness in which "a clash of contest, man against man, and men against fate" took place daily. Every generation has its metaphor of temptation: in mid-Victorian London, that metaphor was the jungle, and the streets were the tangled paths where children were devoured.

In the space of a few short years, Henry had gone from life to death, comfort to want, open heaths to crowded slums—and now, his mother feared, from innocence to moral ruin. His future as a child had spread before him like the fields surrounding Haverstock Hill; his life as a youth now closed around him like the soot-blackened city. London was a jungle, a cliché today but a metaphor that was new, exciting, and frightening for its time. On the streets, as in the wilderness, warned one schoolmaster, "there is a bitter struggle to live." Henry was learning about jungle life, but where a mother saw danger, a boy like Henry saw a New World.

CHAPTER 2

NATURE BELONGS TO MAN

A new hero was taking shape during Henry's childhood: the explorer—the astronaut or rock star of his day.

Throughout the Victorian Age, the Crown funded scientific exploration linked directly to its imperial interests, which the public saw as an expression of culture. Exploration was an act of intervention that altered the fate of one nation while increasing the knowledge, resources, or power of the other. The explorer was the first agent in Western Europe's conquest of "uncivilized" lands, where it was presumed that no law or truth existed. If a missionary, the explorer brought Christian light to the darkness; if a scientist, he spread knowledge among the ignorant. Residents of the periphery, however, did not always share that high-minded vision. They tended to see him as an agent of exploitation, the vanguard of an invasion—a spy.

By the late eighteenth century, exploration was a proven path to power and riches. The Spanish Conquest was the prime example. The Enlightenment's appetite for new facts provided new reason for discovery, and although all Europe participated, Britain maintained the highest level of exploration throughout the nineteenth century. By midcentury, the popular image of the British explorer had begun to evolve. Already a symbol of reason's spread, he now became a celebrity. The great explorations of 1790–1830 had been maritime coastal reconnaissances to produce the celebrated Admiralty Charts, and explorers of this period were seen as self-effacing, duty-driven civil servants whose sole motive was the collection of information for the charts to make sea travel safer. The new explorer who emerged during Henry's childhood was a larger-than-life character whose exploits acted out personal ambitions and public fantasies.

There was no better example of this than Sir Richard Burton, after whom Henry would model his personal style. In 1853, dressed as the fictitious Sufi physician "Sheik Abdullah," Burton joined the Egyptian hajj to Mecca; although he was not the first Christian to enter the holy city in disguise, his 1855–56 *Personal Narrative of a Pilgrimage to al-Medina and Mecca* was the most detailed. Success led him to a legendary quest—the discovery of the source of the White Nile. An initial foray in 1855 ended with his camp ambushed, one companion killed, and a second—John Hanning Speke—wounded, while Burton himself was pierced through the cheek with a javelin. In 1856, Speke and Burton set out again and by February 1858, the two reached Lake Tanganyika, one of the Central African "Great Lakes" thought to be the source of the Nile. With Burton paralyzed by malaria, Speke went on to locate Lake Victoria and proclaim it the Nile's origin. By then the mismatched pair hated one another. They disagreed over Speke's claim, initiating the great controversy that proved irresistible to chroniclers and which ended finally in September 1864 with Speke's death, by accident or suicide, a day before the two were to debate the question at the Royal Geographical Society.

Suddenly the explorer was a central hero in an escape fantasy that gripped the British isles, a champion who trod the earth's wild places and interpreted what he saw through English eyes. Newspapers and journals were filled with the adventures of Burton and Speke, as they were with David Livingstone's expedition up the Zambezi River in 1857–63. A new theme crept in. Since the new explorer was in the vanguard of a culture whose science and technology were clearly superior to that of "unscientific races," his very arrival seemed to prove England's right to rule these newly explored lands. "Natural theology," the Christian belief that nature existed for man's improvement, was already a given; now it was one short step to the conviction that the West was destined to master nature and remake the world.

Henry came of age when explorers' journals were bestsellers. There was something otherworldly about an explorer's description of an unknown land, such as the passage where early explorer MacGregor Laird confessed that the stillness of Sierra Leone's forests "chills the heart, and imparts a feeling of loneliness which can be shaken off only by a strong effort." The quest to fill the spaces on the map became a journey deep inside the mind. Balancing this was an obsessive adherence to the grinding reality of exploration. Meticulous instructions from scientific societies called for daily readings of instruments, regular updates of maps, the careful preservation of specimens, a daily weather log, and observations of native customs, language, population, resources, and trade. The explorer was expected to be leader, emissary, hunter, observer, collector, scientist, mapmaker, and artist. He was told to make frequent notes and leave nothing out, a fail-safe in case he should be murdered. Events less final than homicide could also ruin an expedition: Precious equipment got destroyed, compasses were lost, instruments split from the heat, thermometers calibrated for English weather burst in the desert clime.

There was also a growing realization among Britons that the tropics were the "white man's grave." During Henry's childhood, this meant West and Central Africa, but as the century progressed, it included any equatorial place loaded with "primitive tribes," burning heat, miasmic swamps, swarms of insects, and miles of treacherous jungle. Much of the image was false: The maximum temperature in West Africa and the Amazon Basin rarely surpassed the maximum midsummer heat of the American Midwest, while jungles were often the home of sedentary agrarian people who were far from primitive and who for centuries had slashed and burned the underbrush to carve out farms. The image did have one real basis in fact, and that was disease. European newcomers to the African coast died at a rate of three hundred to seven hundred per thousand during their first year. After that, the mortality rate dropped to 80–120 per thousand, but that was still far higher than any cholera pandemic sweeping through the European capitals. West Africa alone was stocked with sleeping sickness, Guinea worm, yaws, bilharzia, dysentery. The principal culprits were spread by mosquito, and an individual's chance of living a year without being bitten were nearly nonexistent. In some areas, the average number of infected bites per year per person was a hundred or more. There was no escape from malaria or yellow fever.

Of the two, malaria was the oldest, deadliest in number of deaths, and most pervasive. Yellow fever was self-limiting, since the parasite either killed its victim in five to seven days or allowed a complete recovery that brought with it a lifelong immunity. Not so with malaria. It struck, lingered, then returned repeatedly, teasing its victim, slacking off before rolling back like a scalding wave. Anyone who ventured into its domain saw others die around them before they, too, succumbed to debilitating headaches, chills, and fevers. Physicians prescribed general bleeding as a cure: 20–50 ounces of blood taken at the fever's onset, then more for a total that could exceed 100 ounces. Since the body contains about 180 ounces of blood, and anemia is one of malaria's most common symptoms, the loss of so much blood could be fatal. The common treatment with mercurous chloride (calomel), was no better. Malaria is dehydrating; since calomel is a purgative, it intensified the loss of fluid. The two treatments administered together sent many to an early grave.

Only one thing quelled the fever, a bitter alkaloid in the red bark of a tree. That tree grew in only one place on Earth, along the east slope of the Andes near the source of the Amazon—a place that, for most Europeans, seemed like the dark side of the moon.

Although quinine is best known today as a bitter taste in tonic water, during Victoria's reign it was the most powerful antimalarial medicine known to man. It came from the bark of the evergreen cinchona tree, native only to South America. Cinchona belongs to the Rubiaceae, or madder family, which includes gardenia, bluet, bedstraw, coffee, and the ipecac (*ipecacuante*) shrub. There are nearly sixty-five species of cinchona, and most prefer the mountainous forests of Colombia, Peru, Ecuador, and Bolivia. They grow in a narrow belt between 22° S latitude and 10° N latitude at 3,000–6,000 feet above sea level, an area characterized by high humidity and frequent heavy rain. Only four of the species contain an alkaloid content in the bark of their roots or stems high enough to be any use to man: the highest concentrations were thought to exist in the *cascarilla roja,* or "red-bark tree." Quinine's preparation had been the same for centuries: *Cascarillos* uprooted the tree, beat loose the bark, and peeled it off by hand. The dried bark was then ground into powder or infused in water.

In 1859, the London journals began to chronicle Britain's two-year quest to smuggle cinchona from the Andes. That there was something different about this incursion was apparent from the beginning. Most British expeditions claimed discovery and the furtherance of science as their purpose, but the cinchona affair was an unabashed imperial endeavor swaddled in justifications ranging from free trade to the good of mankind. If the single source of this life-giving drug were being mismanaged to the point of extinction, argued colonial authorities, didn't the world's most powerful nation have a moral obligation to plant seeds elsewhere? When *cascarillos* felled cinchona during collection, they pulled up young trees with no thought of replanting or conservation. The world's only defense against malaria might go the way of the dodo. Theft in such a case would be a humanitarian act. Global thievery was a small price to pay if millions were saved. Left unsaid was the fact that malaria's defeat would end the deathwatch of the "white man's grave" and leave the tropics open for colonization.

The midcentury belief that a nation could appropriate for itself another's vegetable resources did not occur in a vacuum, especially if clothed in the garments of religion. Natural theology had long assumed that everything in nature existed for man's use and instruction: Natural riches, scattered to the ends of the earth, were not for one people alone and should be available to all. J. H. Balfour's 1851 *Phytotheology* argued that God's orderly plan was reflected in the structure and function of plants. T. W. Archer's 1853 *Economic Botany* claimed that God had clothed the Earth "with every necessary for men's wants," and even rubber, with its remarkable versatility, was used by the Society for the Propagation of the Gospel as a prime example of God's affection for man. The transfer of plant species from one part of the world to another was nothing new. The Spanish Conquest had opened Europe up to what has since been called the "Colombian exchange." Wheat, grapes, lettuce, cabbage, apples, peaches, mangoes, bananas, alfalfa, and many other crops crossed the Atlantic in one direction or the other. Most important was maize, which thrived in the drier soils of southern Europe, the Balkans, and West Africa, and the white potato, which fed the poor throughout industrial Europe. After the First Opium War of 1839–1842, a plant collector named Robert Fortune brought two thousand tea plants and seventeen thousand seeds out of China and transplanted them in the sprawling Indian plantations around Darjeeling. It was obvious that transplanted crops could make British settlers rich and their colonial administrations solvent: The Ceylon coffee crop, for example, was worth £500,000 in 1850 and would be worth three times that in just another decade. Cocoa showered riches on Trinidad; sugar, on the colonies in the Caribbean. In Burma, officials urged the enclosure and protection of teak; in British Honduras, mahogany companies ran the colony.

The *reality* of plant transfer was not at issue. What emerged with the 1859 cinchona theft was a new rationale. The Earth was a treasure house that had to be managed: Nature belonged to man to harvest and "improve." This was especially true in the tropics, where vegetable riches ran rampant but were also wasted and destroyed. By the Victorian era, when the British Empire had endowed its expansion with a moral flavor, the idea had new urgency. Not only the conquest of nature was at stake but the course of the future. Conservation sometimes meant saving environments from those who lived in them. At best, natives could be educated; at worst, they must be expelled. The "Profligate Native" became a common theme of empire, often rephrased as "the labor problem," used as a justification for the right to intervene. In Cameroon, reservoirs of palm oil were being wasted because the African laborers "waltz through life in a dream with their heads wrapped in clouds too deep to receive the instructions given to them," complained botanist Gustav Mann. To colonial governor Sir Charles Bruce, "the very existence of [tropical] colonies as civilized communities required the intervention of capital and science of European origin." To Benjamin Kidd in his 1898 *The Control of the Tropics,* the native had no right "to prevent the utilization of the immense natural resources which they have in charge." This certainly applied to the profligate Andeans killing off the world's shrinking stock of quinine.

The motive was as old as Babylon. The world's riches were up for grabs. Only the justifications changed.

Every great theft demands a mastermind, and he requires the machine to help launch his schemes. In the case of cinchona and rubber, the usual genesis was reversed. The great machine came first, awaiting its machinator.

The Royal Botanic Gardens, Kew, was that great machine. Formed from two adjoining pleasure gardens of the Hanoverian kings, Kew was given new life in 1841 as a state institution. Funded by Parliament and charged to aid "the Mother Country in everything that is useful in the vegetable kingdom," its mandate was to be the nerve center for "the many gardens in the British colonies and dependencies, such as Calcutta, Bombay, Saharanpur, Mauritius, Sidney, and Trinidad, whose utility is wasted for want of unity and central direction." Although its beauties were open to the public, its true role was that of research and development, providing scientific aid to the empire's vast plantation economy—a mission considered crucial for "the founding of new colonies" and the maintenance of their economies. To succeed in such an endeavor, a walking encyclopedia of botany was needed at the helm. The man selected for this, William Jackson Hooker, was Regis Professor at the University of Glasgow, director of the Glasgow Botanic Garden, founder and editor of several botanical journals, and one of the few professional botanists of the time. He was also politically savvy, transforming the new state institution into a world center for "economic botany," closely tying the young science of botany with the rising fortunes of empire.

Hooker wasted no time in making the Royal Botanic Gardens indispensable. Soon after his appointment on April 1, 1841, he established a Museum of Economic Botany at Kew. This acted as a clearinghouse for global seed transfers, shuffling plants throughout the empire as the economic possibilities presented themselves. By 1854, Hooker could boast that not a day passed without investors, planters, or administrators coming to Kew for information about the useful woods, fibers, gums, resins, drugs, and dyestuffs awaiting exploitation in their chosen wilderness. By 1855, he felt confident enough to claim that Kew was "essential to a great commercial country." By the extent and nature of her power, Great Britain sat in the crossroads of Providence; Kew was there, Hooker said, to organize and improve the bounty from the four corners of the globe.

Hooker was not the first to see rubber's importance to the empire: that honor may have gone to the industrialist Thomas Hancock. By the 1850s, it was becoming obvious that vulcanization was transforming rubber from a natural oddity to a world commodity. In 1830, Britain imported 211 kilograms of raw rubber; by 1857, that figure jumped to 10,000 kilograms, a 4,700-percent increase in a quarter century. Even more telling was the fact that Brazil was becoming the world center for supply. In 1827, the nation exported 69 metric tons of rubber, an amount exploding to 1,544 tons annually from 1851 to 1856. Hancock began to worry that Brazil might someday cease to provide the quantities of rubber that Britain was beginning to require. Rubber was a jungle product, extracted from secret groves in the rain forest by methods no European understood. There was no way to predict annual supply; Hancock knew that the price would increase as the world demanded more. The market needed a prod. In 1850, Hancock proposed creating rubber plantations "in Jamaica and the East Indies" to William Hooker. The director, in turn, promised Kew's resources and vowed to "render any assistance in his power to parties disposed to make the attempt" to move the rubber tree from Brazil to some friendly territory within the British Empire.

Thus, Kew first turned its attention to rubber in the same year that Henry Wickham's father died. Its interest in caoutchouc paralleled Henry's coming-of-age. Unlike cinchona, however, a major mystery prevailed. Industrialists like Hancock had determined that "Pará fine" was the most durable rubber on the market, able to withstand more punishment and shrink less during transport than any other variety. But no one knew what it came from or where. The Amazon Basin was huge and uncharted; accurate maps did not exist; rubber suppliers kept the locations of their rubber stands secret or lied about the source, calling everything "Pará fine" to drive up the price. Rubber went by such names as seringa rubber, India rubber, "Pará fine," and now a new term, *siphonia*, which perplexed everyone. Experts added to the confusion, bringing rubber-producing plants to Kew from throughout Latin America, often in a deteriorated state. They brought *Castilla elastica* and its cousins from Central America; three kinds of *Hevea* from the Amazon and Orinoco valleys; the Ceará rubber tree, *Manihot glaziovii;* and a Brazilian species of *Sapium.* They confused *seringa* with *Castilla* and were lost regarding the ranges of separate species of *Hevea.* There were alternate sources from other continents, since each nation had an interest in proclaiming its rubber the best in the world. Botanists found *Kickxia, Funtumia elastica, Landolphia, Clitandra,* and *Carpodinus* in Africa; *Ficus elastica* in India and Burma; and "gutta-percha" from the leaves of the towering *Isonandra* tree in Borneo, the Malay peninsula, and Ceylon. For a long time, "gutta-percha" and "India rubber" were interchangeable. As rubber became essential, only one thing was certain. Confusion reigned.

Such was not the case with cinchona. Its taxonomy was settled. Collectors knew what they were looking for. Although never parsed so finely, it made a certain sense to smuggle cinchona before attempting the same with rubber: In order to solve the mystery of rubber, one needed to survive the tropics and its great assassin, malaria. In order to do this, one needed a reliable supply of quinine.

There was also imperial demand. The Sepoy Mutiny of 1857 in Bengal and North India catalyzed Great Britain just as the French Revolution transformed France and September 11, 2001, changed the United States. The world was a dangerous place, the homeland surrounded by deadly plots and enemies. In such a world, the best defense was a preemptive offense, and this meant sending British troops, administrators, and their dependents to the empire's malarial possessions around the world. The Dutch were already mounting a campaign to secure and control cinchona. Quinine was more than a drug. It was a fetish: a symbol of the power of science to control an unruly world.

Into this tumult stepped the caper's mastermind. Clements R. Markham was a handsome fellow, with a broad forehead, muttonchops, and pale, distant eyes—one of those eminent Victorians who seemed "irritatingly destined for high office," a biographer said. In time he would be knighted, made president of the Royal Geographical Society, and write forty-four books, most on South America, but in 1859 he was a thirty-year-old junior clerk in the India Office, recently decommissioned from the Royal Navy. Most of his seven-year service was spent sailing off the American coasts, and he still dreamed of the tropics. On April 5, 1859, Markham proposed in a letter to Sir James Hogg, Chairman of the Revenue for the Judicial and Legislative Committee of the Council of India, that cinchona seeds could be stolen by an Englishman and replanted in the plantations of northern India. "My qualifications for the task," he wrote, "consisted in a knowledge of several parts of the chinchona [*sic*] region and of the plants, an acquaintance with the country, the people, and their languages, both Spanish and Quichi." He maintained that past attempts to collect seeds had failed because the work had not been assigned to someone who was "really interested" in the project, like him. Perhaps most importantly, he showed a keen understanding of the parsimonious nature of British bureaucracy by offering to undertake the project for his current salary of £250 annually, plus expenses.

To accomplish the task, he suggested a pairing of the India Office and Kew. The India Office, caretaker of Britain's "crown jewel" among colonies, certainly had the power to see such a project to fruition, and Kew had the know-how. When presented with the idea, William Hooker hopped aboard without any apparent hesitation, laying the groundwork for a successful partnership between Kew and the India Office that would last for decades. Hooker must have immediately sensed the advantages: With cinchona, Kew entered a wider circle of colonial officials, merchants, and planters than had ever been possible. Within a year, the tight-fisted Treasury was granting funds to Kew for such projects as a "double forcing house" for seed germination, money that had not been previously available. Aligning one's fortunes with India was a wise course for the future.

Markham's plan of attack was three-pronged. The cinchona region followed the curve of the Andes for about one thousand miles. Since this was too much for one group to handle, he suggested three. Markham and

Kew gardener John Weir would tackle the southern region—Bolivia's province of Caravaya and southern Peru—where the "yellow-bark" version, or *C. calisaya*, flourished. The "gray-bark" species (*C. nitida*, *C. micrantha*, and *C. peruviana*) grew in the center section, the forests of northern Peru, where collection was entrusted to G. J. Pritchett, an English expatriate. Penetration into the Ecuadorian highlands where the "red-bark" species flourished was entrusted to Kew gardener Robert Cross and Richard Spruce, who'd been collecting new plants in South America since 1849. Except for Spruce, who was still in the jungle, all would train at Kew in the fine points of collecting and preserving cinchona; each group was budgeted £500 and was to complete its mission within a year. The greatest obstacle, Markham feared, would be the "narrow-minded jealousy" of those South American officials who might object to their nation's loss of an important source of revenue.

In December 1859, Markham left London and spent a month in Lima organizing supplies and planning strategy. Revolution was in the air. Seven thousand Peruvians had died in uprisings since 1853, and in Bolivia, coups and civil war had dominated life since that country's independence from Spain in 1809. At least in Peru a British smuggler could call on the British minister or vice consul if he got into trouble. In Bolivia, there were no sympathetic representatives of the Union Jack. Not surprisingly, Markham skipped Bolivia and confined his search to the southern Peruvian province of Caravaya.

On March 6, 1860, Markham and Weir headed into the interior, struggling up the Caravayan Andes into a fantastic country of volcanoes and some of the deepest canyons in the world. Andean condors soared above the clouds and valleys. It seemed prehistoric, and on the last leg of the journey they ascended seven thousand feet in 30 miles to the mountain town of Sandia, where they began collecting seedlings and seeds.

But Sandia was not a friendly town. A certain Don Martel, ex-colonel in the Peruvian army, heard of Markham's intentions and told inhabitants to do what they could to stop him. On May 6, the mayor of Quiaca ordered that any *estranjero inglés* seen in the area should be arrested. Since the mayor had no real authority, he depended on hired guns. Desertions and mutiny plagued Markham's company. He grew desperate to defend his plants if necessary. At one point he brandished a pistol whose damp powder rendered it useless, but the bluff succeeded. He needed to escape with what few seeds he'd already collected, so he sent Weir north as a decoy and struck south over the mountains to the coast. The mules burdened with the seeds seemed determined to do what the Peruvian officials had not. Several times they nearly rolled off the narrow paths into the deep ravines. Yet Markham somehow got the plants to Islay, where he rejoined Weir.

Now he entered the web of Latin bureaucracy. At Islay, the superintendent of the customs house refused Markham's request to ship his four bundles of seeds to England unless he first produced a signed order from the minister of finance in Lima. This was bad news. The previous president had issued a decree forbidding any export of cinchona, but the order had never passed outside Lima. Perhaps the finance minister did not have a copy, he hoped. Since "all the clerks in public offices are changed in every revolution," most officials had no more than two or three years' tenure. If he could bluff his way through the layers of rival officialdom, he might still escape, and if there was one thing Markham understood, it was the tangled ways of politics.

Markham got lucky. The president was "a rough, illiterate, though shrewd and valiant Indian," who cared only for the strength of his army; he'd appointed as finance minister a former colonel of horse who knew little about customs regulations. Markham bribed and threatened the man, and the order was signed. On May 23, 1860, he watched his precious seeds stowed onto a steamer for London and sail away.

Even so, his triumph turned to failure. Although the seeds made it to Kew, one case of seedlings fell into the sea en route to India. The remaining plants baked in temperatures topping 107° F when the ship experienced engine failure on the Red Sea. When the cases arrived in India, the saplings were rushed to the cool Nilgiri hills in the north and quickly replanted, but it was too late. By December 1860, every plant had died.

The bad luck continued. When the gray-bark seeds arrived from the central region, their alkaloid content proved to be too weak for any medical use. England's hopes were pinned to the group in the north, headed by a very sick man.

Richard Spruce was already a legend, at least in the small world of botany. In 1849, he'd traded the safe English life of a mathematics teacher for the risky one of collecting South American plants. A decade later, he'd traveled ten thousand miles by river, identified thousands of plants, compiled the world's most complete collection of mosses, and written a glossary of twenty-one Indian languages. He was restless, gangly, thin, and dark. When it came to Hooker's requests, he didn't seem capable of saying no. In 1846, after spending ten months collecting mosses in the Pyrenees, Spruce was approached by Hooker, who asked if he'd live a similar life of discomfort in the New World in Kew's employ. Spruce readily accepted. He trained at Kew in tropical botany from 1848 to 1849, and on June 7, 1849, embarked from Liverpool to Brazil.

Along with the naturalists Alfred Russel Wallace and Henry W. Bates, whom he met in Santarém, Spruce was a new breed of explorer. Previous wanderers, like la Condamine, had their own wealth to support them or were sponsored by a government, like Charles Darwin aboard the *Beagle*. Spruce lived on a shoestring, earning his slender keep by selling his collections, often for threepence a specimen, often for less—and sometimes for nothing should the dried specimens spoil en route to Kew. His mandate from Hooker was to discover new plants that could be of use to the Empire, and a major part of that commission was to solve the mystery of "Pará fine." There were hundreds of species of rubber-bearing plants in the Amazon Valley, and every region called their rubber "Pará fine." Did it come from the *Maceranduba*, the cow-tree, so called because of its copious secretions of a sweet latex that Spruce called a "drinkable milk"? He spotted several trees near Pará and mixed the latex with his coffee. "Its consistency is that of good cream," he reported, "and its taste perfectly creamy and agreeable." Or was the source a tree yet undiscovered, waiting in the green fastness like some wondrous Sangreal?

It was an impossible task for any one man. Over five hundred species of plant flow with a milky latex that can produce rubber. They belong to several botanical families, in several different genera, distributed throughout the earth. All are united by the milk in their veins, an elastic latex, which, strictly speaking, is any mixture of organic compounds produced in *lactifers*, the cells or strings of cells that form tubes, canals, and networks in various plant organs. The latex flowing in these tubes varies in composition from species to species, but each is an emulsion loaded with hydrocarbons and mixed with other compounds, including alkaloids, resins, phenolics, terpenes, proteins, and sugars. While most have elastic properties, some, like the opium latex of the poppy, are prized for different qualities. No one truly understands the function of latex in the wild. While many botanists believe latex has evolved as a defense against herbivores, since it usually gives the plant a bitter taste, others think that lactifers have developed as a conduit for waste, that lactifers are the sewage system of the plants and latex the liquid manure.

On October 8, 1850—while four-year-old Henry Wickham was just facing life without a father—Spruce began a solitary canoe journey that took four years and covered eight thousand miles of river. Hooker, his mentor, saw the Amazon as a huge warehouse for economic use, a wilderness to be plumbed and tamed. It was a common idea among naturalists, a fresh twist on El Dorado. Their precursor was the German explorer Alexander von Humboldt, whose 1799–1800 journey up the Orinoco River in what today is Venezuela verified a legendary link between the Orinoco and Amazon river systems. Humboldt envisioned thriving cities and a great civilization amidst the intertwined creepers and flooded river plains. Spruce's friend, Alfred Russel Wallace, was just as enthusiastic: "It is a vulgar error that in the tropics the luxuriance of the vegetation overpowers the efforts of man," he wrote in 1853. Just the opposite: The growing season never stopped, the climate was favorable to agriculture, and a man could produce in six hours of work "more of the necessities and comforts of life than by twelve hours' daily labor at home." Two or three families might convert the virgin forest into "rich pasture and meadow land, into cultivated fields, gardens, and orchards," within three years, he asserted.

Spruce did not share Hooker's vision. He preferred an Amazon in its pristine state, an opinion unusual for its time. "How often have I regretted that England did not possess the magnificent Amazon valley instead of India," he wrote in his journal:

> If that booby King James I, instead of putting Raleigh in prison and finally cutting off his head, had persevered in supplying him ships, money and men until he had formed a permanent establishment on one of the great American rivers, I have no doubt that the whole American continent would have been at this moment in the hands of the English race.

He paid dearly for this love of the wild. Soon after setting out, he contracted malaria; its visitations would progressively weaken him. On the Rio Negro, he stumbled into a planned Indian massacre of two Portuguese merchant families: Where the old hatreds lingered, a white skin was the reminder of slavers. He survived the night when the Indians realized they were outgunned—and that *they,* not the Portuguese, would be slaughtered. Farther up the Negro, he fell into a fever: he hired an old "Zamba" woman named Carmen Reja to nurse him. Unfortunately for Spruce, she hated foreigners too. During the malarial attacks, he was plagued by violent sweats, an unquenchable thirst, and difficulty breathing. He was convinced that death was only hours off and gave instructions on what to do with his plants and how to contact Hooker. Then he collapsed in apathy on his hammock, waiting for death to come. During such times, Carmen Reja left the house for hours, hoping to find him dead when she returned. In the evening, after lighting his lamp and leaving a jug of water within reach, she filled the house with friends and spent the night abusing him, calling him names and crying, "Die, you English dog, that we may have a merry watch-night with your dollars!"

Even with such punishment, Spruce tried to find the source of "Pará fine." By 1855, he was nearly convinced that the answer lay in a soaring giant with a peculiar three-lobed seed. He first noticed the tree growing close to the river or rising from the floodplain. It arrested the eye: The lower trunk was thick, dark, and warty, the upper trunk and crown so light that it seemed to shine. The medium-sized leaves were trisected evenly, glowing a beautiful light green. For ten to twelve feet above the ground, the trunk was often striated with the cuttings of the *seringuiero,* or rubber tapper, thin welts from which wept a pale stream of latex like blood. In an 1855 article for *Hooker's Journal of Botany,* Spruce would be the first botanist to accurately describe the techniques of rubber tapping. He mentioned the climbing price of rubber, a sign for him of rising demand. He described a village ball game he'd witnessed two years earlier: The balls were made of India rubber, and he asked the players to save two or three for him after the game. "But during the night," he wrote, "they all got gloriously drunk and burst their balls."

The tree that he suspected is now called *Hevea brasiliensis,* and a cross-section shows why it was so prized. Unlike other rubber sources, its large lactifers lay just underneath the inner bark. Because of this, it could be tapped repeatedly without cutting deep into the cambial layer, producing a steady flow of latex, high in both quality and quantity, for decades. A towering tree, quite common in the steamy Amazon Valley, it seemed an infinitely renewable resource, a well that would never run dry. It healed its cuts and continued to drip latex like the "tree of life" of Indian tales. And when it died, another could always be found down the next jungle path or up the next tributary.

Hevea seemed the logical source of "Pará fine," and when Spruce wrote Hooker of his suspicions, the director asked that he send back some suitable seeds. If germinated in Kew's greenhouses, they could be grown and classified, once and for all solving the taxonomic mystery. Spruce sent several specimens, but the oil in the seedpods turned rancid so quickly that they never survived the sea voyage, a misfortune that doomed every British attempt to procure the valuable seeds for the next two decades.

By 1859, Richard Spruce was tired. He'd collected thirty thousand specimens for Kew, including two thousand new flowering plants. He'd come closer than anyone to solving the enigma of "Pará fine." He'd served his country well, but the decade took its toll. He suffered miserably from malaria and had nearly been killed by natives. That he'd lasted at all spoke to his intelligence, good sense, and good luck, but many wondered how much longer he might hold on. He was nearing the Amazon's headwaters, his final present to himself before leaving South America forever.

Then, just when he thought he had the right to call it quits, he got a letter from London drafting him for Markham's scheme:

> Her Majesty's Secretary of State for India has entrusted the Hon. Richard Spruce, Esq., with the commission to procure seeds and plants of the Red Bark Tree which contains the chemical ingredient known as quinine.

To be so entrusted was an order, and he had no choice: The letter directed him to proceed to Ecuador, collect money at Guayaquil from the British Consul, then head inland to gather cinchona. Then and only then could he come home—with cinchona seeds.

Spruce would be joined by the "very able and painstaking" Kew gardener Robert Cross, a Scotsman who carried an umbrella in the jungle to ward off the sun, rain, and deadly reptiles. They were to meet in Guayaquil. The fact that Ecuador was in the midst of revolution didn't register with the planners. "Matters are in a very unsettled state here, and preparations for war with Peru resound on every hand," Spruce wrote in his journal. "Recruiting—forced contributions of money and horses—people hiding in the forest and mountains to avoid being torn from their families—scarcity and dearness of provisions."

Spruce arrived in Ecuador before Cross and began to climb the Andes from the Amazon Valley below. He followed the Rio Pastaza, which joined the Amazon in Northern Peru; he followed it up the gorges of the Ecuadorian Andes until halting at the town of Ambata, within sight of Mount Chimborazo, at 20,702 feet the highest peak in Ecuador. It was misty and cold, and he was struck by an attack of catarrh, a cough so violent that blood flowed from his mouth and nose. But he suddenly found himself in one of the most moss-laden places on the planet, and mosses, his first and greatest botanical passion, seemed to sooth all aches and pains. The penetrating drizzle created a glade out of some old Celtic romance, where the forest itself was moss-draped and lofty, and every rock and bush was shaggy and green. "I find reason to thank heaven which has enabled me to forget the moment of my troubles in the contemplation of a simple moss," he said.

The idyll didn't last. As he dragged himself up the heights, condors attacked. He sat on his mule above the Rio Pastaza, contemplating the fact that he could no longer feel his hands or feet and wondering whether death had finally caught up to him. He crossed an undulating plain swept by the *paramero*, a wind laden with frost "that withers every thing it meets." Crosses sprouted from the rocks; a pilgrim appeared and said these were the graves of people who had died in the wind. When he was a boy, the pilgrim remembered, he'd crossed this plain with his father. When he spotted a grinning man sitting on an ice-sheathed rock, he told his father, "See how that man is laughing at us?"

"Silence, or say a prayer for his soul!" his father shouted over the *paramero*. "That man is dead."

Robert Cross was having his own problems on the other side of Ecuador. He arrived in Guayaquil in May 1860 but immediately fell victim to fever, then could not proceed until the war abated. In July, he boarded a ship into the interior packed with troops and weapons. Two days south of Ambato, he found Spruce huddled beneath the huge snow-covered hulk of Chimborazo, where glaciers stood out like marble against the cerulean sky. Spruce had found the thickly-matted cinchona forests on the slopes beneath the glaciers. He'd camped at a place called Limón, little more than a collection of huts on stilts. For the next two months, this would be their base as they took thousands of cuttings, collected seedpods, and bought them from locals. Cross sowed a number in case the seeds did not survive the ocean voyage. As soon as they took root, caterpillars attacked, followed by waves of maroon-colored ants. By September, they'd collected one hundred thousand well-dried seeds. They built a sixty-foot square raft from twelve huge balsa logs, then floated down the white-water river until they reached Guayaquil on December 13, 1860. On January 2, 1861, Cross and 637 cases of cinchona embarked for London.

Spruce was ready to go home. He hoped to write his findings in the scientific journals and study his thousands of specimens in the peaceful environs of Kew. But soon after Cross departed, the Guayaquil bank in which he'd placed his savings (about seven hundred pounds, or six thousand dollars in today's currency) promptly failed. He was forced to collect for another three years before earning enough to come home. By then, he was deaf in one ear and suffered partial paralysis of his back and legs. In May 1861, the Ecuadorian government outlawed the export of cinchona, but by then it was too late. England had already planted the seeds in India, a fact remembered bitterly by every Latin nation. The entire venture cost England £857, and by 1880, the government of India would reap thousands of pounds in annual income from Spruce's cinchona, replanted in northern India and Ceylon. Markham, who'd say he'd broken no law by taking the seeds, would eventually become a lion of English society. By 1870, a lifesaving dose of quinine was being sold for "half a farthing" at village post offices throughout India.

The real winner was Kew. Thanks to the success of the cinchona expedition, the scale of Kew's imperial work grew enormously. A new network of gardens was established in far-flung colonies. Colonial officials consulted Kew on an increasing range of questions. Hooker gained powers of patronage far beyond anything he'd dreamed.

Yet everyone touched by cinchona was not equally blessed. Pritchett, who smuggled the gray-bark variety to England in 1866, was too late to matter and is forgotten today. Markham's partner, Weir, was crippled by disease and forced to live on his wife's earnings. Cross suffered from malaria and slept with a gun beneath his pillow.

And Spruce, broken in health and denied a pension until 1877, lived the rest of his life in Yorkshire on one hundred pounds a year.

CHAPTER 3

THE NEW WORLD

Henry Wickham was thirteen when the cinchona mission started and fifteen when Robert Cross straggled back to Kew with the purloined red-bark seeds. Though the triumph was covered in the papers, he could not have cared less about the imperial implications of cinchona's domestication. What mattered was the jungle and the writings of Spruce, Darwin, Bates, Wallace, and others. In the heat and moisture of the Amazon Basin existed a fertility like nothing Western man had experienced, a force of life beyond control. There was something hypnotic about the rain forest: "I have been wandering by myself in a Brazilian forest," Darwin wrote in his *Beagle* journals. "The air is deliciously cool and soft; full of enjoyment, one fervently desires to live in retirement in this new and grander world."

Earthly paradise was the recurring theme. Columbus was the first European to be enchanted by the land where "the good and soft smell of flowers and trees was the sweetest thing in the world." The scientific names for the banana—*Musa paradisiaca* and *M. sapientum*—elicited this notion, the latter associated with the Tree of Knowledge of Good and Evil. The tree was God's dwelling place in sacred mythology: the cypress was sacred, the ash in Scandinavia represented the universe, the fig in India. Just as God spoke from trees in the Bible, He spoke to explorers in the forest. It was in the jungle that Man was closest to the Divine.

But Nature's exuberance could seduce the unwary. In the forests of Yucatán and Belize, there dwelt a lovely but sinister temptress called Xtabay who appeared to hunters who'd spent too much time in the bush.

They glimpsed her through the leaves and could not help themselves; they followed deeper into the forest as the twilight thickened, sometimes drawing so close that they could catch her wild scent or feel the lash of her hair. If they ever awoke, they were lost and disoriented. If they emerged from the forest, they spent the rest of their lives in ruin.

It was irresistible for a dreamer, which by all accounts Henry had become. Although his education was "indifferent," he possessed by his teenage years an intense love of art; when asked at school of his ambitions, he'd said he'd be an artist someday. At seventeen, he began a two-year class in art, and "his many drawings with pen and ink, supplemented by colour washes, reveal considerable ability and technical skill," said Edward Lane. This was not a passing fancy: In the 1871 census, when he was twenty-five and about to embark upon his greatest adventure, he identified himself as a "traveling artist," and throughout his travels he'd keep his sketchbook nearby.

But another spur drove him, too. He'd fallen far in status, and it grated on him. How could he rise in the world? He could not believe, as Lord Palmerston said, that Britain was a nation "in which every class of society accepts with cheerfulness that lot which Providence has assigned to it." He was not content with such orthodoxy. His family had fallen too far, too fast, and if there was no way to rise back in the Old World, maybe there was a way in the New.

Men remade themselves in the Americas, conquering the wilderness, carving out plantations the size of small empires. No other group in Latin American history possessed a glossier patina than planters; through every revolution, they remained at the pinnacle of society, projecting wealth, nobility, and power. Classical and medieval authors echoed the sentiment of Cicero's *De officiis*: "Of all the sources of income, the life of a farmer is the best, pleasantest, most profitable, and most befitting a gentleman."

On August 5, 1866, at age twenty, Henry sailed for the Mosquito Coast of Nicaragua.

It seems an odd destination. Edward Lane believed that Henry's decision was "typical of his generation, when the pioneering spirit, fired by a desire to take part in the development of the Empire, induced so many young men to voyage to the Americas, to Australia, and to Oceania." A studio photo made just before he left showed him already adopting the role of explorer; dressed in khaki jacket and pants, a pith helmet dangling from his right hand, he smiles to himself without the faintest touch of awkwardness. He'd grown into a tall, lean young man, just under six feet. According to Lane, "If anyone stated that he was six-feet tall, he would scrupulously remind them that his exact height was 5 ft. 11¾ in.!" He had jet-black hair, blue-gray eyes, a pencil-thin mustache, and an aquiline nose then called "Wellingtonian." He was a handsome fellow; family attested to his "unbounded energy" and "easy-going indolence."

In some ways, Nicaragua had advantages for a young man hoping to enter the ranks of writer-explorers. The Mosquito Coast was an overlooked stretch of sand and coral, jungle and marsh that fronted the Caribbean Sea. It dropped from Cape Gracias a Dios in the north to the San Juan River in Costa Rica, a four-hundred-mile ribbon of white surf dotted with small inlets and reefs and plagued by treacherous currents and shoals. Blewfields (now Bluefields), in the south, was the largest city and provisional capital. Everywhere else, river mouths and lagoons were plugged by shifting sandbars, and many harbors were announced by the hulk of a steamer rotting on a shoal. The land rose slightly from the coast, covered by dense jungle that continued thirty to sixty miles inland before opening to a broad savannah carpeted by a coarse, wiry grass. This climbed west until meeting the blue mountains of the interior.

Until recently, no one had really wanted the Mosquito Coast except the Indians who lived there. The Spanish preferred the more hospitable elevations of the Pacific side, although a few conquistadors had ventured into the area during the sixteenth century. In 1512, Diego de Nicuesa gave it a shot, but his expedition wrecked near the mouth of the Rio Coco, and later visitors grew discouraged by the inhospitable Indians, unforgiving terrain, torrential downpours, and maddening swarms of mosquitoes and flies. These were so bad, said one adventurer, "that neither Mouth, Nose, Eyes or any part of us was free of them; and whenever they could come at our Skin, they bit and stung us most intolerably." Thus the name of the coast, while others attributed its christening to the presence of the Miskito Indians, a subgroup of Suma, who had migrated from South America.

The informal British presence in Nicaragua mirrored its history throughout much of Latin America. Where actual possessions did not exist, a web of business interests and political meddling secured a foothold, the door kicked open by the 1824 fall of Spanish imperialism. In Peru, the British consul aided the quinine theft. Trinidad and British Guiana provoked conflict with Venezuela. British Honduras galled neighboring Guatemala and was seen as a pivot for British commercial and naval power. Nicaragua had its own British intrigue. From 1655 to 1850, Britain claimed a protectorate over the Miskito, but this was not aggressively pursued and existed primarily as a means to lay claim to the region's potential as a gateway to the Pacific Ocean.

Although Britain officially abandoned its Nicaraguan settlements in 1787, each Miskito king was educated in British schools and the Miskito royal family liked to think of "Mosquitia" as a province of the British Empire. When Henry arrived at Greytown on October 21, 1866, aboard the schooner *Johann,* he entered a land where British subjects were adored. The British in Latin America brought not only guns and money but such ideas as antislavery, capitalism, and the cult of the gentleman. The approach to Greytown was "prepossessing," Henry wrote. The surrounding hills were "crowned with umbrella-shaped trees of great size." *Heliconius galenthus,* a "handsome butterfly," fluttered up on the breeze and into his boat. He kept it as a souvenir.

Almost immediately, the *Johann*'s captain came within a hair of wrecking the ship on a sandbar. This harbor regularly took a mortal toll. During his fourth voyage, Columbus lost a boat's crew there at the mouth of the San Juan River. In 1872, the commander of a U.S. surveying expedition and six sailors drowned while trying to cross the bar. Large sharks swarmed about the river entrance; the crew's fate was known by their mangled remains.

Henry came ashore the next day. Visitors described Greytown as a neat little town of white-painted houses tucked among palm, breadfruit, and other tropical trees, but Henry was an impatient traveler and wished to be on his way. He saw Greytown as "altogether a very uninteresting place," surrounded by forests and water. It was called the second-wettest place on earth, right behind a Himalayan village that endured 128 inches of rain per year. Rainfall didn't pierce Henry's consciousness as much as the inhabitants did. Although friendly and talkative, the women were most un-Victorian, promenading at night with "cigars in their mouths, spitting in the most approved fashion."

Rather than the town, he was drawn to the forest. One morning he entered it for the first time. Those reared on temperate forests are rarely prepared for the tropical alternative. The tropical forest is a "great, wild, untidy, luxuriant hothouse," Darwin wrote, one in which both sound and silence pervaded. The insects are so loud in the evening that they can be heard from offshore vessels, yet inside the forest an absolute stillness reigns. A green murk prevails beneath the canopy, stabbed by canyons of light where the fall of a giant tree has ripped a rent in the gloom. One's shirt and trousers are soon soaked; everything is moist, the tips of leaves dripping from a hundred unseen places, sweat pooling under one's arms. An endless variety of trees sprouts at eye-level, their trunks furry with moss, wrapped in ropy lianas or girdled with spines. The smell of vegetable rot is omnipresent. Brush aside the forest litter, and a web of white threads appears just below the surface, a pallid tangle of tree rootlets and fungal mycelia that pierce everything organic when it drops, leaching out nutrients, transforming the world of the dead back into that of the living.

Henry, like most first-time visitors, was amazed by the insect life swarming around him. Butterflies were "numerous and beautiful, varying from the size of a bat, to that of our very smallest species." The woods were crossed by "beaten roads" of leaf-cutter ants, called wee-wees. Grasshoppers and katydids chirruped and "tinkled like bells" above the background chorus of frogs, some of which shrieked, while some made "a noise like a kettle-drum." Henry's English terrier, Jack, was "perfectly bewildered" by the din.

In such a world, it was easy to forget the one from which he'd sailed. Another cholera epidemic was raging in London, killing 5,600 residents in a matter of days. One can almost hear Henry's mother advising him to get out while he could.

Harriette Wickham looms as the unspoken presence behind this first trip, and for a practical reason: Henry was shooting birds. With a milliner mother, his market for feathers and skins was assured. This was no leisure trip: Henry was always on the lookout for bright plumage, severely disappointed when he missed his aim or ruined the skin. Around Greytown green parrots nuzzled under the foliage; toucans hopped among the branches; beautiful tanagers filled the edge of the forest, velvety-black except for one fiery-red patch above the tail. The trogon lurked near the columns of army ants, diving as insects tried to escape. The male was brightest, a beautiful bronze green on its back and neck, wings speckled white and black, carmine on the belly. There were red-and-yellow headed woodpeckers, jet-black curassows as big as turkeys, and olive-green motmots with their abnormally long tail feathers, naked at the tip. Nicaragua was a milliner's paradise: the demand for "concoctions of feathers, chopped and tortured into abnormal forms" was so great that by 1889 the Society for the Protection of Birds would be founded to combat the craze. The coincidence is too great: That Henry's first expedition would revolve around plumage and his milliner mother not be involved is hard to believe.

Five days after arriving at Greytown, Henry booked passage north to Blewfields aboard the Moravian schooner *Messenger of Peace*. As he passed the *Johann*, the captain waved his hat. "I was glad to see he did not harbour any remembrance of the little difference we had had during our outwards passage," Henry said. There is no other explanation of their "difference" besides a vague mention that he was required to pay duty twice due to a quartermaster's mistake, but it *is* the first sign of Henry's impatience with his fellow Westerner, an impatience that rarely extended to the natives.

If anything, Henry seemed to like the indigenous people he met more than his fellow colonials. The first example of this occurred soon after arriving in Blewfields. He stood on the jetty, gazing out to sea while awaiting his baggage, when "a slight little fellow, who was standing by, asked me my name." Henry answered "satisfactorily," then inquired the youth's in turn.

"William Henry Clarence," said the boy.

Henry secured lodging at a Moravian mission house, followed by his new friend. The Moravians were sprinkled all along the Mosquito Coast. As the first large-scale Protestant missionary movement to go to the world's enslaved and forgotten, they acted as an early Amnesty International, preferring education as a weapon against injustice and choosing indirect rather than direct confrontation with those in power. An apocryphal church story told of a Moravian farmer whose mule refused to plow: In desperation, the farmer finally said, "Brother mule, I cannot curse you. I cannot beat you. I cannot kill you. But I can sell you to a Methodist who can do all these things!"

On the Coast, this approach meant educating the Miskito royalty. "I see you've met our little chief," said a missionary, and when Henry glanced around, there stood William Henry Clarence, beaming at him. Without realizing it, he'd befriended the eleventh hereditary king of the Miskito Nation, and for the next week he'd be shadowed by miniature royalty as he explored the tropical town. "The little chief seemed to take a great fancy to me," Henry wrote, "generally accompanying me when I went on a stroll with my gun. He was about ten years of age, and appeared very intelligent. He lived at the mission-house, and was, I believe, well grounded in his studies."

It was neither easy nor healthy to be king. The first king, known only as Oldman, was taken to England by the Earl of Warwick in 1625 and presented to Charles I. He died in 1687 at a ripe old age, something few others achieved. Two successors died of smallpox, spread by the colonists; one died while attacking the Spanish in Yucatán in 1729. George II Frederic, the seventh king, was assassinated in 1801 by the friends of one of his twenty-two wives, whom he was said to have killed with particular barbarity. His successor, George Frederic Augustus I, was either strangled in 1824 by his wife and his body thrown into the sea or assassinated by a "Captain Peter Le Shaw." On the Mosquito Coast, the lessons of their Moravian tutors did not serve the royal family long, or well.

The "little chief" had been crowned six months before Henry's arrival, on May 23, 1866, after the natural death of his uncle, the tenth hereditary king. William Henry Clarence had been privately educated in Kingston, Jamaica, and according to Wickham's account seemed a happy, unpretentious boy. He would reign in Blewfields under a Council of Regency until he came of age in 1874, but even then he did not gain the power of his predecessors; his would be the first regency in which the Miskitos no longer had control of their fates, due to the 1860 Treaty of Managua. His reign as an adult would be brief: On May 5, 1879, after five years of court intrigue, the young chief who befriended Henry was poisoned and died at age twenty-three.

Despite the attention, Henry was anxious to be away. He hired three men and a large pitpan canoe, stocked it with food, powder, and items for barter, and at dawn on November 5, 1866, paddled across the lagoon and up what he called the Woolwá River. It is hard today to determine exactly which river Wickham ascended, partly because of the changed names of geographic features and partly because of his own confusion. He said they joined the river in the lagoon's northwest corner, but that is the location of the mouth of the Blewfields River, known today as the Escondido. The Escondido is a wide, straight waterway that cuts due west into the interior, nothing like the narrow, winding stream that Henry described. To the south lay a much smaller river; on most maps today, it remains unidentified, a narrow waterway that meanders through the trees, looping north, then west in a huge parabola—the same directions Henry logged. This would appear to be Wickham's river, first in a long string of evidence that suggests Henry never truly knew where he was.

Whatever his exact location, Henry started his journey by raiding someone's field of sugarcane. A hurricane had ravaged the shore the previous year, leveling the forest in every direction. The slash-and-burn cane plantations along the bank were abandoned, their owners ruined or killed. Henry spent the night in a deserted thatch house, dining on iguana and cassava. In the morning he shot and skinned several birds before heading into the interior. After three days of this, the canoe reached a place between high banks where the current was rapid. On the heights above them reared the communal lodges of the Indian village of Kissalala:

Bowing my head, I stepped across the little trench, and passed under the low-hanging thatch. I found myself in what appeared quite another world of manners and customs, which made a strange impression upon me, so totally different was everything that I now saw from all my previous experiences of life. Since that time, I have learnt to feel quite as much at home in an Indian lodge as in any other place.

Thus began Henry's sojourn among the Woolwá. For the next two months he used Kissalala as his base, paddling up the river to the next village or dropping down to unexplored tributaries. He "lay down and rose again with the sun," occasionally working into the night skinning birds by the faint light of his bull's-eye lantern. He settled in the lodge of the tribe's headman, Nash; when Nash left for Blewfields, Henry had the lodge to himself. "Left alone," he wrote, "I soon found that the life of a solitary traveler is not an idle one, for having to be at once master and man, renders his position no sinecure." He hunted birds in the morning, then returned to the lodge for his meal, usually rice boiled with a few drops of coconut butter and mixed with plantains, cassava, and the meat of whatever bird he had bagged. He talked with the Woolwá, who treated him like an amusing if somewhat hapless child. He was not a good naturalist. Many of the birds were "exceedingly difficult" to skin, and when shot, "the feathers usually fly off in a cloud." Those that were not hopelessly mangled were dried in the sun and carefully packed for the monthly mail from Blewfields to Liverpool. The day ended with a "strong cup of tea, brewed in the Australian fashion," straight from the pot like the coffee of American cowboys. He followed this with his pipe, "never a greater source of enjoyment than on such an occasion."

But the tropics have a way of testing those from temperate climes. Life begins to resemble a bizarre hazing ritual; God piles on discomfort to see what you are made of. There comes a point that is very much like a light switch: You either decide you can endure whatever the tropics dole out, or the discomfort grows maddening, and you must leave. That flip of the switch is always evident: The ones who can't stand any more shut themselves up in a permanent sulk or are always on edge, snapping at every irritation. Those who acclimate simply slow down. Enduring the sun is one such tipping point: As it beats down relentlessly, one either bakes or instinctively learns to seek out shade.

The other great test is the insect life, and that drives everyone crazy. "It was a long time before I became used to the ants, crickets, and cockroaches," Henry wrote, "whose crawling, scampering, and buzzing kept me awake for many a long hour." Large glistening cockroaches flew in his hair and entangled their legs: the only consolation was "knowing that they are easily killed." He gave up brushing his hair in the morning when he found that tapping the back of the brush over a fire "caused myriads of minute cockroaches to fall in showers from the hairs, where they had comfortably ensconced themselves during the night." The odor that impregnated the brush "was so unendurable that I had to content myself with only passing a comb through my hair."

Ants were the other insect plague that drove Henry to his limits. One morning he woke to find that a colony had deposited their eggs and larvae in the rolled blanket he used for a pillow. Army ants would march through the lodge, scouring it clean of cockroaches, tarantulas, and other pests. This was a convenience, but the idea of being caught napping by one of the swarms gave him the creeps, and he shivered at tales of the sick who were "much injured, as they lay helplessly in their hammocks, by these ferocious legions." One defense against them was the ant's "peculiar aversion to wet": When the Indians wanted a column to bypass their lodge, they sprayed "mouthfuls of water at the head of the column." Of greater danger was the inch-long *Paraponera clavata*, a glistening black giant sporting a massive hypodermic syringe at the end of its abdomen. The Woolwá called it a "fire-ant" due to the sting's effect: one alighted on Henry's shirt, and an Indian "filliped it off, with the remark that had it bitten me it would most probably have caused a severe fever."

By mid-November, the rainy season set in, the frequent cloudbursts sheeting the land and obliterating all other sound. After a half hour the rain would subside to a steady *drip drip drip*. Fires in the lodges blazed bright, and the Indians took their last meals of the day. Such evenings were lonely for him. The cry of the goatsucker, *Nyctidromus*, would drift across a patch of maize—"Who-are-you? who, who, who-are-you?"—and Henry would ponder that very question. Who was he, a fatherless child, alone in an alien world? What was he trying to prove? "I know of nothing so suggestive of reflection, tinged with a wholesome sadness," he wrote, as "to find oneself alone in a pathless wilderness, associating with a race utterly strange." The moon silvered the forest around him, and a crake rattled in the sedges at the water's edge. He felt face-to-face with the "Great First Cause of all"; the drum of frogs filled the early night, "but later on no sound, except the occasional hoot of an owl, broke the unusual stillness."

One fills silence with talk, and Henry learned the ways of his hosts to pass the time. The Mosquito Shore was populated by several related tribes that had pushed their way north over the centuries from the coast of Colombia into what today is Costa Rica, Honduras, and Nicaragua. They were a handsome people, with fine features, thick hair worn over their foreheads to their eyebrows, and a "warm, reddish-brown skin." The men's arms and chests were well developed but not their legs, probably because they paddled everywhere by canoe, Henry theorized. Nash, the headman, and Teribio, who had two wives and had paddled with Henry from Blewfields, both had a good command of English. Henry passed out tobacco to the village men, and the nightly conversations were long and leisurely. The Woolwá, or Suomoo, as they called themselves, seemed a martial people, and much of their talk was tales of war. The Spaniards of Nicaragua did not recognize the Miskito's claim to self-governance, and so there was anticipation of an invasion. They liked to talk about past engagements between the British Navy and the Spanish, and stories were still told about Horatio Nelson and the attack on Greytown.

The theme of such tales was obvious: Invaders would be repelled. Now, for the first time, Henry heard of the bloodshed accompanying the rubber trade. The Woolwá did not tap *caucho*, as they called it here. That was left to outsiders who understood and profited from the world's need for rubber. A few months before Henry's arrival, some Spaniards from Honduras had ascended the Rusewass, a tributary of the Woolwá, and built a number of thatch houses. They planned to tap *Castilla elastica*, the main source of latex in Nicaragua. These Hondurans were a hard bunch. In the forests south of Lake Nicaragua, they kidnapped Guatuso women and children to sell as household slaves. When the Woolwá demanded payment for the use of their land, a fight broke out; a tapper slashed an Indian with his machete, and that night the tribe returned to club the intruders to death, "leaving none alive to tell the tale."

But on the whole, Henry's hosts seemed a peaceful people, and his time among them slipped away. His attitude to other peoples was a peculiar mix of Victorian racism and an admiration that bordered on the protective. When he first met the Woolwá, he reacted in shock: All were quite naked except for a loincloth reaching from waist to mid-thigh, and of the women he only mentioned "their decidedly light apparel." Like many of his time, he preferred racial purity to "mixing." Unlike others, however, he did not place whites at the pinnacle of creation. Toward the half-black, half-Indian Creoles, he seemed to harbor no ill will, though he thought them a lower race; the mestizos of mixed Spanish blood were sinister or degenerate. But as he grew to know the Indians, he had only the highest praise. They exhibited a "scrupulous honesty" in all their dealings with him, and a praiseworthy etiquette in their relations with each other. Once, through his ignorance, Henry violated a tribal taboo for separation of the sexes: He surprised a woman alone as she washed pots. After the first shock, she recovered her presence of mind, "remembering probably that I was but a stranger from some distant land of barbarism, and therefore unaccustomed to polite society." Though spoken half in jest, he really did like these people and defended them often: "I am sure if some of those who condemn Indians as a lazy race had seen them at their work they would have revoked their judgment."

On November 25, his solitude was broken by the rancorous arrival of the trader Hercules Temple, sitting in a dory filled with men he'd hired to collect rubber. Since Temple was known to the Woolwá, his band was not in the same danger as the slain Hondurans. Each tapper was a Creole, descended from escaped slaves from the West Indies, and "Temple himself was nearly black, with crisp hair, like many of the Blewfields Creoles: he assured me that his mother was an Indian woman of the Toonga tribe." Henry began to realize the strange makeup of the world he'd entered, a world where entire populations were in flux. There were hereditary refugees like Temple and the Creoles; indigenous groups like the Miskitos and Woolwá; and intruders like the Hondurans, who came dreaming of quick riches and slaves. Each was to some extent what sociologist Everett Stonequist in the 1930s called a "marginal man," the individual who suddenly found himself "poised in psychological uncertainty between two or more social worlds." He balanced on a precipice—by leaving one life, he was unable to enter the other, and found himself "on the margin of each but a member of neither."

The New World was not only physical but psychological, and remote places like this were a testing ground for the future. Henry wondered if such different peoples could ever coexist, and if not, whether the jungle was wide and deep enough to hold them all. Temple and the Creoles, though friends with the Woolwá, were much different than their hosts. On the night of their arrival, they fiddled and joked until the early hours, "a contrast to the quiet of the Indian part of the encampment." The next morning, shortly after sunrise, "the whole of the population went off like a flock of birds, some up and some down the river, leaving the Blewfields trader, his son, and myself alone."

Temple was talkative, curious, self-assured, and sometimes dictatorial; if not a king on this forgotten river, he considered himself a merchant prince, secure in the knowledge that he was the Woolwás' sole source of machetes, cast iron pots, and tobacco, items that had started out as luxuries and evolved into necessities. He may also have brought another of civilization's byproducts—disease. Historians believe that malaria followed explorers and slave traders across the Atlantic: Temple's arrival from Blewfields may have duplicated this on a miniature scale, since the female anopheles mosquito, which carries the malaria plasmodium in her gut and passes it to a new host with every blood meal, is an expert stowaway. The egg-shaped parasite attacks the red blood cells; it heads straight for the liver or kidneys, and once it gets in the liver, malaria can linger for years. The bite is actually the most pleasant part of the experience. Three to eight days after being bitten, a victim starts vomiting, runs a fever of 102–104° F, sweats profusely, and shakes uncontrollably. The fever subsides for a day while the parasite changes to a nonpathogenic gamete, which allows it to multiply and spread. Soon the fever returns, and red blood cells are ruptured in huge quantities. At this point, an untreated victim can lapse into coma and die. Death depends upon the number of parasites. In severe cases, there can be as many as 80–100 per microscopic field.

Two weeks after Temple's arrival—about the time for the plasmodium's latency period to end—Henry began to feel the onset of a "slight feverishness and weakness." Every third evening, he was gripped by violent shivers as the sun dipped. He'd build a fire and sit over it, and "being thirsty without hunger, would brew a quantity of strong tea, drinking it as hot as possible," hoping to sweat the fever away. He'd spend the night rolled in his blanket and wake in the morning sweat-soaked and tired. He'd consume great quantities of sugarcane after each attack, this being the only thing for which he had an appetite. On Christmas Day, he lay on his raised bamboo bed and watched the termites "driving their covered ways along the supporting beams just above my head." He missed his home, his friends, his family. For the first time since arriving, he wanted to go back.

Temple had returned to Blewfields before Christmas, but before he left, Henry arranged for him to take a shipment of bird skins and then, on his return, "go with me into the interior." Henry planned to ascend the river to its source, then head north along the edge of the savannah until meeting the Patuca River and tracing it back to the coast. He'd stocked a wide assortment of beads, fishhooks, and knives for trading. There were rumors of never-chronicled tribes in this region, so maybe he'd be the first to discover them.

But when Temple returned on January 22, 1867, he brought something far deadlier than trading beads or the mail. He'd been delayed by an epidemic of cholera, which swept up the coast from the south in a particularly macabre way. In December, it had broken out aboard a river steamer belonging to Cornelius Vanderbilt's Accessory Transit Company, which plied the San Juan River. It hit with such fury that the captain beached the ship in one of the creeks, and everyone abandoned the vessel in fear.

Some Miskito men in Greytown heard of the disaster and boarded the ship for plunder, but before they made it home, one of them was seized by cholera and died. The others threw his body overboard and continued north to Blewfields Lagoon. When another man died, they put into a bluff overlooking the water to bury him. The job was done with such haste that his legs protruded from the grave, but "they continued on their journey north, no doubt sowing there the seeds of the harvest," Henry said. A boy from a nearby village went to the bluff to cut sugarcane, but when he arrived, he smelled the rotting corpse and went to investigate, thinking it might be someone's dead cow. Instead he saw the man's legs sticking from the ground like roots, and rushed back to his village to report his discovery. "As it was Christmas week, he went to a dance in the evening, the custom of these people being to go in a party from house to house, until they have danced in all the houses. . . . While still at one of these houses, he was taken ill, and died before morning." And so the disease spread through the villages, ultimately reaching Blewfields.

Word travels fast on the river, even as far as Kissalala. When Temple returned to Blewfields before Christmas, he'd taken with him a young village boy; now the boy's two sisters set out to fetch him in their family pitpan. The "sickness," as they called it, was feared in every little village up every remote tributary. It had killed thousands during the "filibusters," or invasions, of Nicaragua in 1855–60 by American soldier of fortune William Walker; it came with the ships from the white man's world. The two sisters arrived just as Temple was about to return to Kissalala, and they all traveled back to the village without any apparent harm. But on the afternoon of their return, one sister was hit with diarrhea and grew worse by evening. A kinsman rushed to Henry for medicine, but all he had was some essence of ginger, which he mixed with strong tea.

A grand *mishla* feast had been underway with guests from villages up and down the river when the sisters returned from Blewfields. Because of that, Henry grew truly afraid. *Mishla* was the principal fermented drink at these feasts, the cause of many drunken brawls and squabbles, and as a precaution the women usually hid all weapons until the effects turned to sleep and hangovers. But *mishla* was a far, far greater danger now. Its production was a communal event, a "disgusting process" as Henry called it, in which the women collected a large pile of cassava root and chewed, spitting the juice into a large earthen pot or jar. When their jaws got too tired to continue, they boiled the remaining roots and mixed it all together, stirring and skimming the pot and letting it stand for a day or two until it had fermented. When the sisters returned and still seemed healthy, they added their share to the pot. When the first sister fell sick, Henry went to the lodges and tried to persuade his friends to overturn the pots of *mishla*.

Now the futility of Henry's position became all too clear. He was an outsider, an amusing pet, acting in ways that seemed incomprehensible to the Woolwá, like shooting birds and sending off their skins. This sudden insistence in overturning the very reason for the feast was just another example of the white man's irrationality. What could Henry know of ancient ways? His growing anger and helplessness shows through his journal: He is a boy of four again, his father dying in the next room, and there is nothing he can do. For the first time since his stay in Kissalala, he grows so frustrated as to think of them as ignorant savages, but he likes these people, he has become their friend, and there is a moment's relief when his ginger tea seems to soothe the girl's pain. Maybe they will be reprieved; maybe *this* Paradise will not be destroyed. But that same night, the girl's family gave her another draught of *mishla,* and "just before dawn," Henry wrote, "I heard the crying of the women, by which I knew she was dead."

Panic set in. Each lodge threw fuel on their fires, believing that heavy smoke would act as a disinfectant. Guests from other villages loaded their canoes, "and I heard the rattle of their paddles while it was yet dark." At daybreak, the second sister began retching and, "creeping down the steep bank to the water's edge with great difficulty, with the aid of a staff, died there in about two hours." Another woman who'd been drinking from the same *mishla* pot began vomiting; *mishla* pots everywhere were dumped, but too late. Teribio, Henry's companion since his arrival, "looked quite pale and complained that he felt very sick." The panic so overtook them that flight was their only thought. Henry had already packed, and Temple and he jumped in a canoe and followed others north. Friends were screaming, cutting their hair in grief, staggering to the water's edge and retching; others scattered to their canoes, leaving the accursed place for good. As Henry looked back, he saw the body of the second sister draped across the sand. Above her wailed her mother and the younger brother whose innocent visit to Blewfields had brought death and ruin upon them all.

Henry, Hercules Temple, and the Woolwá paddled for their lives deep into the interior. The wildness seemed to calm them. They were absorbed into a primeval world of immense and towering trees joined together by a "wild, matted tangle of flowering vines, and by the multitude of other parasites, which blend the whole into one gorgeous mass of flowers and leaves." The profusion of greenery was so great that it overwhelmed one's senses. Henry made a rare aesthetic comparison and wondered whether Nature was not more pleasing if carefully tended as in England, where "our oaks, elms, and beeches stand out in individual completeness and beauty of form."

By January 28, they were five days upriver from Kissalala and no one else had died. The rest of the village split off up different little creeks; the community of Kissalala was no more. Wickham, Temple, Teribio, and two other Woolwá continued deep in-country, and that night they saw for the first time another inhabited Indian encampment, the first people they'd seen since their flight. For those five days, the desolation had been complete. "[A]t all the other places we passed the Indians had fled far up the little creeks at news of the 'sickness,' generally leaving at the mouth of the creek a wand, with a piece of white rag fluttering at the end, to indicate the direction they had taken."

For the next twelve days, they pressed north up the steadily narrowing river, spending more time portaging over falls and twisted, rocky rapids than paddling their pitpan. The world around them was silent except for the flight of startled bats, which shot like arrows from crevices in the riverbank, only to disappear into the shade of dark caverns. The abundance of game diminished. Temple complained about the lack of meat, and Henry suffered physically. He developed a ring of suppurating boils around his ankles, and the pain in his feet and legs made travel tormenting.

In Kaka, the last settlement on the river, they heard that the mining town of Consuelo lay beyond the forest in the nearby hills. On February 9, Henry limped with Temple's aid along a faint track up the side of a steep hill. They broke from the woods at the summit, and before them lay "a view of great extent and beauty: the plain beneath, diversified by hills of different elevation, stretched far away to the foot of the distant mountains." He'd made it to the savannah as he'd hoped, and near the opposite hill they passed through a narrow valley strewn with wooden shacks, workshops, and machinery on every side. Henry knocked on a door and asked if any Englishmen were around. The resident led them through the town, and "Temple and I saw enough to convince us that we were in a mining settlement of considerable importance." On the right, fifty yards from the road on a grass-covered slope, stood a whitewashed wooden house surrounded by a veranda and connected to a rear kitchen by a covered walkway. They limped to the house and their guide pointed to a man sitting in a hammock. Two women, looking pale and English, "were cooking at a stove what looked more like beefsteak than anything I had seen for a long time." Henry's mouth watered. He turned to the man in the hammock and tried to stand straight. "Are you English?" he inquired.

The man leaped up in shock. "I should rather think so!"

So ended Henry's first expedition to the New World. He made it no farther than the little gold mining town of Consuelo, part of the celebrated Chontales Mining Company. The pain from his sores was just too great to bear. His surprised host was Captain Hill, R.N., a transplanted Cornish miner, part of that wave of British investment that started in the 1820s with the collapse of Spanish mercantilism. When the new Latin republics opened their arms to Old World money, few investments were more enticing than gold. This hilly Santo Domingo country was loaded with the precious metal, running in parallel veins of auriferous quartz so numerous that in a mile-wide band a new vein could be found every fifty yards. The value of ore treated by Chontales averaged seven pennyweight per ton, enough that several company towns sprang up to stay for about fifty years. Even today, the Santo Domingo country retains a legacy of blue-eyed Latin children and that local delicacy, the Cornish *pasti*.

Captain Hill was both floored and delighted to shelter this unexpected Englishman. The large, square house was roomy and comfortable, commanding a view of the mine sections across the valley. The company doctor arrived, examined Henry's lesions, and pronounced him unfit to continue until they healed. Captain Hill "took me to his room," Henry said, "where a dinner of beefsteak and bread was already on the table, of which I was very glad to partake." The steak dinner made the prospect of ending his journey more palatable. He paid off his men and accepted the offer to stay. "I was not able therefore to take many birds among the surrounding hills or see as much of the country as I desired," Henry lamented, but considering the speed with which he settled into this new life, he did not seem overly disappointed.

He stayed at Consuelo a month and a half, until March 23, 1867, when he backtracked down the jungle river to Blewfields. His time in the mining camp was peaceful, strange, and sad. Captain Hill regaled his young guest with tales of his voyages in the South Pacific and the cannibal coasts of New Guinea while serving in the Royal Navy, information Henry stored away in his mind.

One evening, as the two compared adventures, a mine worker ran up panting and said that a man had been stabbed while gambling in the carpenter's workshop, a regular hangout on payday. They grabbed their revolvers and hurried off. The street was thronged with people, the workshop so crowded they could barely elbow their way to the workbench where the wounded man lay. "I recognized him at once," Henry wrote:

a tall, gaunt Spaniard whom I had seen in conversation with the native miners as they came from receiving their pay. . . . [A] single glance at his livid face was sufficient to show that his minutes were numbered; and so it proved, for before we left the shed he was a corpse. When we entered, the doctor was doing all he could for him; but though his wound between the ribs looked wonderfully small, and there was very little blood to be seen, the internal hæmorrhage must have been great, for he was very soon choked. I could not help thinking, whilst looking on the powerful frame before me, laid so low, and by so small a thing (a pocket clasp-knife, afterwards found in the shavings in the shop), how easily the "silver cord" is loosened. It seemed that none had witnessed the fatal blow, though the Cornish Captain, on hearing the disturbance, had gone in with his revolver to disperse the disputants in time to see the Spaniard fall. The next day, an officer, with some Nicaraguan soldiers, arrived, and made inquiries into the murder; in consequence of which about a dozen men, witnesses and petty offenders, were put into the stocks. They did not secure the murderer, who, of course, had made his escape into the bush.

Where previously his journal had been full of wonder, now it was tinged with mortality. Hercules Temple came back for him when his sores had healed, and they retraced their route down the Woolwá River. On April 2, "we passed Kissalala, which place presented a most desolate appearance. The thatch had already been partly blown off the houses, and the whole land was choked up with weeds." The villagers had never returned to their old home, but treated it as a cursed place. Where his Indian friends had seemed so vital and full of life, now his melancholy observations were laced with thoughts of "what might have been." He returned to Blewfields on April 5, and from then until mid-July visited Miskito villages around the lagoon while staying with his old friends the Moravians. On May 10, he watched a prayer meeting in the Miskito settlement of Haulover: "The missionary standing, book in hand, in the centre of the roughly-thatched hut, surrounded by a circle of dusky listeners, the men on one side and the women on the other, speaking to them pleadingly in the sonorous Moskito language." Outside, in the sunlight, girls peeped through the cracks between palmetto stems or drooped against the posts of doorways. It was a peaceful scene, but one drained of vigor. "Although, no doubt, in the old times the Moskito men were very superior in war . . . yet they do not appear to me at present to bear a very favourable comparison," he observed.

Disease and creeping corruption had been the unspoken theme of Henry's journey among the Woolwá, and now the picture was painted explicitly:

I was surprised to meet one day, near Temple's lodge, a handsome young Woolwa, who had been one of the crew of my *pit-pan* on the river. He had a heavy axe . . . and was engaged in cutting some logs of wood, to be used, I believe, in building Temple's new house. I was shocked to see how altered he had become; his skin, once as clear as bronze, was covered with rough blotches, the perspiration was running down in streams, and he seemed much exhausted. When with me he could not speak a word of Moskito; and I fear he must have had a hard time since then, for the Creoles are inclined to be tyrannical, and make perfect drudges of the Indians when they have the chance. Had I known that Temple would have brought him to Blewfields to make a servant of him, I should have seen that he returned to his home up among the rapids.

On July 12, 1867, Henry left Blewfields forever, departing on the *Messenger of Peace,* the same Moravian schooner on which he had arrived eight and a half months earlier. The captain was Temple's brother, and conversation centered on the threat of Nicaragua annexing the Shore:

[T]he captain loudly deplored the falling off of the warlike spirit of the Moskitos. They were once sole masters of the coast as far south as the San Blas Indians, who alone were able to withstand the onset of their dorys of war. But he expressed a hope that, if the hated Spaniards did come, they would again clean and sharpen their rusty old lances, and arise from the drunkenness caused by the villainous stuff sold them as rum by the traders, which, with the aid of their own *mishla,* caused such demoralization. . . . As we sailed along, he pointed out several places where these Indians had fought . . . and related how the king used to go in his large dory to take tribute of the Spaniards of Grey Town.

But all the old glories, conflicts, and empires had passed, turned instead into a parody of life and death, like the funeral of a Spanish child Henry watched on July 14 when he landed in Greytown:

It was a strange sight:—first came a lad with a spade, after him followed two men bearing between them the coffin, which was gaily painted and dressed with flowers; on one side walked the men, and on the other the women, some of whom were smoking cigars; behind, were men playing on fiddles and guitars; and lastly, a number of people, who were throwing crackers about, and amusing themselves in various ways. One of these combustibles fell amongst a flock of guinea-fowl, which seemed to cause much merriment for the company.

On July 15, he left, booking passage on the Royal West Indian Mail ship *Tamar* bound for Colón in Panama and then on to the island of St. Thomas. His description of Colón regains some of the romance he felt nine months earlier: "The mountains behind Porto Bello looked very beautiful: They were the deepest blue imaginable—here and there intercepted by dense rain-clouds and showers." The town was surrounded by mangrove swamps, and the Panama railroad crossing the isthmus ran through town, the small-gauge engines "racing to and fro." At the wharves, Indians sold canoeloads of plantains and seashells to visitors. It all seemed too postcard perfect, somehow unreal.

Then, again, death intruded. He heard that Captain Hill, his savior and host at Consuelo, had died in Colón while the *Tamar* lay in harbor. This occurred during Henry's final layup in Blewfields, when the old Cornish mine captain was returning to England to visit family. He'd been buried at a spot called Monkey Hill, a little distance out of town along the railway.

Before he left, he tried to find the grave:

> I walked along this line one day for some distance, and discovered that the first part runs through a dark mangrove swamp; as far as I went there was very little elevated land. The amount of human life sacrificed in laying this line must have been immense, in consequence of the workmen turning up the slimy deposit of ages under a fierce sun: they say . . . that one man died for every sleeper.

But he never found his old friend's grave. When he turned back, it was intensely hot, hotter than he'd ever experienced, and though his skin was tanned and dark, the back of his neck became scorched and blistered. Somehow it was all wrong, so different than what he'd expected. Nature had seemed transcendent at the beginning of his travels, a peaceful world in which man could flourish and prevail. But as he continued deeper into the wilderness, that world changed. Nature ran amok, its luxuriance a trap, a brightly colored cover for a hymn of strangulation. The tranquil dream of Eden turned to chaos without end.

CHAPTER 4

THE MORTAL RIVER

Henry's voyage to the New World had not been a rousing success. His packages of skins and feathers to his mother were not abundant; his drawings and journals were haphazard and confused. He even tried to buy a parakeet for his sister and failed. He came home weary and dispirited, with a lingering case of malaria and ugly scars on his ankles and shins. He made friends in Nicaragua and lost them; he'd seen the diminishment of a people he loved. He gazed at beauty and found death instead.

Given the tone of his closing journal passages, it is surprising that he chose to return to the tropics. Most young men would chalk the affair up to sad experience and announce their wanderlust permanently cured. Not Henry. There were times when he seemed dazzled, as if the tropics transcended reality, like an opium dream. The gateway to his journey had been Castries, the naval base and capital of the island of St. Lucia. The cliffs plunged from overhead into the ocean, and "the very rocks are robed in the deepest green." He chose to remember this, not the horrible deaths of the Woolwá sisters, the stabbing of the man in Consuelo, or his fruitless search for the grave of Captain Hill.

From autumn 1867 to autumn 1868, Henry stayed at home in Marylebone and probably helped his mother in her Sackville Street shop. He worked on his journals and sketches and probably tried, without success, to get them published. That in itself would have made it hard for him to forget the tropics, but popular art and entertainment also seemed smitten. In February 1868, the Crystal Palace opened a Great Show of

Singing and Talking Birds. Macaws, mynah birds, parakeets, and other winged tropical wonders were showcased in the newly built Tropical Department, "the most pleasing place for such a great exhibition."

Henry's choice for a second expedition was just as dangerous as Nicaragua, if not more so. When Venezuela gained her independence in 1830, she initiated a seventy-four-year conflict between liberals and conservatives, including the feudal landlords called *caudillos*, which resulted in the death of three hundred thousand combatants and seven hundred thousand civilians. Venezuela's civil strife was unusually virulent, even for South America. Operating in a sparsely populated country separated by huge geographical obstacles, the *caudillos* developed a rare skill in tearing the republic apart to increase their own power, and the creation of twenty federated states gave them the opportunity to do as they pleased. Now a new war had started. By 1871, another three thousand would be dead.

Henry's new route was more ambitious than the wide circular track he'd planned in Nicaragua. For the first leg of his trip, he proposed traveling the length of the Orinoco River, the massive waterway that stretches 1,500–1,700 miles and includes 436 tributaries, whose total basin covers 340,000–470,000 square miles. The river delta itself extends 165 miles, with 50 mouths opening to the sea. The Orinoco was the principal highway into wild country and had only been open for navigation since 1817. It was also the principal escape route for draftees avoiding military service and for blacks and Creoles fleeing slavery in the West Indies. But such freedom was precarious, and one observer commented that "all that many Creoles enjoyed along its banks was a gun, a hammock, a woman, and a fever." The wars of liberation left large tracts of land abandoned and uncultivated. Between 1810 and 1860, millions of people left Europe for new homes, but only 12,978 settled in Venezuela. This was a harsh land that few people wanted, but it *was* up for grabs.

When he left England in December 1868, Henry hoped to become the first Englishman to duplicate the route of German explorer Alexander von Humboldt and his companion Aimé Bonpland—and survive to tell his adventures. Von Humboldt was probably the greatest European explorer of South America of his age: he and Bonpland did nothing half-heartedly. When they set out for the Orinoco's source in February 1800, they began by walking across the long flat grassland, or llano, that stretched between Caracas and the River Apure, the Orinoco's largest tributary. From there they embarked on a long canoe journey, and by May 1800, accompanied by a Jesuit priest, had completed an overland portage to the Rio Negro, a major tributary of the Amazon. Near San Carlos de Rio Negro, on the disputed frontier between Brazil and Venezuela, they found the legendary Casiquiare "canal."

The Casiquiare is actually a two hundred-mile tributary of the upper Orinoco that also flows southwest through level marsh into the Rio Negro, forming a natural link between the Orinoco and Amazon that makes it the largest "bifurcation" on the planet. Until Humboldt's account, it was just another jungle legend. In 1639, the Jesuit priest Father Acuna reported rumors of the canal, and in 1744, another Jesuit, Father Roman, accompanied some Portuguese slave traders when they returned from Venezuela to Brazil via the Casiquiare. No other European explorer had plied it since von Humboldt, though not for lack of trying. Alfred Russel Wallace tried from the Brazilian side but fell victim to a bout of malaria that nearly killed him. Spruce had contemplated the ascent until he found himself in the armed standoff in San Carlos between Portuguese merchants and Indians. Henry intended to drop down to San Carlos via the Casiquiare, then float down the Rio Negro to the Amazon and from there to the Atlantic, a journey of approximately five thousand miles.

He began in typical fashion, with little preparation, carrying with him his fowling piece, trade goods, and sketchbook. There was no mention of a map or bottle of quinine: He'd depend on a mental sketch of the landscape and the good will of strangers. Historian Edward V. Lane believed "the main purpose of his journey was to study the rubber trade," but in truth he didn't mention rubber until he ran out of money and cast about for a quick source of funds. If there was a change in his method, it was that he took a companion, Rogers, "a young Englishman who accompanied me." Henry seemed to know very little about him: Lane called him a sailor, which suggests they met on the voyage over from England. Their journey would be a litany of hardship, and through it all, poor Rogers rarely got a kind word. Why Henry invited him remains a mystery.

Something had changed in Henry this second time around, and it wasn't always attractive. He was more impatient with others than in Nicaragua, at times downright disdainful, a trait that led one historian to dub him a "self-respecting British prig." Part of the change might lie in different circumstances. In Nicaragua, he'd been a tourist, happy to let new experiences wash over him, thinking perhaps his drawings of the tropics would bring some recognition. This time he was a seeker: His craving had blossomed during his year's wait in London, and he sought, like others before him, an unspecified El Dorado.

Henry was also more solidly "English" now than he'd been a year earlier. In Nicaragua he could easily criticize his homeland for abandoning the Mosquito Shore and their protection of his Indian friends. This time, on the way to St. Lucia, when they were delayed by the three-masted British warship *Royal Alfred* for a routine check of documents, he proclaimed, "What a difference there is in the appearance of the boat's crew from an English man-of-war, on a foreign station, to the sailors belonging to any other power." On board the *Tamar* with him was a recruiter for the 4th West Indian Regiment, "a very fine specimen of a West Indian soldier," Henry enthused.

Finding that no steamer would leave for the Orinoco for another three weeks, Henry spent two days securing passage on a smaller vessel. On January 11, 1869, Rogers and he hitched a ride with a boatload of smugglers. "Our little craft, about the size of a Margate lugger, was well manned; the crew were all excellent fellows in their way, although confirmed smugglers; indeed, the boat was afterward confiscated by the authorities." The Creole master from Trinidad had been to England and seen the Crystal Palace; a man from Guadeloupe steered, while the Trinidadian cook "had traversed the Spanish main ever since he was twelve years old." Pedro, a mestizo from the island of Margarita, wrote down the native names of local fish and birds: "One hardly expects to find such a pitch of education in fellows with shirts like the rags remaining to us as relics of the Waterloo standards!" They crossed a well-defined line "where the greener water of the sea is borne back by the yellow tide of the Orinoco." Sand spits around the mouth were thick with egrets and scarlet ibis. They penetrated the delta's narrow, mangrove-lined channels and were engulfed by clouds of mosquitoes and flies.

For two days they drifted up the labyrinthine channels of the delta, the sun broiling the inhabitants of the open boat, the still air sweltering and thick by midday. Sometimes they'd spot in a side channel the small canoes, or curiaras, of the indigenous Guarani, but they'd always disappear among the dense thickets, "paddling as for dear life," Henry said. The tribe's members were kidnapped and sold into slavery by the Spaniards, and smugglers like Henry's hosts were not above participating. These Indians were said to "possess the knowledge of an ointment that is obnoxious to mosquitoes, which cease to torment them after they have anointed their bodies with the valuable charm." After two days of this torment, they sailed into the main channel of the Orinoco and were driven upstream by a cooling coastal breeze. At sundown on January 22, nine days after entering the delta, they pulled into Ciudad Bolívar.

The chief city on the Orinoco, Ciudad Bolívar was picturesque, built on a low hill overlooking the river. It was flanked by a deep lagoon and the dry, sandy llanos, which stretched for miles in every direction. Simon Bolívar hoped the city, then called Angostura, would be *the* great port of South America. This didn't happen, but it did become the gateway to a frontier that in many ways resembled the American West. The llano was filled with cactus and tough, thorny hedges; cattle ranches stretched for thousands of miles. This prairie was not the rolling kind of the American West but a flat, treeless expanse covered in short brown grass, extending from the Orinoco delta to the spurs of the Andes, where the sun beat down with an average temperature of 90° F. Rumors of gold beyond the eastern hills brought streams of prospectors for supplies. Beyond the horizon, hostile Indians still threatened. The city was an oasis, a neat Spanish town of "rough-paved, but clean streets" dominated by a German merchant class who'd grown rich off local salt. In the harbor lay German, Dutch, and American ships that had come for salt and beef to trade in the Caribbean.

Ciudad Bolívar was also home to the governor, Antonio Della Costa, whose permission Henry needed to continue in-country. Henry and Rogers slung their hammocks in a large stone-flagged room in the town's only hotel, and the next day Henry went to the Government House, where he was graciously received. Della Costa was proud of his district: The gold near the hills of Utapa would bring riches like the California gold rush, he claimed. His nation might be in turmoil, but in this region there was law and order; there was opportunity. He gave Henry a signed letter allowing free passage up the river and wished him Godspeed.

Henry soon witnessed frontier justice in action. Prison labor maintained the roads and public works. Each morning, a chain gang straggled from the district jail under guard, "and a more villainous-looking collection of different types of men I think I never beheld:

> Among them was a low-class Frenchman, who, associated with a repulsive visaged Negro, had long been in the habit of robbing and murdering travelers on the road from the Caratel mines. On the quay, one day, some soldiers ran past, loading their old flint-lock muskets as they went, halting occasionally to level a shot at a man who had just left a *pulperia,* and was making up the road. The object of pursuit (but that day released from a term of imprisonment) had found his former sweetheart at the *pulperia* in the company of a rival, and overpowered by jealousy, had run her through with a machete. The miscreant was a tall Negro, who had been notorious as a bully among his fellow convicts; he was ultimately severely wounded and captured.

Obtaining the governor's passage was just the first step in leaving Ciudad Bolívar. Henry had to find a boat, and this proved difficult. Although Della Costa gave him an iron lifeboat salvaged from a wrecked river steamer, the boat was rusted through at the seams. For the first time, Henry's habitual lack of planning had consequence: The daily two dollar hotel charge ate into his meager savings, so the travelers moved their hammocks to the house of an American woman, "one of the last of the southern settlers who came two years before." She hailed from that peculiar and desperate species of American exile that Henry would find sprinkled throughout the tropics, a refugee group that would have such an impact on his destiny.

In the first four years following the end of the American Civil War, 8,000–10,000 former Confederates left the South and sought new homes in Latin America. Called the *confederados,* they thought they could escape the wreck and ruin that covered the former Confederacy from Texas to Virginia. Although the majority of Southerners stayed home, a few thousand sailed from southern ports to colonize land grants in the wilds of Latin America. The Confederate dream of spreading south to form a new slave society predated the Civil War. Its main spokesman was Matthew Fontaine Maury, one of the most renowned American scientists of his day, famous internationally for his achievements in astronomy, geography, and hydrography. Maury was superintendent of the U.S. Hydrographical Office and astronomer of the Naval Observatory in Washington; the fact that he was also a Southern expansionist gave the idea of colonizing the tropics in the name of the South an air of legitimacy. From 1849 to 1855, Maury wrote unceasingly about turning the great rivers of South America into the South's own slave colony. He was as enthusiastic as Clements Markham in England, painting Latin America as a land of unlimited resource ripe for the taking. Maury was, in fact, sounding the same imperialist trumpets as Britain, but with an overlay of slavery; his real focus, however, was the Amazon Valley. It was "a gold and diamond country" awaiting the cultivation of cotton, coffee, sugar, "and numerous other commercial agricultural products," but kindred colonizers used his arguments for the Orinoco, too. It only seemed logical, said Maury, that "the fingers of Manifest Destiny pointed southward as frequently as westward."

Maury was such a believer that in 1851–52 he used his influence to promote a scientific survey of the Amazon Valley by a relative, William Lewis Herndon, and Midshipman Lardner Gibbon, an officer assigned to accompany him. In letters to the two, he confided that, although the expedition's stated purpose was to expand "the sphere of human knowledge," they must remember that this was "merely incidental." The real object was to investigate the possibility of transporting a large portion of the South's slave population to the jungle and to gain land grants for future Southern planters.

Such schemes did not bear fruit until after the Civil War. The Confederate colony in Venezuela, chartered on February 5, 1866, was one of the first to start, and one of the first to fail. Called the Price Grant after the colony's leader, Dr. Henry M. Price, the land given to the ex-Confederates for settlement was a vast, 240,000-square-mile tract extending along the Orinoco's right bank to Colombia, then stretching to the common lands between Venezuela, British Guiana, and Brazil. The first fifty-one colonists arrived in Ciudad Bolívar on March 14, 1867, and were greeted by Governor Della Costa. They settled in Borbon, a small town twenty miles upriver, but they didn't stay long. Fourteen left to dig for gold 125 miles away, and all fell ill or died. Back in Borbon, the leader, Frederick A. Johnson, woke one morning to find himself alone. At noon the others appeared, said they'd found gold in a nearby bluff, and held forth a bucket filled with glistening black sand. Johnson took the sand to be assayed in Ciudad Bolívar. The test identified the sparkly stuff as mica, and the only use for the yawning hole outside Borbon was a grave in which to bury their dreams. That drained the heart from the settlers, and most went home. Johnson remained until April 12, 1867, when he left to recruit more colonists. His personal resources had dwindled to sixty-five cents, and as far as is known, he never returned.

Henry did not sympathize with his American hostess or her bitter plight. She "was not blessed with a particularly amiable temper," he said. In addition, she kept a raucous "stock of some dozen parrots in readiness for a Yankee skipper, who traded with New York." To escape the din, he and Rogers explored the surrounding llanos and made a practice of swimming in the river near town. The daily constitutional was a ritual that Henry prized: "I believe exercise is even more essential to health in a tropical than in a cold country," he said, a nod to the Victorian belief that regular habits prevented malaria, and so he took a dip every sunrise and sunset. It was also tempting fate: A local man warned them of the dangers of *tembladors,* or electric eels. "A shock from an eel would send a bather to the bottom without reprieve," Henry said.

La Condamine was the first to bring *Electrophorus electricus* to Europe's attention, but von Humboldt was the first to determine the nature of its deadly charge. When his guides drove a herd of mules and horses into a marsh where eels were known to rest, the disturbance caused them to deliver a shock. The frightened animals stampeded from the water, and the guides drove them back in. After several repetitions, during which a few horses died, the eels exhausted their batteries and could be flipped onto land with dry lengths of bamboo. Humboldt dissected the fish and discovered that its fibers of electrical generation were equal in weight to its muscular tissue. He was the first to detail the Hunter organs and Sacks bundles—but not before accidentally stepping on an eel. "I cannot remember ever having received a more terrible shock," he wrote, realizing he would have been killed if not for the sacrifice of the horses. As it was, he had violent pains in his knees and other joints that lasted the rest of the day.

Forgoing a bath was "a tremendous denial in such a climate," so Henry bathed with a calabash, or *tutuma,* in a shallow creek past town in the cool morning. During one of these dips "poor Rogers was stuck by a *raya,* [or] 'stingray,' whilst wading in the shallow water at the brink of the river, and suffered considerably; his leg swelled and he was rendered almost incapable of walking for some twenty-four hours." This was one of the twenty-five to thirty species of mottled freshwater *Potamotrygon* stingray, more feared on the Amazon and Orinoco than the piranha or eel. Fishermen were stung while spreading their nets. The large venomous sting at the base of the tail was known to kill children, and even adults if it hit a major vein or artery, while others told of affected limbs feeling numb for years. One can't help but feel that Henry was partly responsible: He'd been in the tropics before and should have been aware of its dangers. Since he described Rogers as a "young acquaintance," he probably was younger and less experienced than Henry. From this point on, Rogers's health began to deteriorate, and they pushed on before the young man recovered. After this, Henry grew increasingly irritated with his companion and seemed to have little sympathy for those who, for whatever reason, slowed him down.

Besides further delaying their leavetaking, Rogers's injury had additional consequences. They moved their quarters again to an even cheaper location, the house of an old Barbados woman named Mother Saidy "who was quite motherly to Rogers" in his pain. Mother Saidy seemed to realize that cockiness and spirit were not enough in this harsh environment. She looked out for all wayward young men, black and white, whose tumbling fortunes led them to her door. Such kindness extended to children. She had a weakness "for picking up, and caring for stray chicks of doubtful pedigree," Henry observed, apparently unaware that he was a "stray chick" too. Expatriate blacks from throughout the British West Indies rendezvoused at Mother Saidy's: "It was most amusing to see what pride they took in being British subjects, and the contempt in which they held their dark brothers" who were not so blessed, he said.

They lingered a month, a delay that grated on Henry's nerves. They'd steadily lost money, and when they finally cast off on February 22 in a "fast little native-built lancha" with a pilot named Ventura, Henry also felt they'd lost valuable time. The rains would come in April, and with them flood, fever, and disease. His impatience included himself and others. Something valuable *had* to come from this journey, but what could it be?

He may have caught a glimmer three days before leaving town. On that day, he considered tapping rubber for the first time. "I proposed exploring the Caura, a major tributary on the south bank which joined the Orinoco about 100 miles upstream from Ciudad Bolivar, in search of India-rubber." Nowhere in his journal did he seem swept away by the gold fever then consuming young men of all nations, but commercial plants of all stripes—that was a different story. At this point, any valuable species would do. As he ascended the river, he searched for the sarrapia tree, *Dipteryx odorata,* source of the tonka bean, a black-skinned, aromatic seed that smelled like vanilla and was used as its substitute in soaps, perfumes, and pipe tobacco. Antonio Della Costa may have alerted Henry to such possibilities, either he or an English trader named Derbyshire, for whom Ventura worked. The most likely source of his newfound commercial interest, however, was Mother Saidy, a walking encyclopedia of botanical lore. Rogers and Henry both would be nursed through virulent fevers by her knowledge of jungle medicine. She knew plants that were helpful and harmful, and where they could be found.

The voyage upriver seemed cursed from the beginning. The river was low; they constantly ran aground. The Orinoco's bends were compressed between high banks; the wind worked against them and the lancha was too bulky to steer. They abandoned it in the village of Maripe for more nimble canoes, one for each man and his gear. They planned to paddle up the nearby Caura River, but just before they left, Ventura got roaring drunk, so Henry stowed him away in the bottom of a canoe and pressed on. Forest surrounded them at the tributary's mouth. Cicadas sang in the trees, one sounding like a train whistle, another like "the jingling of little bells." Masses of cloud rolled up behind the forest, and their camp was swamped in a deluge. They paddled deeper up the river in search of better luck. The left bank was held by hostile Taparitos, the right by agricultural Arigua awed by their wild enemies. "There does not appear to be much actual fighting between them, however, as an alarm of their approach occasions a general bolt into the bush," Henry observed. "The Taparitos then content themselves with planting a few arrows in the deserted houses." Nevertheless, Henry carried his double-barreled shotgun and Rogers a Snider rifle wherever they went. When Henry tried to persuade several Arigua to search with him for rubber, no one volunteered. They continued on alone.

On April 2, fever arrived with the rainy season. Several species of the anopheles mosquito that can transmit malaria flit around the Amazon and Orinoco basins, each as ubiquitous as rain: *A. darlingii* breeds in swamps; *A. aquasalis* breeds in salt water along the coast; *A. cruzi* prefers the dirty pools of water in bromeliads. Whichever species attacked bit Henry first. That morning he'd gone into the forest to shoot some birds when "a feeling of giddy faintness came over me, accompanied by a disagreeable sensation of doubt as to whether I should be able to get back to the camp." He struggled on, "and succeeded in reaching that destination just before the fever obtained mastery over my limbs." He tried to tell himself that he'd succumbed to the "effluvia" of a tree that he examined, but by April 15, all three companions were touched by fever. They abandoned their quest and turned downstream.

It took nine days to return to Maripe, a journey they'd made upstream in six. They lashed the canoes together and drifted with the current. The sun beat upon them in the open boats, and Ventura alone had strength to steer. Sometimes they couldn't even do that but beached the canoes and lay in the water. They were in the same position now as Alfred Russel Wallace, who'd been attacked by fever on the Rio Negro in 1851: "I began taking doses of quinine and drinking plentifully cream of tartar water, though I was so weak and apathetic that at times I could hardly muster resolution to move myself to prepare them," Wallace recalled. "During two days and nights I hardly cared if we sank or swam. . . . I was constantly half-thinking, half-dreaming, of all my past life and future hopes, and that they were all doomed to end here." Henry wasn't even taking quinine, his only relief a wet towel wrapped around his head. After each spasm of fever, he'd drag himself "to the brink of the river and lay myself down in the rippling water."

They reached Maripe on April 24, only to find the lancha ruined. The hawser that made her fast had been stolen; she'd rolled on her side in the current, and her planks were warped and sprung. Henry was deathly ill, and his companions were probably as pitiful as he. Their only choice was to float back to Ciudad Bolívar in the canoes. On May 8, they arrived and were put up by Mother Saidy. The "good natured Barbados woman" nursed them back to health and probably saved their lives.

Henry had to face facts: All his plans had fallen through. He had no cash. His sailboat was ruined. His only possessions were the three canoes and what he could carry. A return to Trinidad would be an admission of failure, but he couldn't even do that, since he lacked sufficient funds for the fare.

While Henry mulled his dim future, he was approached by a young *confederado* named Watkins—and Watkins had a plan. He wanted to get out of this damned country, he said, and he intended to do so by ascending the cataracts of the Orinoco and gaining the Amazon via the Casiquiare canal. Any alternative seemed better to Henry than rotting in Ciudad Bolívar. "Watkins and I were in good condition," he said, a bit of wishful thinking since his fever never fully abated. Watkins "had seen much rough service in the late American war and in New Mexico," and had just arrived in town after walking across the llanos from the coast. Henry perked up: He was with a man again, not just that whiner Rogers. He agreed wholeheartedly to Watkins's project and "did not fear as to the result."

On August 6, three months after returning in tatters, Henry set out once more into the interior, this time with Rogers, Watkins, and an Indian named Ramón, a replacement for Ventura. The rain was at its highest, but all were overjoyed to finally be away. The governor had given Henry letters of safe passage to the governor of Amazonas, his opposite upriver. They were as ready as they'd ever be.

Four days into the journey, things again went awry. They'd only made it upriver about twenty-five miles, close to the first *confederado* settlement of Borbon, when "declaring himself unwell, [Watkins] took leave of the expedition, and returned to Bolivar." This was a blow for Henry: "As I had rather relied upon him, I was much disappointed." It was as if every plan was doomed from the onset. They entered Borbon in hopes of hiring someone to take Watkins's place, but no one could be tempted, and they moved on. By now, the woods were flooded. "At this, the height of the rainy season, little or no land is to be met with . . . for several days' journey," so they made fast beneath a huge floating tree trunk and ended their day with a meal of jerked meat and tea.

This upriver journey grew worse as they paddled deeper into the inundated world. Rogers's malaria returned full-force. By midday, all steering and paddling lay with Henry and Ramón. Massive tree trunks reared from the water. The only animal life they encountered were unforgiving insects and *perros de agua,* packs of huge river otters. When the strange beasts heard the stroke of their paddles, they "gave forth their peculiar mewing cry . . . or suddenly lifted their heads and shoulders out of the water, in order to reconnoiter us, at the same time displaying a goodly set of sharp white fangs." When Henry found land on a small island, the dank smell of earth was replaced by the musty odor of otter droppings and urine. Henry watched as the dark-brown, slickly furred heads, flattened on top, thrust from the water; the pack chattered at the intruders: "Uh! Uh! Uff! Uh!" Ramón imitated the chatter, and the otters surfaced beside them, six feet in length. Drops of water slid from their side-whiskers. They resembled murderous representatives of some Ministry of Madness. This was an alien world he'd entered, so different from Nicaragua, where life at least had some semblance to the known world.

One day they entered a stretch of forest dominated by *matapalo,* or "tree-killer," the local name for the strangler fig. Henry was reminded of a "vegetable Anagonda (*sic*):

> [O]nce it has embraced the trunk of a forest tree, it mounts higher and higher, till its glowing foliage mingles with, and then tops that of its

supporter: its supple limbs, now tightly compressed, flatten out, and gradually spread over the whole trunk of its victim, so enclosing it as to deprive it of life.

In actuality, the pressure exerted by the *matapalo's* aerial roots reduce the host's ability to move nutrients to and from its crown. In effect, it starves. The host dies off, and then the *matapalo* "stands self-supported, a great tree, bearing aloft a dark green dome."

Resting in its shade, Henry studied the parasite fig. If the tree were in fruit, raucous green parrots fed noisily in its upper limits. A troop of spider monkeys passed on their daily rounds. There was a wealth of life on its trunk, every niche occupied. Geckos and anolis lizards fought for territory across a no-man's-land of bark; army ants and *Polistes* paper wasps battled for nesting space in the large cavities. The meandering creatures of the forest floor climbed up the surface in the absence of dry ground. Land snails left silver mucus trails behind them; centipedes hunted constantly. Around the roots the men cast their hooks and caught a "caribee," no longer than a perch, whose sharp, powerful teeth protruded from its jaws and took off the top of Henry's finger when he spitted it for the grill. Another day Ramón gave a whoop as a cayman steered straight for the canoe prow where Rogers lay dreaming. There was a splash; the cayman moved on.

This pitiless river was like the *matapalo*, spreading over the visible world. It drowned men in its molten waters, snipped at their hopes like the crazed abominations that mewled and chattered at the edge of the gunnels, or tore off the tip of one's finger when by all rights it should be dead. This was the modern world; Henry was a citizen of one of the most prodigious empires the world had ever known. Yet out here that fact counted for nothing, as if Nature herself harbored an animus against civilization and took special pleasure in grinding her representatives into the loam.

By August 20, they'd passed the mouth of the Caura; by September 1, the Rio Cuchivero, and by September 4, the Apure, the huge tributary heading west into the plains. They rose from the water-soaked world to dry land. Sandy beaches lined the banks; villages dotted the shore. Their mackintoshes were ripped to pieces by the thorns through which they'd forced passage. All was rotten or damp, and they called a halt to dry their clothes. With land came a new plague: *zancudos,* or sand flies, and clouds of mosquitoes. But there were glories, too. One night as they camped on an island to escape insects, an immense flock of snowy egrets roosted close. "They were very noisy," Henry wrote, "as the sun went down in a glorious sky." As they approached the end of each day's journey, Henry would ask, "*Bostante lejos, Ramón?*"—Far enough, Ramón?

And Ramón, whom he trusted, grunted back, "*Lejos.*" Far enough.

He'd come far, indeed, and now he was about to step into the Abyss. On September 4, Henry's fever returned, his *calentura* as he called it, and that left him vulnerable to the most repulsive thing he'd ever encountered—the *zamora* vulture, the ubiquitous "turkey buzzard" that glides in great gyres over the sick and the dead. Henry was sick, all right, and he wrote, "When one is unwell, it is especially unpleasant to have a mob of these disgusting birds fluttering and croaking disagreeably in the trees you select for shelter and rest." In great enough numbers, they rendered a place unendurable with their odor. Sometimes he could no longer stand their presence and, grabbing his rifle, sent a bullet "at one more intolerably impertinent than the others." They were "half-weird . . . when skulking about a camp, dodging behind stones and bushes, or peering down from the boughs over-head." If he hit one, its death was sinister and absurd: "If a bird was struck whilst walking on the ground, he appeared simply to lie down suddenly on his side; there was no kicking; the ball drilled a hole through the body, and continued its way." Like an old man in a stained overcoat, it fell over and lay still.

With this second bout of malaria, things went a little mad. By September 5 and 6, the fever was so bad that Henry could no longer paddle. He fell into deep sleep, only to be jerked awake by a squall that threatened to drown them all. The night was "pitchy-dark," and he could only see Rogers and Ramón by the "rapid flashes of vivid lightning. Ramon looked really terrified and was yelling out something," which Henry could not hear above the tempest. Everything in the canoe was lost, drenched, or in shreds. They landed at a village, and an old Indian named Cumane was so alarmed by Henry's condition that he "wanted to take me home with him to be doctored by his women for a few days."

But Henry refused, insisting that they push upriver. When they left the next day, they passed Cumane's rancho high on the riverbank. Cumane called out, begging him to stay, but Henry refused, conceding to accept a drink of sugarcane juice before pressing on. Toward noon, the fever hit again in full force. He pulled to a shady spot, but the *calentura* felt as if it would burn him to cinders, and he tried to walk back to Cumane. "I managed to control my legs just long enough to stumble into the rancho, where there were two pretty Spanish-looking girls." He fell at their feet and they screamed for their father. "I did not remember anything until the fever lessened," he said.

He stayed with Cumane six days, until September 15, the fever recurring the same time each day. He drank hot fluids when in a sweat and cold when the fever broke. The girls fed him soup and a decoction of a medicinal plant he called *frigosa*. This was apparently the *mimosa frigosa*, or *mimosa pudica*, the so-called sensitive plant native to South America that, when touched, closed up on itself. "Stay longer, do not be in a hurry," Cumane urged, pointing out the richness of the soil of his farm, the abundance of game in the forest and manatee in the river. "All of this, but I am old," hinting also that he'd saved much silver. Henry realized what he suggested: A young fellow like him might take a fancy to one of his daughters and stay with him forever, working and inheriting this corner of Eden.

The next day Henry said good-bye. In two days, his band reached Urubana, sheltered beneath a rocky hill overgrown with trees. Several rocks had faded inscriptions etched upon their face; they were entering a land of cultures that were ancient or dead. Secure at the dock was a lancha large enough to house twenty-four men and women under the palm-thatched *toldo* (awning) built on-deck. It belonged to the Governor of Amazonas, the state beyond the cataracts, and was bound for the Rio Negro. Henry arranged with the captain to take him and the supplies beyond the falls, while Ramón and Rogers took the canoe on ahead.

The land took on an ominous air. The falls at Atures and Maipures denote a major declination between the llanos of northern Venezuela and the forests of the south. The llanos were the home of over two million cattle, the feudal ranchers, and traditional Spanish culture as represented by Acting Governor Della Costa; the forests were unknown, and outposts of civilization seemed fewer and more fragile. On September 23, they crossed the mouth of the Meta River, flowing into the Orinoco from the west; in the grasslands to the horizon "myriads of fireflies sparkled like gems low over the surface, seeming to give an undulating motion to the misty plain." But as they drew closer to the Atures and its brutal three-mile portage, the river around them grew brutal, too. On the 25th, they made a landing at which the vultures rustled around them in a great flock. An oarsman baited a fishhook and caught one, and what happened next was very strange. The man plucked out two of the vulture's long tail feathers and thrust them through the bird's nostrils in its beak, forming "a pair of ferocious moustaches."

> Released with a kick, the unfortunate bird . . . endeavoured to regain the mob of his companions, which had been gravely looking on; but the more he tried to do so, the more they edged away from him, and at last took themselves off altogether.

Part of the brutality was inspired by fear. This was the haunt of the "dreaded Guahibos [or "Guaharibos"], who appear to be perfect Ishmaelites, whose hand is against every man." Today the tribe is known as the Yanomami, dubbed the most violent people on earth. Now they've been pushed back into the upper Orinoco and the forest around the Siapa, a tributary of the Casiquiare canal, but in Henry's day their range extended to the cataracts, and Europeans and local Indians alike were terrified of them. They were fair-skinned and green-eyed, and some anthropologists think their ancestors were the first Indians to reach South America from the north, linking them to the warlike island Carib. Although they cultivated plantain, they were primarily hunter-gatherers, and there was not always enough food to go around. During lean times they killed newborn girls, which created a vicious cycle—since there were never enough women, the men incessantly fought for them. Within the tribes, these fights were ritualized—combatants hit each other over the head with ten-foot poles. But between tribes, they raided for women, killing rivals with six-foot arrows tipped with curare. They had no concept of natural death, attributing it instead to black magic—which had to be avenged.

Henry heard them described as a kind of human army ant, which nothing could stop. They spread out on march, their sole purpose to pillage, building bridges of palm leaves when they swarmed across a stream. In 1853, Richard Spruce suggested they had good reason for ferocity, considering the depredations of the slavers:

Shortly after the separation of Venezuela from the mother country . . . the Commandant of San Fernando was sent with a considerable body of armed men to endeavour to open amicable relations with the Guaharibos. He reached the Raudal de los Guaharibos with his little fleet of fifteen piragoas, and as the river was full, the whole of them might have passed the raudal, but it was not considered necessary, and his own piragoa alone was dragged up, the rest being left below to await their return. A very little way above they encountered a large encampment of Guaharibos, by whom they were received amicably, in return for which they rose on the Indians by night, killed as many of the men as they could, and carried off the children. Treatment such as this of course, is calculated to confirm, and perhaps it was the original cause, of the hostility of these Indians. . . .

Henry soon witnessed such casual enslavement first hand. On September 27, they reached the deserted pueblo of Atures, an abandoned Spanish mission over which hung "an unpleasant air of mortality." The ancient footpath connecting the Lower and Upper Orinoco passed through this ruin. The last four inhabited houses surrounded an overgrown square, their sides fallen in and rafters exposed. The villagers reminded Henry of London chimney sweeps, "their faces being covered with black spots, that are left after the attack of the mosquitoes." The bones of the tribe, for which the town was named, were interred in a nearby cave. Once stored in large baskets, or *mapiri,* the bones lay scattered on the floor, along with a horse's skull. Henry gazed at the plains below, stretching out of sight, and noted the silence. "Whilst I gazed into the tomb," he wrote, "a beautiful little humming-bird flitted by, and hovered over the white bed of bones."

That night, their journey took an even coarser turn when they met

"Señor Castro," the Governor of Amazonas. The governor's rum flowed freely as he waited for his lancha to take him above the falls. Soon everyone in Atures was drunk, singing his praises in inebriated falsettos. Castro sang back, his own voice shrill. The festivities continued into the next night, even after they left Atures, and annoyed Henry as much as anything he'd encountered. "It is singular," he groused, "that these people endeavour to render their voices as ridiculously effeminate as possible whenever they attempt to sing!"

The besotted revelry continued, around the Upper Falls at Maipures, through the nearby pueblo and beyond. Whether from boredom, fear, or discomfort, Castro's behavior assumed a desperate tone. "Drunken carousels continued without intermission," Henry wrote. "Wishing . . . to make a forced march, El Governador plied his men with so much rum" that three paddles were lost, and they returned to Maipures for more. "Castro at length reduced himself to such a pitch of nervous . . . phrenzy, that I thought it advisable to give him an opiate, which had the desired effect."

The next morning, they tried to leave at sunup, but the men were hung over. The governor exploded. "I was sorry to see Castro bend his bright toledo in thrashing the first offender that appeared," Henry said.

They rowed through groves of giant thorny palms, past the mouth of the Vichada River, on to San Fernando de Atabapo, the last sizable outpost of civilization, situated at the confluence of three tributaries where the Orinoco veered east into the forested mountains. There was no stop to rest. It was the first wakeful night for the rowers, and Castro fueled them from a demijohn of Málaga by his side. "Row!" he screamed. They dropped asleep in midstroke, still keeping time to the chants, though sometimes "the stroke might be taken in air." One man dropped his paddle and kept rowing until he woke from his coma and looked "round with a stupid grin."

A canoe filled with cassava appeared in the river; a "good-looking matron" sat beside the produce while her two sons rowed. Castro pulled aside to buy cassava, but suddenly pressed the two boys into service as oarsmen and stole the cassava for his own. The mother began to cry, but to no avail—she was left in the curiara alone.

Later in the morning, Castro ordered a stop for breakfast. When they put to shore, the boys disappeared into the trees. "It is a wonder that these simple people do not even more seclude themselves in their mountain forests," Henry concluded. Whole tribes would disappear like the boys: "No one had seen them go, but they were nowhere to be found."

On the morning of October 9, the lancha rounded a bend and, nestled in a grove of cocoa palms, lay the thatched roofs of San Fernando de Atabapo. The air was heavy and close. Without warning, the sky turned overcast, and a torrent came down. Castro was home and installed Henry in the empty house next to his. The door and windows were riddled with bullets, evidence that even here, in what the Venezuelans called *Ravo de Venezuela*, the tail of the country, the endless faction fights could erupt at any time.

Henry stayed in San Fernando for two and a half weeks, reduced to debility by the heavy atmosphere and a constant diet of "fish, fish, fish!" he said. He longed to return to his tiny curiara on a breezy river, potting with his rifle the contents of a simmering stew. Birds flitted across his window— a lively finch of a deep russet color and sky-blue finches that perched in guava bushes and orange trees. He paid them little mind. He was in a funk, a deep, incapacitating depression that had not appeared in his journals but would be repeated in his history. He'd considered dropping south along the Rio Atabapo until reaching the nine-mile Pissuchan portage to the Rio Negro, but what would that accomplish? He wouldn't duplicate von Humboldt's route; he wouldn't do anything but survive. He was too broke to pay the passage back to London; his clothes were in tatters, his future more so. In the South Pacific, he'd be labeled an "island bum." Here there weren't even balmy beach breezes to cool him or tropical women for consolation. A bad case of malaria was the only thing he could call his own.

Then another stranger arrived, like Watkins in Ciudad Bolívar, this one bearing fantastic tales. Although Henry never seemed tempted by mineral wealth, we now see the buds of a lifelong pattern: He'd rush headlong into any scheme that might turn him into a planter, with all the prestige and power that implied. A "Venezuelan Spaniard" named Andreas Level, a young friend of the governor, regaled Henry with sagas of the Upper Orinoco. Level had traveled three days above the supposedly impassable cataracts of Maguaca. He'd seen Indians with skin as white as Wickham's, and red hair. He'd married the daughter of a local chief and hoped to corner the market in balsam and Indian hammocks, which he'd sell to export houses at Ciudad Bolívar and on the Amazon. He'd start a trading empire in this corner of the jungle. He'd grow rich before his thirtieth birthday.

But more than anything else, Level described rubber, miles of untapped rubber trees growing along the banks of the Upper Orinoco between the old Spanish mission of Santa Barbara and Esmeralda, deep in Yanomami country. The Indians had tapped trees for centuries, but they were lazy, not like a European, a man of daring whose vision stretched beyond this fevered river to the commercial heart of the world.

Henry was sold. On October 24, Ramón and he left to go scouting, leaving Rogers behind. "The air was heavy with the odour of the flowers of the water-loving *gica* tree, when the sun rose over the rolling bank of mist," he recalled. They headed east toward the Serra Yapacini, the mountain range that lay across the river like an immense blue bar. They passed the ruined mission of Santa Barbara, overgrown with guava, a lonely wooden cross standing at the water's edge.

By November 3, the river curved into the mountains' shadow, splitting into the Ventuari, which angled north, and the Orinoco, which angled southeast toward Brazil. Several people were already at work in the area, collecting rubber, or *goma,* as it was locally called. This was not the *Hevea* of the Amazon basin, but *Siphonia elastica,* a similar tree that produced latex of an inferior grade. Two days upstream they found the camp of a "Señor Hernandez," a white Creole from the coast who'd fled to the forest to avoid the wars. He was glad to see a friendly face and showed Henry how to tap the rubber trees that dotted the island where he made his camp. The trickle of latex was slow, he said, apologetic, but he lived in hope of better things when the dry season set in.

Henry was convinced, and they headed back to San Fernando for supplies. They arrived on November 12, at noon. The town was deserted: "We found that the whole of the inhabitants had been seized by a kind of mania for *goma*, and were gone *al monto* in search of it," he said. A "rubber madness" had them in its grip: "The idea appeared to have struck them that this really must be a good thing, if an Englishman, like myself, coming from so far, desired to go in for it."

They struck out ten days later, on the morning of November 22. Joining Henry, Ramón, and Rogers were two boys they hired in San Fernando. Manuel was very bright, "somewhat approximating in character to a London street boy," intelligent in a canny, watchful way. Henry made him a kind of personal servant, but like street urchins, he was "given to pilfering." Narciso was twice the size of his companion, but "decidedly stupid; he did not seem to comprehend Spanish very well, so it was *trabajoso* [difficult] with him." Along the way they picked up two others: Mateo, "a queer, wizened-looking old fellow, who, whenever he glanced at me, assumed a most insinuating grin, making me feel as if my own features were involuntarily taking the same expression," and Benacio, "a stolid old man" whose most distinguishing characteristic was the way he took his time in getting anywhere. Henry dubbed Ramón the headman, since he had tapped rubber before; he did this despite Rogers, who by now had taken a dislike to Ramón.

They went past the meeting of the Orinoco and the Ventuari, where Señor Hernandez had been joined by friends from San Fernando, deep into the forest until, after five days of paddling, they reached a tributary called the Caricia, or Chirari, a stream so small it rarely shows on maps. This is difficult country, a landscape alternating so suddenly between choked forest and swamp that travel by foot is almost impossible. There were few footpaths, and most travel was by curiara. Birds and insects dominated the wildlife. Large squawking swarms of parrots and macaws glided over the treetops, while clouds of mosquitoes explained why no other rubber tappers were near.

But rubber trees were everywhere. On the first day, he let Mateo and Benacio off on a small dry spit, and soon they returned with fifty-seven notches carved on their counting sticks—that many trees in a small space of land. Farther upriver he found an island dominated by rubber; two streams within sight pierced deep into the shadows, and they found more rubber trees. Henry's depression seemed ages in the past; he'd found the cure he wanted, and it showed by his plans. On December 1, he built his rancho on a bluff overlooking dark water and settled his workers on tiny creeks at various rubber stands. He'd collect rubber in a triangle formed by his creek and the Orinoco, clearing paths between the trees with machetes while Rogers went for supplies. He estimated that by the start of the dry season he'd have a thousand trees for tapping—a grandiose plan for a beginner, but one he didn't consider unrealistic. He was cutting a plantation out of the jungle, a tiny empire "in this little creek in the very core of the continent."

❋

Henry's typical day as a rubber tapper began at 5:30 or 6:00 A.M. He ate a frugal breakfast—black coffee and a handful of *chibéh*, or farina with cold water—then strolled into the forest with his gun, shot pouch, powder flask, and machete at his side. He followed the path for two or three hours as it meandered between the rubber trees, always alert for any dangerous snake or unexpected game. He stopped before a rubber tree and seized one of the small tin cups piled at its foot. He ran his finger around the rim to clean out dirt, and with a small curved *faca*, or rubber knife, he examined the trunk of the tree. He selected a spot above the last day's cutting and with the blade of the *faca* struck one sharp blow sideways at a rising angle, letting the blade almost bite into the cambium, or formative layer of the tree. This is the real art of the tapper: Too shallow, and he doesn't reach the latex; too deep, he injures the tree. After this stroke, he stuck the tin cup into the bark under the lower angle of the cut so that the slowly exuding latex would ooze down into it. He made three or four such gashes in the tree, and if the tree were newly tapped, as many as five to seven per day, a lattice of cuts that extended 10 to 12 feet high. Greedy *seringueiros* made more cuts; the prudent tapper rarely exceeded five.

What exactly had he done? He'd severed several bundles of vascular ducts that carried the latex through the tree's outer cell layers, causing it to bleed. His little ax had cut a gash of about one and a half inches in length and no more than three eighths of an inch in width. If he exceeded these dimensions or cut deeper into the trunk he'd destroy his livelihood,

Ramón explained. The wood bled for three or four hours. It was very much like dragging a razor blade across the skin. Severed blood vessels bleed freely, but then begin to heal, the edges of the wound meeting and closing, leaving only a slight scar. Cut too deep and infection gains a foothold; the surrounding tissue becomes diseased. So with a tree. The *faca's* gash is often attacked by ants and other insects, and by fungus: Cut too deeply and rot sets in. Some tappers twist the blade of the ax to increase the flow of the milk, not knowing or caring that by doing so the vessels are widely separated, thus creating a permanent wound. Treat the tree with respect and the riches keep flowing, Ramón preached. One must be a good husbandman and give his trees time to heal.

On a good day an experienced tapper could visit 150–200 trees during his morning round. This was impressive enough, considering the torturous route of the *estrada,* the path, doubling back on itself, crossing fallen logs over gaps in the hillside or stream. By noon, he'd finish tapping and return to his rancho for the *balde,* or hollow gourd, used to collect latex. He repeated his morning route, emptying each tin cup of white milk into the *balde,* then placing it bottom up in the pile beneath the tree. A tin cup filled with two ounces of latex was a good average per tree: Those trees that were young and healthy could yield three to four ounces, while those that were old or abused might yield less than one.

Henry ate lunch—usually a repeat of his breakfast—then devoted the afternoon to smoking the latex over a fire. Every tapper had a *defumador,* or rubber-smoking hut, near his rancho. The best fuel for smoking rubber was old nutshells scattered at the base of the *cucurito* palm. When the smoke rose steadily, Henry settled down for the most tedious part of his day. With a wooden paddle alongside the blaze and the basin containing latex at his side, he dipped the paddle into the milk and twirled it over the smoking fire. In a few seconds, the latex coagulated and turned the color of cream. After a dozen turns of the paddle, he'd created a thin layer of cured rubber. Again he dipped the paddle, repeating the performance until all the milk was gone. He might attain a thickness of one or two inches, depending upon the amount of latex collected that day. It might not look like much, but the operation took at least three hours and required some 1,200–1,500 movements of his hands and arms. When all the milk was smoked, the fire was extinguished and the paddle supported on sticks so that the blackened, pliable ball of rubber retained its shape as it hardened and was not pulled out of shape by gravity.

Henry sometimes felt "shut out from the rest of the world." He'd pass through the mouth of the creek and all trace of humanity was lost, even those in his band. It was silent as a grave. A chief feature of the forest around him was the corded vines, or *bejucas,* that bound the forest together—they tied the canopy into bundles, wound around the trunks, and coiled upon the ground. Cutting through them wore him out, but that exhaustion gave him peace. In his rancho at night, he would "watch the cold shadows of night gradually creep up from the water on the opposite side of the creek, and when the topmost boughs of the forest trees were alone tipped with golden light, I had the fires built for supper."

Psychologists have suggested that *place,* as a force, can impress itself upon the psyche and serve as a crucible for creating a sense of the holy. They speak of "peak" and "flow" sensations, where the individual experiences a loss of self; where the normal distinctions between subject and object, "I" and "everything else," break down. When that happens, the observer becomes immersed in the present: He or she *transcends.* This is when people feel the touch of God—the connection to all things.

Now Henry felt touched by that hand. "Sometimes, during the time for rest, I would sit down and look up into the leafy arches above and, as I gazed, become lost in the wonderful beauty of that upper system—a world of life complete within itself." Above him existed a world of strangely plumaged birds and the "elvish little *ti-ti* monkeys, which never descend to the dark, damp, soil throughout their lives, but sing and gambol in the aerial gardens of dainty ferns and sweet-smelling orchids."

He felt heavy and earthbound, while all above him was light. "All above overhead seemed the very exuberance of animal and vegetable existence," he wrote, "and below, its contrast—decay and darkness."

INSTRUMENTS OF THE ELASTIC GOD

Darkness and decay. Henry dwelled on these words as the New Year passed and his body began to rot. The first hint that he'd be smart to leave his thousand-tree utopia were the infected mosquito bites covering his hands and legs. Unable to control his scratching, he ripped apart his skin. "The constant irritation," he complained, "caused my hands and feet to swell." The sores grew so inflamed that his knuckles and the backs of his hands became red, runny patches. "My feet especially were so inflamed, that I was confined to my hammock for some days, whilst Ramon and the two boys were putting up the lodge." He was stung by a scorpion, but he'd learned to rate degrees of torment: It was "not so painful as I had anticipated . . . [and] the smarting and accompanying feeling of numbness was not so great as that caused by the sting of a forest wasp." His camp was raided by a giant vulture, *uruba-tinga,* which bore off a large piece of salted fish when Henry returned from the forest. Their ammo was running low, and they were all slowly starving, so he couldn't waste shells on anything but food. His only consolation was to imagine "the pangs of thirst he will suffer after such a gorge of salt fish."

Then Henry became another creature's home and meal. The culprit was the human botfly, *Dermatobia hominis.* Flies were a plague. First he bemoaned the fact that uncovered troughs of liquid rubber would fill in an instant with "self-immolated blue bottles," ruining the day's effort. Meat was honeycombed within hours with masses of eggs. On January 2, 1870, a maggot fed on him. It started as a simple itch, the same as a mosquito bite, but a small mound formed on his skin. Other mounds rose, like volcanoes: "The first time I felt them, I could not imagine what on earth was the matter with me: it seemed as if some one was making a succession of thrusts into my side with a red-hot needle." Ramón checked his back where Henry couldn't see and gave his verdict: Several large-headed worms were wriggling in his skin.

"Edible," Ambrose Bierce would write eleven years later in his *Devil's Dictionary*: "Good to eat, and wholesome to digest, as a worm to a toad, a toad to a snake, a snake to a pig, a pig to a man, and a man to a worm." Once again, mosquitoes caused his woe. Botflies are too large and slow to parasitize targets like monkeys and men, so they've opted for stealth: The female fly captures a female mosquito, glues an egg to its hairs, and then releases it. When one such tagged mosquito landed on Henry, his body heat triggered the egg to hatch; the tiny larva fell from its original carrier and burrowed into the bigger and more palatable Henry.

Ramón delivered the disgusting truth: His back was home to several botfly larvae, all happily wiggling in their pustules. The larva has evolved two anal hooks to hold it firmly in place; pull it out and the maggot bursts, filling the cavity with toxins and loosing an infection more dangerous than the original larva. It breathes through snorkel-like spiracles poking from the skin. Henry and a botfly could have coexisted peacefully, except for when it started feeding, and that's when the pain began. There were two cures, Ramón said. The first was the "meat cure," sandwiching a piece of raw, soft steak over the spiracle tightly enough that the botfly was forced to burrow up for air. But they were low on steak, so Henry chose the painful alternative: killing the maggot with a dollop of latex or tobacco juice, after which Ramón sliced out each corpse with a sterilized blade.

By January 8, Henry had tapped one hundred trees, learning the *seringueiros'* secrets by trial and error. Never again did he speak of tapping a thousand trees; the latex yield was small, the smoked rubber polluted and disappointing. The trees were loaded with green fruit; some trees were too young to tap, others between seasons. Ramón and the others were having the same bad luck; the latex flow was weak, the trees green.

There'd be no quick riches from this triangular plot of *goma*. Their disappointment was made worse by the sickness creeping over each man. Mateo and Benacio were pale and weak, Rogers so frail from fever that he could barely steer the curiara down the river for supplies. Ramón had a game leg that swelled so badly that he couldn't leave his hammock. Narciso refused to work and ran into the forest when Henry caught him loafing. Manuel turned to full-scale thievery. The only man healthy and honest was not even one of the party, a young Spanish Creole named Rojas Gil, who, with his two young wives, had followed Henry from San Fernando and built a rancho up the river, around the bend. At first, Henry resented the intrusion, but Gil proved a lifesaver. What little cassava they procured was shared among all; what little fresh fish they trapped was parceled out. Rojas shot a sloth one morning as it swam across the stream. The meat was badly cooked but delightful, since at least it was a change.

On February 8, the recurrent fever Henry had dreaded so long struck again, without warning, flattening him in a wave of nausea as he was out tapping trees. He crawled back to his curiara: "[E]ach time the fit of nausea returned, I became quite powerless, and had to drop to the damp earth and wait until the paroxysm was over." When he staggered to his feet the machete tripped him up; he was lucky he didn't gash himself, which would have been the end. He reached the canoe and tried to paddle to his rancho, "but the sun was too powerful for me" and he scrambled to the cool dirt of shore. His skin burned from within. He crawled on, "the remainder of my strength fast failing." He reached camp and pulled himself up on the raised bed. "I remember little of what passed during the four days that the constant nausea and vomiting lasted," he said.

These days merged into a hallucinatory blur. He recalled the torment of being eaten alive by mosquitoes and sandflies, of being too weak to do anything but lie on his bed and provide a meal. He was close to death, and none of the others in his group could help him because they were either away, like Rogers, or fighting for life themselves.

Algot Lange, an American adventurer who wrote several books about his explorations of the Upper Amazon, documented the sufferer's psychological state in excruciating detail. Lange traveled to the rubber lands of the Javari River, a major Amazon tributary forming the border between Peru and Brazil. He visited a rubber camp where every tapper was prostrate, and in a few days the sickness hit him, too. "For five days I was delirious," Lange wrote, "listening to the mysterious noises of the forest and seeing in my dreams visions of juicy steaks, great loaves of bread, and cups of creamy coffee." Soon the hunger turned into a "warm, drugged sensation" that "pervaded his system," and he listened to "the voice of the forest"—the nighttime call of tree frogs. He imagined "the murmuring crowd of a large city" and was soothed.

But soon his visions changed. The jungle was no longer a gentle place. "I saw myself engulfed in a sea of poisonous green, caught by living creepers that dragged us down in a deadly octopus embrace." His dreams and actions became indistinguishable. Did he bolt from the hammock, crying and babbling, or was he hallucinating? He fled from the jungle's "impenetrable wall of vegetation, its dark shadows, and moist, treacherous ground." Lange and Henry had reached the same point: They found the jungle no longer transcendent but claustrophobic, a parasite world of shadows and green. It was "a place of terror and death," filled with savagery.

Jungle tales almost always dwell on this savagery. It is a trope, a literary motif, and the question in each survivor's account is often the same: Is savagery an omnipresent part of the forest or one that man brought with him? In the 1920s, the Capuchin father Francisco de Vilanova addressed the same problem. Vilanova and his party made a famous *excursión apostoliá* to the Putumayo River in northern Peru, a place that had become synonymous with rubber and murder. They went to save the native Huitoto tribesmen from rapacious rubber barons but in the end could barely save themselves. De Vilanova watched as one guide fell sick and insisted upon a shaman's cure that killed him. He lost heart when treating the wounds of Indians flogged by rubber bosses.

It is almost something unbelievable to those who do not know the jungle. It is an irrational fact that enslaves those who go there. It is a whirlwind of savage passions that dominates the civilized person who has too much confidence in himself. It is a degeneration of the spirit in a drunkenness of improbable but real circumstances. The rational and civilized man loses respect for himself and his domestic place. He throws his heritage into the mire from where who knows when it will be retrieved. One's heart fills with morbidity and the sentiment of savagery. It becomes insensible to the most pure and great things of humanity. Even cultivated spirits, finely formed and well-educated, have succumbed.

Soon, Henry's reality merged with strange dreams. His memories mixed with hallucinatory incidents that are not recorded in his journal, but they were apparently passed down to family. He'd been unconscious, how long he did not know, and when he awoke "he found several vultures calmly awaiting his death," Edward Lane recorded. One was especially close, bolder than its fellows. Henry stared back for an instant at the pale, unblinking eyes, the crepuscular, warty head. How he loathed these birds—their shuffling, rustling walk, their croaks and carrion smell. The memory of the Indian catching one on a hook flashed before him. He laughed and lunged convulsively, grabbing its beak and greasy feathers in his paws. His laughter carried across the silent rancho as he ripped out feathers and thrust them through the bird's nostrils as the Indian had done. The vulture resembled a music hall comedian, the feathers sticking from the holes in its beak like outlandish mustachios. He flung the bird from him and collapsed on the bed. The unlucky vulture limped to its companions, who were frightened by this strange, demented apparition and flew croaking into the air.

He was saved by Rojas Gil and his wives. "I recollect one afternoon, as I lay prostrate and incapable of moving, and part of my back bared to the swarms of sand-flies . . . a woman of Rojas' entered, and seeing my condition, she passed her cool soft hands gently over my burning brow and back, brushing away the plagues." He never recorded her name, but he always remembered the moment: "It is singular what an impression the slightest mark of kindness and human sympathy makes on one in such an extremity. . . . Although unable to thank her, I think I never felt so grateful for anything."

The anonymous savior reported Wickham's condition, for soon Rojas took Henry to his rancho, where the air wasn't so close and there was a slight breeze. Another neighbor helped, tapping Henry's trees while he was unable, feeding him *gaurapo,* the heated juice of sugarcane, the only thing he could keep down. Although they bathed and cleaned his wounds, they already considered him doomed, for he'd seen the *curupira,* "the little pale man of the forest" who appeared to those foolish enough to wander alone under the green canopy. The *curupira* rose from the dark, twisted roots lacing the undergrowth. He was the deep sound in the forest that had no explanation, the shadow at the edge of one's vision that was "the sure precursor of evil to the unlucky beholder, if not of his death." The souls of those who died at the hands of the *curupira* wandered forever in the forest. Survivors left part of themselves beneath the canopy and were never the same.

There was a belief among the Indians of the Upper Rio Negro and Orinoco that the best cure for the curse of the *curupira* was the Prayer of the Dry Toad. One day Henry heard some strange incantations near him, but he drifted in and out of consciousness and would never know more. The Prayer of the Dry Toad was attempted only as a last resort, and few Europeans were ever allowed to witness its power. A large *cururu* toad was buried in the ground up to its neck and forced to eat glowing coals. "Peace will come to me in the dust of the earth," the shaman muttered as he buried the squirming creature in the dirt. "You guardian angels, stay with me always, and Satan will not have the strength to seize me." A week later, the toad had vanished, burned from inside by the fire. In its place a three-branched tree was said to sprout, each branch a different color: white for love, black for mourning, red for despair.

Henry survived but was left weak and helpless and spent all of March recuperating. The fever waxed and waned, but never as virulently as in those first few days, when he feared he'd lost his mind. He tried to spare others the same travail. One day he returned from a short walk in the forest and found "a French gentleman in my lodge." The Frenchman was journeying up the river to La Esmeralda, the last large mission on the Orinoco and gateway to the Casiquiare. He wanted to go past the tiny outposts of Ocamo and Manaco to the great peak of Duida in the Serra Parima range. "In so remote a situation all Europeans appear as countrymen," Henry said, trying to dissuade this apparition from his suicidal plan. He was heading alone into Yanomami country, something few travelers lived to describe. But the Frenchman was adamant, "having determined in his mind that there must be a mine of gold in that direction." Henry had nearly lost his life and knew what such dreams of El Dorado did to a man. Henry begged him to turn back, tried to tell him of the disease, vipers, and starvation that awaited him. But the Frenchman laughed and said he would return in a month and they would "have a raid" on the bars together. Henry's visitor paddled up the river, waving one last time as he slipped out of sight. He never returned in a "month or less," and as far as anyone knows, never came back at all.

The doomed Frenchman was the last straw. By Good Friday, Henry admitted it was time to end their plans. Ramón's leg was so bad that he'd been "unable to work for some time past." They had no need to fast for Lent, since they'd been starving for so long. The waters rose in April. They caught electric eels on the submerged *estradas,* which meant tapping the trees was certain suicide. By the end of April, he decided to abandon the camp and push upriver to the Casiquiare, but when he tried on April 28, only little Manuel and Ramón remained. All the others had left or drifted into the forest, and Rogers never returned from his last trip for supplies. They were all desperately weak and "helplessly sick." The best they could do was drift back on the current to San Fernando de Atapabo.

He returned to that village, took the portage south through the jungle, and, sometime in May 1870, finally reached the Rio Negro. He lingered there, a shell of his former self, recuperating in the village of Maroa "on the cool and limpid water of the Black River" that would take him to the sea.

That August he drifted down the Rio Negro in a trading lancha owned by his friend Andreas Level. The journey took most of the month. The deep, lifeless river winds nearly 1,550 miles through stunted *campana* forests, past hundreds of islands and white sand beaches that emerge in low water. The water itself is a black tea brewed through the sandy soil. Henry's journals lack detail here. He was exhausted and felt lucky to be alive.

The Rio Negro ends where it meets the Amazon at Manaus. Two English-built steamers were anchored there, looking "very suggestive of a return to civilization." It was September 3, 1870, and he'd been on some tropical river in this vast continent for one year, eight months, and thirteen days.

Henry gasped at the size of the city. He'd forgotten that Manaus, once the fabled site of El Dorado, had become the modern equivalent. When it came to black rubber, this was the center of the world.

In Manaus, Henry entered a land where people were growing rich, fabulously rich, off rubber. It was the beginning of what investors in New York and London called the Rubber Age. Henry arrived four years before rubber began to be applied to telegraph wires and transatlantic cables, but even as early as 1860, it had become obvious to Thomas Hancock that with the discovery of vulcanization, rubber would be the world's most useful plastic: "Nothing has been discovered which would even be a substitute," he said. A threat to its free access would be a threat to Great Britain's national interests; its control was "the ultimate hard currency of exchange."

In the decade after Hancock's pronouncement, rubber had become essential for war. In addition to its many uses in railroads and steam engines, military catalogues of the era show new designs using rubber for shoes and boots, blankets, hats, coats, pontoon boats, bayonet guards, tents, ground sheets, canteens, powder flasks, haversacks, and buttons. Rubberized silk was used for military balloons. War also created a boom in reconstructive surgery using hard rubber teeth, nose pieces, and custom-molded prosthetics. Now the 1870–71 Franco-Prussian War was in its early stages, and all of Europe watched. Even in the early campaigns it became evident that victory would depend on rapid mobilization of troops by the fastest means possible. That was the railroad, which depended on rubber.

By the time Henry drifted from the wilderness, *Hevea brasiliensis* was the Holy Grail. Like the Grail, no one knew exactly where it could be found. Did it grow along rivers, where the first trees were encountered, since the seeds had the ability to float, or further back in the rain forest, on higher ground? Hevea did not grow in pure stands, like trees in temperate climates, but was sown through the forest at the rate of two or three trees per acre, as if someone had scattered them from cloud level like a giant Johnny Appleseed. Locating the trees and working the twisting *estrada* to collect that daily cup of latex was too slow and inefficient for modern demands. Why had no one tried to domesticate something this valuable, as Hancock had suggested in 1850?

Actually, someone had, but his early efforts were ignored. In 1861 and 1863, the Brazilian explorer and botanist João Martins da Silva Coutinho wandered through the rubber lands. By the end of his journeys, he recommended to the governor of Pará that, instead of depending upon the whim of nature, rubber plantations should be grown. His suggestion was considered foolhardy. The forest was bountiful. Why mess with a good thing? Soon, a package of rubber seeds was sent anonymously (probably by Silva Coutinho) to the National Museum in Rio de Janeiro, where they were sown. Federal authorities recognized a windfall, if their provincial representatives did not; nevertheless, their efforts seemed focused on attracting foreign investors.

In 1867, Brazil sent Silva Coutinho to the Universal Exposition in Paris, where he was made chairman of the jury evaluating rubber samples from around the world. He demonstrated that Brazilian hevea was in all ways superior, and he estimated the costs of plantation production in a report published the next year.

One person who noticed was James Collins, curator of London's Museum of the Pharmaceutical Society. Collins had been reading reports by Amazon travelers and corresponding with them for years. He haunted the London docks, inspecting rubber shipments to learn more about the several species from which it was derived. In 1868, after reading Silva Coutinho's report, Collins published an article in *Hooker's Journal* revealing what he'd learned. He followed this next year in the prestigious *Journal of the Society of Arts* with the more detailed "On India-Rubber, Its History, Commerce and Supply." This article painted a history of all things rubber and quoted everyone from Juan de Torquemada, chronicler of the Spanish Conquest, to la Condamine, Bates, and Richard Spruce.

Two things impressed Collins. As Silva Coutinho proved, the world's best source of rubber was the elusive *Hevea brasiliensis*. And shouldn't it be possible to grow rubber in the empire's extensive East Asian plantations? For the first time in print, he made the obvious connection: "The introduction of the invaluable *cinchonas* into India has been attended with marvelous success." If that smuggling expedition had been a success, why not do the same with rubber?

The years had been kind to Clements Markham since his adventures in Peru. In 1870, he was forty, married, in charge of the India Office's Geographical Department, and secretary of the Royal Geographical Society. He was one year away from being knighted, three years away from his election as a Fellow of the Royal Society of London. He'd attained everything envisioned by his driving ambition, and more was on the way.

However, the cinchona affair had not proved the unqualified triumph that it was popularly portrayed as. In 1865, Charles Ledger, a British adventurer who'd smuggled alpacas from Peru to Australia, showed up in India with the best variety of cinchona ever seen. Ledger had collected his plants in Bolivia, the very place avoided by Markham, and he brought with him *C. ledgeriana,* a hitherto unknown yellow-bark variety whose quinine content proved more concentrated than any other species—in some cases, as high as 13.7 percent. But Markham was absent from the India Office when Ledger arrived, and no other British official would touch the new plants, unwilling to rock the boat. Ledger left in disgust and sold his superior seeds to the rival Dutch, who bought all twenty thousand and planted them in Java. By 1870, the Dutch East Indies had begun to compete with Great Britain, and by the end of the 1880s, the Dutch would corner the world market in quinine, limiting England's variety to local sale.

In the decade since he'd smuggled red-bark cinchona from the Andes, one gets the impression that Clements Markham had grown restless and bored. It is hard to imagine his not rising to the challenge of Collins's article. In 1870, Markham decided "it was necessary to do for the india-rubber and caoutchouc-yielding trees what had already been done with such happy results for the cinchona tree." The empire had only one choice if it wanted to remain a Great Power:

> When it is considered that every steam vessel afloat, every train and
> every factory on shore employing steam power, must of necessity use

India-rubber, it is hardly possible to overrate the importance of securing a permanent supply, in connection with the industry of the world.

Markham envisioned vast rubber plantations stretching across India, and he had more influence now than a decade earlier, when he'd been an obscure functionary. He knew that the India Office was always on the lookout for potential export crops of significant commercial value. And he had a highly placed friend.

Joseph Dalton Hooker had been appointed Kew's director five years earlier, effective immediately after his father's death in 1865. Few seemed bred for the sinecure like the younger Hooker: his father's home at Woodside Crescent, within walking distance of the Gardens, had a museum and herbarium; he grew up in an environment where plants arrived for study from all corners of the globe. As a student, his first thesis concerned three new species of moss. His first scientific voyage, in 1839, was as assistant surgeon and botanist aboard the HMS *Erebus* in Sir John Clark Ross's four-year voyage to Antarctica.

It was with his second expedition, however, that Joseph Hooker began to make a name in botany. From 1848 to 1850, he trekked through the Himalayas in his spectacles and tartan jacket; he had no mountaineering equipment except for some woolen stockings and an "antiglare eyeshade" made from a veil by the wife of a friend. Like Robert Cross and his umbrella, there was something very English about Hooker: He dined on biscuits, tea, and brandy. His porters carried his solid-oak traveling desk and brass-bound ditty bag around moraines and glaciers. Neither he nor his companion, Dr. Thomas Thomson, ever got much sleep, since the yaks they used for pack animals would stick their heads into the tents and snort until they woke. By the time the journey was over, he and Thomson had traveled farther on Kanchenjunga, the world's third-highest mountain, than any European before them, and they returned with seven thousand specimens for Kew.

The son's career seemed blessed. In 1851, he married Frances Harriet, eldest daughter of Darwin's professor at Cambridge. Hooker and Darwin grew close: in addition to championing Darwin's 1859 *On the Origin of Species* he followed Darwin's suggestion about giving his wife chloroform during labor, as Darwin had done for his own wife, Emma. When Hooker's six-year-old daughter died in 1867, he remembered that Darwin had suffered similar sadness a dozen years earlier with the death of his own daughter Annie and wrote, "It will be long before I cease to hear her voice in the garden, or feel her little hand stealing into mine." His scientific reputation and personal connections were impeccable, and his father's influence made his ascension to the directorship certain. William Hooker convinced the government that Kew had grown so much in scope that he needed an assistant, and on June 5, 1855, Joseph had been appointed the garden's assistant director. As William approached the end of his life, he offered to leave his vast private herbarium to the nation if Joseph were appointed his successor.

But even with such advantages, Joseph Hooker was notoriously "nervous and high-strung," said his son-in-law; "impulsive and somewhat peppery in temper," his good friend Darwin conceded. He was famous for hard work: by the end of his life, his list of publications would fill 20 pages; the gardens under father and son increased from 11 to 300 acres; there were 20 "glasshouses" containing over 4,500 living plants; the herbarium contained 150,000 species, allowing global comparisons between old and new species unequaled elsewhere. Even so, Joseph Hooker looked upon the world and saw enemies. At five feet eleven inches, he was tall, spare, and wiry; he scowls from under heavy brows in every photograph, frown lines carved deep in his face. He took offense at so many things. He insisted that the garden's main purpose was scientific and utilitarian, "not recreational," and despised those holiday-goers entering his empire whose intentions were "mere pleasure" and "rude romping and games." He resisted all attempts to extend visiting hours for the public and worried about the scientific standing of Kew and of botany in general. Botany was enormously popular among nonprofessionals, especially women, at a time when Victorian science was almost entirely inhabited by men. All these gardeners, flower painters, "rude rompers" diminished the status of his field.

And there was a very real danger of losing it all. By 1870, Hooker had clashed several times with Acton Smee Ayrton, the first commissioner of the Office of Works, over attempts to cut public spending on scientific institutions. By 1872, their private war would become public when Ayrton proposed transferring Kew's herbarium to a newer facility owned by Richard Owen, keeper of the natural history collection at the British Museum. What became known as the Ayrton Controversy would be extremely nasty and political: Ayrton wanted more control of Kew; Hooker was enraged by a bald threat; both sides marshaled their forces. By the general election of 1874, Hooker and his supporters succeeded in getting Ayrton voted out of office permanently.

But even in 1870, Hooker could sniff what was in the wind. When Clements Markham appeared with his proposal, Hooker remembered well his father's historic partnership with him, which had consolidated Kew's place in the sciences and the empire. The transfer of rubber could be as defining for Joseph Hooker as the transfer of cinchona had been for his father. It would be a political coup.

And so, as a broken Henry Wickham returned to civilization, the word went out in the small world of botanists and diplomats that somehow, by any means possible, the British Empire wanted mastery of the rubber stream.

Manaus was the logical place to begin. Located in the center of the Amazon Valley, the jungle city was the collection point for rubber from the interior, and thus its dealers set prices for the world. In 1870, visitors still thought of Manaus as a shabby little place. In 1865, British visitor William Scully estimated its population at five thousand, with 350 houses and government buildings, most "in a very dilapidated condition." Manaus was like an American gold mining town just taking root in the Rockies. There was a hardscrabble feel and a distant scent of possibilities, but for the moment, all money was devoted to making more money, not advertising its success to the world.

In 1853, when Richard Spruce had drifted into Manaus after a four-year journey upriver, he had been as awed as Henry was now. When Spruce had left for the wilderness, the sleepy little village was still called Barra, a place where small-time merchants loaded boats every few days with rubber pallets for Europe and the United States. When he returned,

Barra was Manaus, and he counted twenty boats in port. The harbor swarmed with canoes, lanchas, and steamships from other nations. "What has happened?" Spruce shouted.

"Rubber has happened!" a friend shouted back.

Spruce witnessed the start of the first Rubber Boom, generally considered that period between 1850 and 1880, before the invention of the pneumatic tire and the endless maw of the auto industry. While he'd been gone, Brazil had opened the river to international trade. A stampede of townsmen and farmers abandoned their holdings for rubber, dropping everything to live in a forest rancho and walk a daily *estrada*. In the province of Pará alone, twenty-five thousand people dropped everything to steam up jungle rivers. Local agriculture was abandoned; local sources of farina, sugar, and rum dried up, and the commodities had to be imported. As shopkeepers in Pará and Manaus advanced money and goods to new rubber prospectors, they began a debt-peonage system that still exists today.

Although the *seringuiero* occupied the bottom rung of rubber's ladder, his was an autonomous existence that many found appealing, at least in the early years. He'd voyage to an unclaimed rubber site, return with a canoe laden with rubber, and trade directly with the dealer, often in goods. This was what Henry experienced, even as the business was changing into a hierarchy with more levels of middlemen. Manaus was both the center and the embodiment of that change. As the more ambitious gatherers established wider rights of ownership in virgin rubber lands, they became intermediaries, receiving trade goods on credit from wholesale dealers, who in turn financed more tappers in a classic pyramid scheme. Newly hired *seringuieros* were granted the right to tap on an owner's grove; they recognized him as their patron, or *patrão*. The *patrão*'s real profit came from the sale of trade goods to his hired help; the mark-up was so high that a tapper could end his season in the red. Yet the *seringuiero* still had some independence: The tapping season was limited to six months, and the manner and intensity of work were his own affair.

As the world rubber trade expanded, a second kind of middleman emerged. Henry saw them throughout Manaus in their new rubber warehouses. Although the wholesale rubber dealer—known as the *aviador*, or "forwarder"—was the top of the chain in Brazil, he too was a middle-man. He sold to the exporter, the direct source of overseas capital, who was usually British. As more money poured into an *aviador*'s coffers, he'd launch a fleet of riverboats that pierced deeper into the Amazon Valley. There the trees were said to be thicker in girth, less scattered, and, most important, virgin.

Although it would be the *aviadors* who grew fantastically rich, it was the *patrão* who owned or managed the "rubber estates" in the jungle, who formed the most vital link in the chain. He was the interface between labor and capital. Since he grew rich from the number of trees that were tapped, his success depended upon the number of tappers he could employ. In theory, his control was paternalistic, hence his title. He solved problems, offered guidance, provided resources—as a father provided daily bread. In return he expected obedience, hard work, and smoked balls of rubber. Those considered lazy were corrected, first through debt and diplomacy, later, as huge sums of money flowed unchecked, through intimidation, beatings, enslavement, and torture.

Rubber served an imperial purpose as well. As the *seringuieros* fanned out to the most distant parts of the interior, they carried with them Brazil's territorial claims. By settling in the wilderness, a tapper validated Brazil's claims of *uti possidetis*, "as you possess," the right of owning land simply by taking it—a common practice worldwide. Settlers in the American West and Australian outback claimed homesteads under the same principle; European nations annexed guano islands in the Pacific and assumed dominion over vast swaths of Africa. The entire Earth was up for grabs.

At Manaus, Henry witnessed a side of the trade far different from any-thing he'd encountered on the Orinoco. On the streets, merchants in white linen suits mixed with the upriver traders who had machetes swinging by their sides. One could dine on pâté de foie gras, Cross & Blackwell's jams, and imported wine. Rubber flowed into Manaus down the Rio Negro from Venezuela, the Içá and Japurá rivers from Colombia, the Solimões from Ecuador, the Acre and Madeira from Bolivia, and the Juruá, Purús, and Javari from Peru. The city was built on four hills. The Jesuit cathedral was the grandest structure in town, while on the opposite hill stood the governor's mansion. When the zoologist Louis Agassiz and his wife, Elizabeth, visited Manaus in 1865–66, the governor's mansion was the social center for the elite. There were innumerable balls, in which the women wore dresses "of every variety, from silks and satins to stuff gowns, and the complexions of all tints, from the genuine Negro through paler shades of Indian and Negro to white. . . ."

But it was not in the governor's mansion that the city's true power resided. Power reigned in the warehouses of the *aviadors,* each a huge cavern for receiving, examining, and boxing rubber, with offices, clerks, and assistants on the floor above. Older or more prosperous firms built their caverns on the Rio Preto that flowed beside the city and into the Rio Negro. A lancha filled with rubber could be docked and unloaded a few steps from the firm. Rubber was stacked in balls and pellets and lined in rows for grading; wagons loaded with rubber trundled past outside.

Each warehouse was an oasis of quiet in the clamorous city. Buyers inspected lots and made their choice in silence. The seller drew up a contract and handed it over; the buyer's continued silence confirmed his acceptance of the deal. It was outside that one heard the din of lesser commerce—the wooden clappers of the haberdasher, the panpipes of the cake seller, the clang of the tinsmith's iron spoon. Behind mud walls, roosters crowed. Piercing the din, the wandering sellers of lottery tickets shouted their promises.

"Get rich, get rich!" they cried.

PART II

THE SOURCE

Stolen apples are always sweetest.

—Medieval proverb

CHAPTER 6

THE RETURN OF THE PLANTER

Rubber was many things to many people. For Henry, it had nearly been his death, but it had also become an addiction. For Joseph Hooker, it was a means to maintain power; for Clements Markham, a door to past glories. To Charles Goodyear, a religious calling; to Thomas Hancock, an international commodity to be bought and sold. The governors in Manaus and Pará, and the emperor in Rio, saw it as a future source of greatness. To the *seringuieros* scattered through the Basin, it was the escape from a miserable existence; to their *patrãos*, a dream of wealth and ease. The white milk that dribbled like blood became a mirror: In rubber's slick, obsidian surface, each man saw his need.

Strange tales surfaced of those who disappeared into the forest in search of rubber, then emerged reborn. The most famous was that of Crisóstavo Hernández, a Colombian mulatto who fled into the dense Putumayo region separating Colombia and Peru. Information about him is sketchy at best. In the same decade that Henry floated into Manaus, Crisóstavo wrenched a rubber empire from the forest by force and maintained it into the 1880s. With the aid of a tribe of Huitoto Indians as his personal army, he enslaved a whole region to tap rubber. In the mid-1900s, a Huitoto oral history emerged told by a man whose mother was Huitoto and father a white rubber tapper: according to this, Crisóstavo was a *cauchero* in Southern Colombia who killed a man in a fight and fled to escape imprisonment. He rafted down the Caquetá River until spotting thatch roofs; when he entered the village, the Indians froze. Was this black creature a person? A spirit? They'd never seen a black man before.

He was taken to the village chief, who decided that Crisóstavo was a harmless refugee. He learned the tribe's language, was honored with a wife, and became a member of the tribe. In time, however, he fell for another man's wife; they consummated their affair and fled into the jungle.

After six days of travel, the couple came to another Huitoto village. The pattern repeated: the villagers were frightened but decided that he was, after all, a man. This time, however, Crisóstavo had brought with him an iron ax, shotgun, and machete, wonders they'd never seen. These, with his singular blackness, made Crisóstavo unique and powerful, and an idea dawned in him. He'd move again, but this time with two hundred Huitoto allies from his new tribe; he continued like this, building an army, until finally coming to the large and powerful village of Chief Iferenanuique. He settled there, increasing the chief's power and prestige, but after four years told Iferenanuique that he wanted to go home and visit friends. But he could not do so empty-handed. The chief asked what he wanted. Crisóstavo marked out a space three meters long that reached over his head—he said he wished this filled with rubber. In three months it was done, and when Crisóstavo returned from civilization he brought with him a similar mountain of magical gifts: metal axes, knives, clothing, machetes and shotguns, needles, beads, combs, salt, mirrors, and cane liquor. There was something for everyone, and thus his power grew. Although he would be known as a conqueror, feared throughout the Putumayo, the core of his power lay in temptation and trade.

Crisóstavo's was the British way, using rubber as a fulcrum for wider influence. Great Britain didn't have the political or military muscle in South America that it had in Africa, Asia, and the Middle East: early attempts at expansion in Argentina threw the people into civil war; clumsy attempts to bully Rio de Janeiro to abolish slavery weakened Britain's influence in Brazil. But it did possess economic clout, what has since been called "informal empire." After 1856, that economic clout became considerable when Brazil's railways—built and operated by British companies—gained generous concessions from the emperor. The São Paulo Railway Company, Ltd. operated a railway down the mountain from its namesake city to the coffee port of Santos: the company lasted from 1876 to 1930, climbed from £2 million to £3 million in total shares, and paid an average dividend of 10.6 percent per year. Business was good. Although the railroad was Britain's most lauded investment, Englishmen bought stock in Brazilian mines, banks, coffee exporters, public utilities, and natural gas; from 1825 to 1890, they preferred Brazil as a field of investment to all other Latin American countries because of its political stability. In 1856, Great Britain got its foot in the rubber trade when it bought the Amazon Steam Navigation Company and a million acres of Amazon real estate; Goodyear's discovery and the opening of the river to international trade made rubber cheap *and* accessible. It was as if the world had suddenly discovered rubber; with a solid stake in its trade, there might be no limit to Britain's power in the region.

One person who saw this clearly was James de Vismes Drummond-Hay, the British Consul in Pará. The office of "consul" was ancient: It was natural that merchants in foreign lands, trading in alien and sometimes hostile cities, would appoint a spokesman to conduct affairs with the local authorities. For their part, the local authorities found it easier to deal with one official than a mob of businessmen. In a seaport like Pará, a consul's first responsibility was the protection and regulation of British ships and seamen, but in Latin America especially he often found himself thrust into the role of the "man on the spot," that strategic British character who rose above himself when Britain's interests were at stake and gave his all for the empire. For consuls in particular this meant watching for new opportunities and acting as agents for expanding British trade. The role sounds important, but in practice consuls were the forgotten stepchildren of the Foreign Office. They were considered "lower in dignity" than diplomats since their mercantile duties belonged "rather to individual interests than those of the state." The uniform was embroidered in silver and not the diplomat's gold; the pay could be below subsistence level. Diplomats did not call on them; they did not belong to the same club.

Few of these restrictions applied to the family Drummond-Hay. James Drummond-Hay came from a long line of distinguished consul-generals, all posted in Morocco. Because of its interests in Gibraltar, Britain was anxious for that North African country to remain independent, and a remarkably close relationship sprang up between the British Empire and

Morocco's sultan, due in large part to two successive consuls, Edward Drummond-Hay (1829–45) and his son John (1845–86). The latter was so effective that he was knighted in 1862. Between them, they arranged for the Royal Navy to take the sultan's sons to Mecca for the hajj, smoothed the way for the English education of Moroccan princes, and arranged the military training of Moroccan officers in Gibraltar and England. James was the younger brother of Edward, and in 1856, the vice-consulship in Tetuan passed to him. He stayed in Morocco for ten years, married, had a son, then in 1866 was appointed vice consul in Pará with the understanding that it would lead to a consul-generalship somewhere in the New World.

The Amazon must have seemed like another planet to someone from the Levant. Tangier was dry and exotic, blown by sea breezes; Pará was humid and crumbling, plagued by periodic bouts of yellow fever. In Morocco, he was days away from London on a fast clipper ship; from Pará the journey took weeks, and the fare was an astounding £60 or £70, a sizable portion of his annual salary. Although the approach to Brazil itself from the sea was lovely, running from deep blue to light green water, Pará lay another eighty miles up a wide muddy estuary afloat with uprooted trees and vast islands of grass that had drifted from thousands of miles inland. The city fought a constant struggle with vegetable life; the jungle was like an invading army, and from roofs and cornices grew plants and small trees that stormed the ramparts and waved their tops like enemy flags. Every consul found himself the inheritor of at least one Distressed British Subject: faced with such a pathetic case at his doorstep, the consul had no choice but to come to his aid. One consul in Siam found such a case asleep in his bed, full of the contents of his liquor cabinet. Here they floated from the jungle filled with madness and parasites. They looked more animal than British, as penniless as church mice, demanding the fare back home.

But now a particularly interesting straggler had materialized: he called himself Wickham, and Drummond-Hay was intrigued. In September, the consul had drafted his "Report on the Industrial Classes in the Provinces of Pará and Amazonas, Brazil," in which he concluded that the average English worker could do well for himself out here, if he were willing to apply himself scientifically to rubber's domestication. Yet none of the natives seemed to consider its possibility: "The labour of extracting rubber is so small, and yet so remunerative," he wrote, "that it is only natural" to continue tapping rubber for a few months each year. In a good district a man could extract thirty-two pounds of rubber per day, yet no real attempt had been made to tap rubber on a large scale. The investment potential was phenomenal: "The rubber-bearing country is so vast," yet locals had not considered "the idea of planting the rubber-tree or caring for its growth."

Now into his office came this twenty-four-year-old, burned-out Britisher named Henry Wickham, who'd hopped a steamer from Manaus and floated downriver. He seemed like any other wanderer down on his luck until Drummond-Hay flipped through his journal and spotted the first accurate taxonomic drawing of hevea he'd ever seen, something even the vaunted Richard Spruce hadn't sketched. This was tucked away among other drawings—an Indian standing by a rubber tree, a sunset on the Orinoco, a rancho in the jungle. What's more, the boy had learned to tap rubber, had actually made enough money from the effort to cover his passage back to Pará and Liverpool. This was another first—the first time, to the consul's knowledge, that an Englishman had gone native and tapped rubber himself. Richard Spruce had described the process, as had Wallace and Bates, but none had spent a season as a *seringuiero*. As far as he knew, it had never been done.

So Drummond-Hay did something unusual for a consul. He took Wickham into his confidence, bucked him up, and turned his life around. He told Wickham that a young man willing to sweat and suffer a bit could remake himself out here. Rubber was the key. If a man arrived with some capital or fellow hopefuls, was willing to work hard, he'd become in two or three years the *only* rubber planter in the entire Amazon Basin. His fortune would be as great as the sugar planters of the West Indies, the coffee men of Malaya, the tea planters of Assam. The place to do it was five hundred miles back upriver in a region rumored to have the best hevea in the Amazon Basin. Somewhere in the unmapped jungle near the town of Santarém.

In fact, Wickham already seemed to know the place. It was one of the few excursions he'd made off the boat, though he was apparently more interested in the town's dwindling group of American Southerners than in its bordering jungle or environs. The consul urged the young man to publish his "rough notes" as a travel diary, and he did something unprecedented. He gave him his own report to include in his book, as a kind of afterword.

Wickham seemed inspired by the confidences of this important man. After a narrative of privation, failure, disease, and near death, he would unexpectedly write:

> I have come to the conclusion that the valley of the Amazon is the great and best field for any of my countrymen who have energy and a spirit of enterprise as well as a desire for independence, and a home where there is at least breathing room, and every man is not compelled to tread on his neighbour's toes. I purpose to make the tablelands in the triangle betwixt the Tapajos and the Amazon, behind the town of Santarem, in future the base of my operations.

We know who set him on this course—one that would change so much and affect so many. For when his notes were published, Henry dedicated the book to James de Vismes Drummond-Hay, C.B., "in remembrance of the many kindnesses by which the author was indebted for a pleasant ending to a somewhat arduous journey."

Henry arrived home in London in autumn 1870 a changed man. His mother had never seen him this way. When he'd returned from Nicaragua, he'd seemed lost. Now he was talking as if a new life awaited them all in the jungle, and the more he talked, the more convincing he sounded. More than that, he got things done with a drive that had never been part of his makeup before. Within the space of a few months after his return, he wrote up his notes and found a publisher. And he got engaged.

Her name was Violet Case Carter, the daughter of a bookseller on Regent Street whose shop was only a few short blocks from Harriette's on Sackville. She was four years younger than Henry, and it is possible they could have known each other before he left for the Orinoco. Although booksellers were not the most fashionable shopkeepers on Regent Street, they were the most interesting and controversial. By displaying in their windows lithographs and etchings of everything from tranquil landscapes to current artwork and classic nudes, they were a draw for window-shoppers and a good indicator of popular and intellectual tastes. Violet's father, William H. J. Carter, set up his shop at 12 Regent Street, just off Picca-dilly Circus. Carter lived on the premises with his wife Patty and daughter Violet. Like most booksellers of the time, he was a literary jack-of-all-trades. He printed books, sold artwork, and gave speech lessons.

William Carter also published Henry's journals of his Nicaraguan and South American adventures, combined in one volume: *Rough Notes of a Journey through the Wilderness from Trinidad to Pará, Brazil, by Way of the Great Cataracts of the Orinoco, Atabapo, and Rio Negro*. The book contained sixteen line drawings by Henry, including sketches of a rubber tapper in the jungle and the leaf and nut of *Hevea brasiliensis,* drawings that would have a greater impact on his future than the daily narrative.

The production seems rough and hurried. The journals are unedited and unchronological. The reader suffers with Henry through the Orinoco before he gets to meet the younger Henry Wickham in Nicaragua. People come and go without introduction or description, most notably the poor, suffering Rogers, whose origins are never given and whose fate we never learn. If ever an authorial voice is shackled to immediate sensation, it is Henry's. One suspects that this was his entire approach to life, and for all his bonhomie and physical courage, he seemed unable to empathize with others. It's hard to call this selfishness, for that implies awareness of another's wants and desires. Henry's lack of empathy is more elemen-tal. Victorians called it manly force of will. Literary naturalists like Jack London or Joseph Conrad cast it as a "force of nature," usually destruc-tive. Freud would christen it the id.

People like Henry are on fire, and he must have burned like a nova. During those nine months, from his arrival in London in autumn 1870 to his return to the Amazon in summer 1871, he convinced his mother, sister Harriette Jane and her fiancé Frank Pilditch, brother John, and an unspecified number of English laborers to return with him to Santarém to start new lives as planters. John Wickham was so filled with Henry's vision that in the 1871 census he listed his occupation as "farmer." Henry convinced Violet's father to publish his *Rough Notes* even though he would be gone when the book came out. Family lore suggests that Carter also subsidized many of Henry's future adventures. He burst into the life of a girl who'd lived around travelogues and exotic prints and no doubt dreamed of far-off places. He swept her up in a whirlwind courtship, and they married on May 29, 1871, Henry's twenty-fifth birthday.

"Born within the sound of Bow Bells, and therefore a thorough Cockney," she said of herself, Violet had never left England. A group photo taken four years later in Santarém shows a small woman, with a sharp jaw and high forehead, the most wan and sickly of the six posed. By then, three of their original party had died; within another year, two more would succumb to malaria or yellow fever. Her sharp features and mousy brown hair made her the least attractive of the three women seated for the photo, but she would prove tougher than all of them, more adaptable to the wild places that Henry dragged her into than even Henry himself, able to face hardship with a straightforward humor that Henry never had.

As a working-class girl, marriage had been an expectation since she could remember. "To be married is, with perhaps the majority of women, the entrance into life," proclaimed the article "Old Maids" in the July 1872 *Blackwood's Edinburgh Magazine*. Marriage was "the point they assume for carrying out their ideas and aims." There was also a belief that behind every successful man an intelligent and sensitive woman oiled the gears— that, in fact, a man *needed* such a woman to succeed. "The general aim of English wives is practically to convince their husbands how much happier they are married," assured the author of an 1852 women's health manual, but Henry didn't need convincing. So great by now was his need to rise in the world, so wholeheartedly did he accept the Victorian formula for success, that the "good wife" of Proverbs was part of his internal equation—she was identified with him, and he with her. She deserved the benefits of his success. Left unsaid was that she shared the reality of his failures.

Unlike Americans today who subscribe to ideas of empire, Victorians felt that they must not only civilize and dominate but also populate the world. Young men were not the only ones recruited to inhabit the world's wild places; family groups were encouraged to establish little islands of English values. A lot was vested in these domestic outposts of civilization, wrote Herman Merivale in his popular *Lectures on Colonization and Colonies:* "The sense of national honour, . . . pride of blood, tenacious spirit of self-defense, the sympathies of kindred communities, the instincts of a dominant race, the vague but generous desire to spread our civilization and our religion over the world." All traits, it was believed, that made the British an imperial people and ensured an ever-expanding empire. The men would clear the land, brave the dangers, and dominate or convert those who were not British, white, or Christian. A woman's role was subtler. Childbearing had obvious imperial functions, but women were needed to civilize males. It was feared that men without women went native and indulged in local mistresses. A wife embodied the moral standards of empire, and husbands must "tenderly preserve them, as the plantation of mankind."

Given this, it is understandable but still remarkable how many women Henry convinced to accompany him to the Amazon. Henry's mother would go, his sister Harriette Jane, and his brother John Joseph. Harriette Jane was accompanied by her fiancé, twenty-five-year-old London solicitor Frank Slater Pilditch. John Wickham brought *his* fiancé, Christine Francis Pedley, who also brought her mother, Anna Pedley, age fifty-two.

The older women may have felt they had no option but to go. An 1875 group photo taken in Santarém shows Christine Pedley in profile. With long, curled tresses and dark, doe eyes, she was the most classically beautiful of the bunch. According to the 1871 census, she was the daughter of fifty-nine-year-old architect James Pedley, who made his home in St. Georges Square, Marylebone, close to the communal residence of the Wickhams. This was a comfortable address, a life not easily cast off, and the fact that Anna Pedley left it for the Amazon suggests either an incredible devotion to her fairylike daughter or the sudden death of James Pedley in the interim. If the latter was the case, and her only child was headed to the jungle, she faced the prospect of a widow's life, alone.

Henry's mother had faced that prospect for a long time. By now she was in her sixties, and her eldest had convinced everyone in the house to start life anew in the Amazon. What was Harriette to do? The milliner's shop on Sackville Street had never been a great success. There's the suggestion that she gave it to her daughter and lived in semiretirement with her children when not needed at the store. There was no safety net for Harriette if her children left—no Social Security, pension, or 401K. She would have faced a life in the workhouse as she grew old and feeble.

However, Henry's sister left the most behind. The 1871 census listed Harriette Jane as "head of household" in the Wickham residence, a woman "of independent means." In all likelihood she'd inherited her mother's millinery shop. She was headed for a comfortable middle-class existence, the same she'd forfeited to cholera as a child. She was picking up where her mother left off: Frank Pilditch, her fiancé, was a solid, conventional choice, much like her father. The 1875 photo shows a watch chain hanging from his vest pocket, a cigar balanced confidently between the index and middle fingers of his right hand. He is safe, uninteresting, a flat, forgettable face that fit nicely in court, not with the band of swarthy pirates that the Wickhams resembled. Harriette Jane stares straight at the camera, the rock-solid center of the group, a level-headed counterbalance to her brother Henry's flights of fancy.

And what did Henry offer in exchange for this future of privilege? Adventure, riches, exotic climes. A chance to work together as a family for a new beginning. They could build a world on their own terms. Henry promised Paradise, and they bought it hook, line, and sinker.

It was a paradise they thought they would recognize. Paradise is almost always tropical; the vegetation is lush and filled with succulent fruit; the men and women are beautiful and wear little or no clothing. Columbus in his first letters to Ferdinand and Isabella described his New World as a paradise; Gauguin and Rousseau peopled Paradise with fantastic vegetation, sensuous inhabitants, peaceful lions and tigers. To be a denizen of such a place meant a life that was happy and languid.

They left Liverpool by steamer at the end of summer 1871 and a month later were in Pará, now Belém. The capital of this rain-forest state is not actually on the Amazon but lies sheltered in the Bay of Marajó and connects to the continent-crossing waterway by the Pará River and a series of natural canals. Heading for the mouth of the Amazon itself was a tricky business for an oceangoing steamer. New shoals and sandbars were always forming and dissolving; a powerful eastward-flowing current threw a ship from side to side.

It was almost exactly a year since Henry had departed this place, and he hurried to the British consulate to announce his return. Except for letters and a legal suit, Henry would write no more about himself until 1908. Henceforth, the story is picked up by public records, scant correspondence, the chance observations of others, and Violet's unpublished memoir. Her observations are sharp, short, and to the point; she spares no one, least of all her husband. And now, her first contact with the tropics as she strolled through the narrow streets and tree-lined avenues of Pará with this congeries of Wickhams "was very like being dropped into deep water never having learned to swim."

Pará was a gateway to and preview of all the surreal beauty and mortal absurdities a tropical newcomer would encounter. There was the intermixture of Portuguese, Indian, and Negro races; the nobles dressed in their scratchy woolen suits and high stiff collars; the imposing cathedral and customs house; and the slow construction of the Teatro do Paz, one of the largest theaters in South America, begun in 1869. Perched in rows along the rooftops were Henry's old friends, the Urubu turkey buzzards. Henry muttered under his breath when he saw them. Violet knew the story and understood her husband's loathing. After a tropical shower they stood erect with their wings outstretched, a blasphemous imitation of a crucifix, feathers clacking in the breeze. Sometimes they seemed to commit suicide, flying across the open square to smash against a large white building. Examination showed them to be swarming with parasites, "a singular winged parasitical insect of a disgusting appearance," one observer later noted, "resembling a flattened house-fly."

They transferred to a river steamer in a matter of days. Steam navigation began on the Amazon in 1853, and by 1871, forty steamers owned by the Amazon Steam Company plied the waterway. Each was a multideck affair, open on all sides, popularly called a *gaiola,* or "bird cage." The upper deck was reserved for officers and first-class passengers, the lower for the engine, cargo, second- and third-class passengers, animals, and crew. The cargo was loaded in layers: merchandise, mules, and dogs on the bottom, passengers swinging in their hammocks above them, with space in the rafters for monkeys, boxes of insects, and squawking birds.

Violet had never seen anything like this mode of travel. Hammocks were strung from every rafter; she didn't know what to think of the gently swaying *rede,* or "net," but soon found them delightful in hot climates and discovered she couldn't sleep in anything else. "They should be nearly square," she wrote, "and you lie diagonally across them instead of length-wise." They should also be high enough to place one's foot on the ground to keep up the swinging, "the rocking and the rhythmical ring of the swinging cords soon lulling you off."

One hundred miles above Pará they entered a channel called the Narrows. This is ninety miles long, seldom one hundred yards across, threading a maze of one thousand forested islands. The main channel twists and curves to the west of Marajó, the central alluvial island about the size of Switzerland. Once past that, they were in the Amazon itself, which at the mouth was like entering an inland sea. Violet saw a vast expanse of water, a milky, yellow-olive concoction that stretched three to six miles from shore to shore. Great beds of aquatic grass lined the banks or broke loose to form floating islands. These had to be avoided at all cost, lest the propeller foul and they became part of the island themselves, borne backward to the sea. Fruit, leaves, and giant tree trunks floated past them in such quantity that it seemed the interior must be denuded of plant life. The level banks were lined with lofty, unbroken forest, the dark, straight tree trunks forming a living green wall right to the river's edge. A range of low hills that connected to the mountains of Guiana stalked back from the north for about two hundred miles. All was covered by forest, with frequent flocks of parrots and great red-and-yellow macaws screaming overhead.

She'd entered one of the largest rivers and most massive river basins in the world. The best current estimate is that the Amazon stretches approximately 4,000–4,200 miles, if the river's source is considered near

Balique in the Andes, just east of 50° W longitude. The length of the Nile is usually set at 4,100 miles. The Amazon's flow is five times that of the Congo and twelve times that of the Mississippi; it disgorges as much water every day into the Atlantic as the Thames moves past London in a year. Brazilian hydrologists believe that the river's annual discharge is about 57 million gallons per second: by such an estimate, the Amazon could supply in two hours all the water used by New York City's 7.5 million residents each year.

The valley itself is fan-shaped. In the delta, Violet bobbed in its 200-mile-wide apex. From there, it broadens until reaching the Andes, where the valley's width exceeds 1,500 miles. It drains an area of 2.4–2.72 million square miles—nearly all of northern and central Brazil, half of Bolivia and Colombia, two thirds of Peru, three quarters of Ecuador, and part of southern Venezuela. Until the 1970s, rain forest covered two thirds of that expanse. Since then, 10–15 percent has been cut back, though some has returned as secondary forest. There is no discernable slope: The main river drops a mere 213 feet between Peru and the Atlantic, a distance of 1,860 miles, while eleven of its main tributaries flow more than 1,000 miles uninterrupted by a single rapids or waterfall. Yet the current maintains an average velocity of 1.55 mph during the dry season and more than double that during high water, and in some places the water level rises and falls 50 feet or more. So much water flows below Manaus and Santarém that the river is a plow, cutting a channel below Manaus that has been sounded to depths of 330 feet. So much sediment is deposited that the lowlands have begun to sink under their own weight, and sediment depths of more than 16,400 feet have been recorded. Violet would hear the river described as the basin's excretory system, the means by which the waste products of the huge forest around her were eliminated, as if the endless riverine landscape were not so much an ecosystem but a creature itself, a huge, relentless, uncomprehending *thing*.

Richard Spruce saw the forest as a monstrous tree. Its tributaries were the boughs; its streams and creeks, the branches and twigs. He imagined in his journals a dark region where the river branched without end, losing itself in an impenetrable wall of green. One entered this verdancy and vanished; you stepped off the path and density closed behind you, as if civilization had never been considered, not even a gleam in God's eye. Only the forest mattered, the sole reality since the beginning of time.

But time had no real meaning here. Hours flowed into hours, and at first domestic details made the strongest impression on Violet. The bread, for example, was sliced and rebaked so that it could be carried into the interior, and its stale taste was her first foreboding that trouble lay ahead. "We had a few people traveling with us 2nd class intended as laborers for our new venture," she wrote. "One of them came up as a spokesman for the rest to show us the bread they gave them to eat—however, on hearing we had the same, they were somewhat quieter."

But even stale bread was a luxury. The "real bread stuff," she soon learned, was farinha, made from the manioc root—tiny dry pellets of starch that the locals picked up between the fingertips and popped into the mouth like popcorn. This took a little getting used to. Violet compared farinha to sawdust and was able to get it down only after soaking it in soup or gravy.

In her father's store, Violet had read about the wild tribes who inhabited the country's thousand streams. In 1541, Francisco Pizarro, secure in his defeat of the Incas on the Pacific coast, "had received tidings that beyond the city of Quito there was a wide region where cinnamon grew." He ordered his brother Gonzalo and his lieutenant Francisco de Orellana to find this cinnamon land.

They crossed the mountains with five hundred Spaniards, four thousand Indian porters, and herds of llamas and pigs. They descended the Andes into the jungle along the Rio Napo, and soon it was obvious there was no cinnamon, much less provisions for an army. The expedition was plagued by desertion, starvation, and sickness, and this river plunged on forever. Orellana was ordered to build a brigantine and scout ahead. Soon it became easier to continue downstream than return, so Orellana took the "flowing road" that carried him across the continent and months later into the ocean beyond.

It was during this accidental voyage that the rain forest became the haunt for fantastic Western dreams. Soon after they took off down the Napo, an Indian chief told Orellana of El Dorado, a fantastic city of gold just a few miles off—probably nothing more than an attempt to get rid of these mad strangers. "It was here that they informed us of the existence of the Amazons and of the wealth farther down the river," wrote Gaspar de Caraval, the Jesuit priest with Orellana's band. A certain tribe lived by a lake whose banks were lined with gold. Each morning the Indians coated their chief with a thin film of gold; each evening, he washed it off in the lake in preparation for the next day. *He* was El Dorado—the gilded one. Although gold was not discovered, other delusions prevailed. The most famous was Orellana's "encounter" with female warriors near the mouth of the Rio Trombetas, or River of Trumpets. These long-haired natives—more probably men than women—were responsible for the naming of what the Indians themselves called the *Paranáquausú,* or "Great River."

Three hundred years had passed, and the Amazon was still a mysterious river of dreams. An estimated 332,000 people lived in this region, up from 272,000 a decade ago. Brazil's boundaries were legalized in 1750; in 1807, when Napoleon invaded Portugal, Queen Maria de Braganza and Regent Dom João transferred the entire court to Rio de Janeiro with British naval assistance rather than surrender and abandon its alliance with Britain. After Napoleon's defeat, Dom João returned to Lisbon, but in 1822, his son, Dom Pedro, refused to depart and declared Brazil independent. In 1840, Dom Pedro II was installed as emperor at age fourteen; he still ruled today, his government a constitutional monarchy. Talk was underway of progressive policies, like outlawing slavery.

But in the remote Amazon, tides of history had little meaning. History was measured by the riches pulled from the forest instead. The first such "extraction cycle" was that of brazilwood, which began in 1503 and lasted to the nineteenth century, halted by the near-extinction of the species and discovery of synthetic dyes. Also known as dyewood, the tree was abundant along the rainforest that lined the Atlantic coast. To extract the brick-red dye, the heartwood was crushed before it was cooked. As the trees gave out in one area, they were felled at increasing distances inland. During the first three decades of Portuguese settlement, the harvest was estimated at three thousand metric tons each year. For the next two centuries, Brazil was considered an inexhaustible mine of dyewood, a cornucopia of deep red trees.

The next extractive cycle, involving sugarcane, began in the sixteenth century and reached its zenith in the first half of the seventeenth. Tea, chocolate, and coffee were becoming fashionable in Europe, and between 1600 and 1700, Brazilian sugar to sweeten them dominated the markets of the world. To break the monopoly, the Dutch occupied northeastern Brazil from 1630 to 1661, but they were eventually ousted and the Portuguese returned. Huge sugar plantations, or *engenhos,* spread along the coast because of more fertile land and easier access to Europe. Firewood was needed to maintain the sugarcane furnaces: They ran around the clock each year for seven to eight months, each *engenho* using up to 2,560 wagonloads of lumber per harvest. The industry created a wasteland. One historian lamented that sugarcane cultivation "left us a north-east despoiled of its very rich forests, [with] impoverished soils, and a miserable and servile population, possessing only its popular culture and the humility and humanity which only many generations of suffering can teach."

As Brazil's sugar production faltered, mines were dug. The most important gold strike came in 1693 in what is now the state of Minais Gerais. Later discoveries produced gold rushes as intense as those in California and the Klondike. Workers abandoned the cane fields for the gold mines, delivering a killing blow to the sugar industry. This flow of gold allowed Portugal to live well beyond her means. Between 1500 and 1800, the Americas sent to Europe £300 million in gold. Of that amount, £200 million came from Brazil. An estimated three hundred thousand Portuguese citizens joined the gold rush, so depopulating the nation that travel restrictions were enacted. Famine and disease hit the mining camps, forcing prospectors to seek new gold fields. They burned down forests and enslaved Indians. In 1700, Brazil's gold production reached 2,750 tons a year; by 1760, 14,600 tons. When the gold gave out, the depression that hit south central Brazil was as great as in the northeastern region, now the waste dump of sugarcane.

The cotton cycle came next, in the southern half of the country, but in the forested north, livestock was king. An estimated 1.3 million head of cattle grazed in and around the Amazon in 1711. Trees were cut down and savannahs burned to "strengthen" the range. Deserts formed along the São Francisco River, in Minais Gerais, Goims, and Mato Grosso.

Now the rubber cycle had begun. On one chilly morning, Violet awoke to see that their boat had anchored beside the small jungle port where the best rubber in the Amazon was said to be found. Santarém looked pleasant enough, resting on a slope at the meeting of the Amazon and Tapajós rivers, with a fine sandy beach, a handsome church with two towers, and houses painted yellow or white, their doors and windows bright green. A mist rose from the forest in the hills above town. *So this is my new home,* she thought, and hoped she'd grow to love it. She prayed that life would be kind to them, the innocent prayer of every young bride.

CHAPTER 7

THE JUNGLE

Violet liked Santarém from the start. It was a pretty little river port, with rows of white houses surrounded by green gardens, standing on a slope near the mouth of the Tapajós. A white sandy beach ran its length and a small, rocky knoll marked its western edge. Set at the juncture of the two massive rivers, the air was not stagnant, but pleasant and breezy. Rows of canoes lined the shore, and beyond them, a score of larger vessels. Since there was no pier in Santarém, barges ferried passengers and freight, or the boat pulled close to the beach and passengers hopped into the shallow water. The shore was lined with washerwomen of all colors: The occasional promenader picked his or her way through the drying clothes.

Nearly every Amazon town was divided like Santarém into *cidade* and *aldeira*, the city and village—the former the modern town, the latter the Indian village from which it sprang. The avenues of the *cidade* paralleled the river but were overgrown with grass, since there were few vehicles and everyone walked. Three long streets ran the length of town, and after that a fringe of jungle, from which the shrieks of the jaguar could still be heard. At the far edges of town, in the *aldeira*, people still locked their doors at night in fear of the cat. Beyond that, a grassy plain, or campo, rose to a high, forested plateau. Although the Wickhams rented an empty house, Henry indicated that in a few days they would be heading into that high interior.

Santarém was a trading town of about five thousand inhabitants whose livelihood depended on the Tapajós, not the Amazon. The locals called it the Rio Preto, or Black River. In fact, its waters were not black, but a deep, impenetrable blue. The Tapajós began 1,650 miles to the south in the jungles of the Mato Grosso, and although its drainage basin covered 188,800 square miles, it was only the fifth-largest tributary of the Great River. Where the rivers met, the Tapajós formed a huge lake: called *rias* by geologists, such "mouth-bays" are common in the Amazon. The mouth-bay of the Tapajós was so incredibly deep that oceangoing steamers could pass up the river to just below the trading villages of Aveiro and Boim. In high water they could go beyond that, past the limestone quarry at tiny Boa Vista and sometimes as far as 150 miles to the falls at Itaituba. Some Santarém trading houses had branches that distant. Canoes and lanchas brought rubber downriver, along with jungle drugs, cacao, small boxes of gold dust, and the occasional shrunken head.

The very spirit of peace seemed to rest over the village, something Violet suspected she'd miss once they struck off for the interior. But Santarém had not always been peaceful, and there were many ghosts here. The area once supported one of the highest populations in the preconquest Americas, with the towns and villages of the Tapajós, or Tupaya, civilization stretching for miles along the riverbanks. According to tradition, the Tupaya descended from tribes that had emigrated from Peru or Venezuela. They subsisted off the rich stocks of fish in the river and corn grown in the rich alluvial soils, replenished each year when the Amazon flooded. A number of flat-topped hills surrounded Santarém, and on them could be found *terra preta do Indio*, the "Indian black earth," a rich compost built up over the generations by Indian farmers. It was this black soil that Henry heard about on his first trip. He made the mistake, as did other newcomers, of confusing this manmade buildup with the natural state of the forest soil. The first European accounts, dating from the early sixteenth century, told of swarms of war canoes coming out to do battle and of Indian longhouses lining the banks. "Sixty thousand bows can be sent forth from these villages alone," wrote one Jesuit chronicler, "and because the number of Tapajós Indians is so great, they are feared by other Indians and nations, and thus they have made themselves sovereigns of this district." They were "corpulent" warriors, very "large and strong"; the tips of their arrows were poisoned, and "that is the cause why they are feared of the other Indians," the Jesuit Father said.

The modern town began when the Jesuits gathered the Indians to convert them to Christianity. The stone fort was built, and Santarém became a mission village. In 1773, the town was overrun by warrior Mundurucú Indians. The citizens and soldiers gathered in the fort and held off the attackers with concentrated musket fire. The Mundurucú fought hard, their wives carrying their arrows, but their greater numbers were repulsed by modern firepower.

The first decades of independence were no more peaceful. A civil war raged along the Amazon from 1835 to 1841. Known as the *Cabanagem*, or War of the Cabanas, what started as a slave uprising soon turned into a race war. By the time it was brutally suppressed, 30,000–40,000 were dead. Those unable to speak the *língua geral*—or who had any vestige of facial hair—were put to death. The Cabanas plucked all hair from their faces so as to not be mistaken for Europeans. Attacks of Cabanas on the homes of colonizers at night were preceded by harsh, jarring blasts on the *ture,* a horn made of long, thick bamboo. Now, once a year, their descendants, called *cablanos,* would assemble for carnival and march by torchlight to Santarém's European quarter, blowing the *ture* as they came. They danced before the doors of the principal citizens, but those old enough to remember shuddered at the sound, as if death descended from the forest once again.

Even now, racial matters continued to shape life in Santarém. It has been estimated that one third of all African slaves shipped to the New World came to Brazil; by the nineteenth century, that figure approached 60 percent. Although the nation's slave trade was banned in 1817, it didn't truly end until 1850, when Britain, motivated by a newfound morality in foreign policy, threatened Rio with a naval blockade. Even then, slavery didn't die. In reality, Violet observed, it still existed on the Amazon: "All children were born free but still the household had a large proportion of slaves one way and another." She was amazed by the resilience of the slave women: They'd perfected the art of multitasking when carrying loads around town, "a large jar of water on their heads, a plate of meal in one hand, palm uppermost just over the shoulder, a child sitting on their hip on the other side."

The laxity of the slave laws allowed Santarém to be home to two diametrically opposed groups of exiles. Escaped slaves from the sugar plantations of Barbados had settled along the Trombetas and Maicuru tributaries across the river on the Amazon's north shore. And there was a dwindling remnant from an original two hundred *confederado* refugees who'd come to Santarém in hopes of growing cotton and founding a new slave society. The *confederados* were scattered everywhere: on the Amazon's north bank, so that their trading boats passed those of the jet-black Barbadians when they came to town; in Santarém itself; and in a farming community south across the campo in the forested plateau. "That is where we are headed," Henry told her. Their new neighbors would not be the shopkeepers of Regent Street but backwoodsmen from the defeated Confederacy.

The last and largest group drawn into the current form of slavery was the Indians. Old animosities still lingered after the War of the Cabanas and the depredations of the slavers. Indian slavery did not always go by that name. An orphanage of Indian children existed near the Wickhams. Children were "drafted out to the different families to be brought up," a form of adoption that still takes place along the river, although along the Tapajós the practice seems more informal, between neighbors and extended families. In Henry's time, however, these "adoptions" had an economic motive. Housewives in Santarém "all made their house keeping (*sic*) money by sending out their slaves with things to sell," Violet wrote, "one day vegetables, another with drinks made from various kinds of palm fruits, another with pillow lace which these Indian girls made under their mistresses' tuition." The children seemed happy enough, "no cases of gross cruelty tho' you could often hear the *palmatore* going, a piece of wood shaped like a bat, with which the mistress of the house kept order by striking the erring one on the palm of the hand." Henry and Violet may have adopted a child themselves at this early date, for Violet later mentioned "a little Indian boy who had been given to H to bring up." There is no hint of the *palmatore* in her memoir, and the boy seemed devoted, following Henry everywhere.

Violet, too, followed where Henry led. This was to the campo. The path from town ascended a mile or two, passing through the narrow belt of woods, then entered grassy land that sloped gradually to a broad valley watered by rivulets. The vegetation of the campo reminded Spruce, who'd visited twenty-two years earlier, "of an English pleasure ground" like Hampstead Heath. Henry may have made the same connection, for there was a spring in his step, which Violet now recognized as a sign of his more ebullient moods. While the others tried to maintain some vestige of London fashion, Henry had gone native, or at least his interpretation— linen shirt, khaki pants and bush jacket, low-slung leather belt, and sheath knife hanging down his leg. He always looked the roughest of their group, with his black, tousled hair, black walrus mustache and per- manent three-day's growth of beard. He delighted in the image, and Violet seemed amused that the cheerful eccentric she'd married in London had become an even greater character out here. The plain they rode across was scattered with low trees, rarely exceeding thirty feet, and here and there gaily flowering shrubs. Beyond that, the deeply wooded hills extended as far as the eye could see. Some isolated hills were pyramidal peaks, the Indian black earth and pottery shards on top the last vestige of the huge Tupaya civilization. Complete solitude reigned over the campos as they rode south. The residents of town had little interest in a land they considered the haunt of jaguars and forest demons. A few tracks from town led to poor farms. Except for those, there were no roads, no signs of civilization.

The ride from town took a day, women on horseback, men afoot. Violet never gives a complete listing of their party, but the English workers would have been with them to clear the farm, while the older women— Henry's mother and John's mother-in-law—probably stayed behind. Violet soon got a rough reminder that she was no longer home. The saddles they brought were built for sturdy English horses, not the stunted Amazon variety: "I had not gone far when my English saddle turned around the horse's belly," dumping Violet unceremoniously on the ground. It was a preview of things to come: Henry had cautioned the sad- dler to shorten the girth, but things never turned out exactly as Henry planned.

They left the campos for a jungle path that seemed to Violet mostly "bush ropes and tangle," then ascended a hill. At the top, Henry turned with a triumphant grin. "Here we are," he said. They stood on an escarpment of the forested plateau six miles south of town in a place called *Piqui-á Tuba*, or "place of the Piqui-á trees." The naturalist Henry Walter Bates, who sojourned in Santarém at the same time as Richard Spruce, called these namesakes "Pikia" trees, colossal things as tall as the gigantic Brazil nut trees. They bore an edible fruit surrounding a seed, "beset with hard spines which produce serious wounds if they enter the skin." Although the fruit tasted to Bates like raw potato, the townspeople loved it and "undertook the most toilsome journeys on foot to gather a basketful."

They'd stopped in front of what Violet took to be a thatch-covered barn, and before that barn an old bearded man in high boots leaned upon a rake and smiled. She took him for a farmhand: It was not the last time she'd be fooled by appearances. The "farmhand" was Judge J. B. Mendenhall, leader of the *confederados,* and the barn was his home. Henry had decided to settle among these crazy-looking people until he could build his own house and clear his own land.

The strange tale of the Amazon Confederates began with the man some blamed for the tragedy of the Donner Party, the California-bound group of pioneers whose name became synonymous with survival cannibalism. Lansford Warren Hastings dreamed of an empire. Born in 1819 in Ohio, he traveled to Independence, Missouri, in 1842, joined one of the first wagon trains to the Oregon Territory, and was elected one of its leaders. The next year found him in California, where he envisioned a new republic with him at its helm. This required followers, so Hastings wrote *The Emigrant's Guide to Oregon and California,* in which he extolled the virtues of the new land and encouraged wagon trains to head to California along a desert trail he'd scouted and called the Hastings Cutoff. Hundreds of families took the cutoff. One was the eighty-seven-member Donner Party, trapped in the Sierra Nevada mountains during the winter of 1846–47 and forced in the end to dine on the recently dead. Barely half their number survived. One had to be restrained from killing Hastings after their rescue.

The Donner Party tragedy was infamous and ended Hastings's dreams of empire in the West. But his ambition was adaptable. He married the daughter of Judge Mendenhall and became an adopted Confederate. After the war, he wrote a new book, *The Emigrant's Guide to Brazil*, in which he painted an elysian Amazon, "a world of eternal verdure and perennial spring." Like Matthew Fontaine Maury, he imagined a Southern empire in the rain forest, and before his death of yellow fever in 1868, he had organized and shuttled two hundred Confederate hopefuls to a land grant in the jungle highlands behind Santarém.

But the empire wasn't turning out as planned. Most of the newcomers were like Henry, people endowed with more hope than practical experience. When Wickham arrived, only fifty of the original two hundred remained. The others had returned home or died. Jungle farming proved harder than anyone anticipated, and the band's more restless spirits returned to Santarém to drink and loaf. They were so raucous that the government withdrew the financial support promised for all. Those who stayed in the forest discovered that the soil was not as fertile as they thought. When they scratched away the topsoil, all that remained was clay. They could have stayed in Georgia for that, and at least in Georgia one could plow the land. Here it was a hopeless task, since the forest was laced with stubborn vines and shrubs. The *confederados* were too poor to buy tools or draft animals, so they pulled their own plows. The promise of growing cotton turned into a joke, so they hit upon sugarcane, distilling the juice into a rum sold in Santarém. Their wives and daughters toiled beside them until they dropped from exhaustion and succumbed to fever, and every year saw more wooden crosses planted in the Confederate cemetery.

The *confederados* were still clearly struggling. The last fifty were grouped atop a five-hundred-foot bluff overlooking the Tapajós River. Only one of their number prospered, and he'd settled elsewhere: R. J. Rhome had arrived with lots of money and parlayed that into a managing partnership at Taperhina, the plantation owned by the Baron of Santarém, which lay twenty miles downriver on the Amazon. The others were backwoodsmen from Tennessee, Alabama, and Mississippi, "lean, hard men with their wives and children . . . who had come to stay," one observer said. Judge Mendenhall had assumed the colony's leadership after the death of his son-in-law Lanford Hastings. R. H. Riker, a railroad president back in South Carolina, had established, with Clements

Jennings, James Vaughan, and others, a satellite colony at Diamantino, a stream feeding into the Tapajós where diamonds had been found years earlier. There were no instant riches now.

These were Henry's friends. For the first time, Violet doubted her new husband's wisdom, if not his sanity. She'd left her life in London for *this*? She could have dissolved into tears or started screaming at Henry, but instead she took a deep breath and determined that "we would never grow quite as they were." At least, she told herself, she'd wear her best dress on Sunday, unlike these backwoods women, but no sooner had she put it on than it was suggested that they go see a field of sugarcane atop a nearby plateau. After a half hour's climb, she looked down and saw her "white stockings and legs and feet about the color of the rich black soil we had been exploring." She reassessed her fashion sense and admitted, "This put rather a damper on my ideas of finery."

It would be a good metaphor for her coming life. "Alas," she later wrote, "I have grown as utterly careless of such things as any born backwoods woman anywhere."

❧

This would be their home for the next three years.

Piquiá-tuba sat on the edge of the escarpment, with vistas of the town to the north and the Tapajós to the west. The plateau was known as the *Montanha*, though there was nothing mountainous about it; it was just a spur of the higher region in central Brazil. The approximate location can still be found. Until 1997 and the advent of clear-cutting for rice and soybean farms, the plateau was much as it had been in Henry's day, a rise of primary forest, home to hundreds of species of gigantic trees. Where, at most, twenty-five kinds of tree might be found in a temperate-zone forest, in the tropics the same area can hold as many as four hundred. The difference is that they grow singly, widely spaced, as protection against parasites, whereas in temperate zones a single species can grow in stands of hundreds or thousands. On Piquiá-tuba, in addition to rubber and the indigenous piquiá, some of the more common trees included the towering, golden-crowned ipe, or ironwood tree, now known in America as the wood used for the boardwalk in Bill Gates's coastal mansion; the purple ipe, whose bark has been suggested as a possible cure for cancer; the giant Brazil nut tree, soaring at least 200 feet, with no lower branch less than 120 feet above the ground; and the rosewood tree, an essential ingredient in Chanel No. 5. Everywhere Violet looked she would see splendid masses of green—cacao trees, limes, great pale banana plants, and coffee bushes straying up from the campos into the woods.

Why *were* there so many trees? The forty-square-mile Reserva Florestal Adolfo Duche, a rain-forest preserve northeast of Manaus, is currently estimated to have 1,300 species; while the preserve itself is two thousand times smaller than the land mass of England, it has forty times more native trees. About 120 theories grapple with this problem, but two seem to prevail. First, the ice ages that scoured the temperate zones repeatedly wiped out vegetation, leaving tough terrain in a tough climate that assured that only the hardiest plants survived. Second, in the tropics, the stresses are biological, not physical or climatic. A species of fewer, more widely spaced trees would stand a better chance against parasites and insects, which helps explain the rampant diversity.

The rain forest was alien to Violet in so many ways. She had never experienced a silence like that found there. It was a force in itself, almost hypnotic. She might follow a large morpho butterfly into the woods, entranced by its wings of glistening cerulean, when suddenly she'd realize that even her footsteps did not make a sound. The ground was not ground as she'd known it; it was a dark compost that gave off a thick, bitter smell. The trees seemed as silent as centuries; no air stirred the leaves, not even a bird called. The morning light was strange and diffused, like light in a dark cathedral. There was a verdure in the air itself that seemed so thick she could crush it between her fingers. The forest's real life swarmed overhead, where birds nagged and clicked and howler monkeys bawled with a strange plaintive loneliness that echoed for miles. It seemed like the cries of madmen. Once she spotted the white face and whiskers of a capuchin money, clutching its squashed fruit to its chest, leering at her like a gargoyle.

Night was a different world too. After sunset, the forest's silence was replaced by cacophony—the roar of the howler monkey, "the deafening

Henry Wickham, 1899.

Henry Ford, 1934.

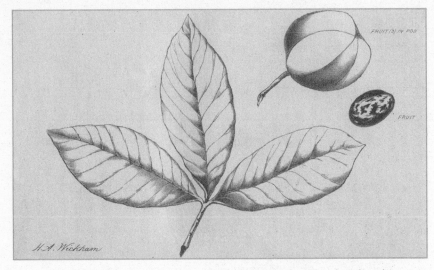

Henry Wickham's sketch of *Hevea brasiliensis,* leaf, seed pod, and seed.

Wickham's sketch of the graves outside Santarém.

Santarém, 1875. *Standing (left to right):* Henry Wickham; his brother-in-law Frank Pilditch; and his younger brother John Joseph Wickham. *Seated (left to right):* Violet Wickham; Wickham's sister Harriette Jane; and Christine Wickham, holding her son Harry, who was born in Santarém.

Wickham's sketch of Violet at the first camp outside Santarém.

Wickham's sketch of tapping a rubber tree on the Orinoco.

Smoking rubber in Brazil.

Wickham's sketch of his rancho on the Orinoco.

Charles Goodyear,
discoverer of the vulcanization process.

Joseph Dalton Hooker,
second director of Kew Gardens.

Sir Clements Markham.

A rubber tree in Belterra, with tappers' scars.

Taperhina plantation house where Spruce, Wallace, and Bates met in 1849, and where Henry Wickham may have recuperated from a near-fatal wound in 1873–1874.

Taperhina, from the heights.

Henry Wickham standing by the oldest tree in Ceylon, 1905.

Portrait of Sir Henry Wickham, after his knighthood and shortly before his death.

One of the original seeds brought from the Amazon.

clamour of frogs of all varieties of hoarseness," the swoop of the giant bat among the trees. Strange forms of life came out at night—foot-long katy-dids with spines like armor plating and mandibles that cut through leather and vampire bats that crawled up at night and nipped a fold of flesh, lapping the drops of blood. These were particularly troublesome at Piquiá-tuba. Though Violet and Henry were never bitten, others were, and suffering most were the domestic animals. The bats would bite her poor horses night after night in the same spot, "not only weakening them but raising such a sore you cannot saddle them," while her chickens emerged from the henhouse each morning "staggering from weakness and their combs as white as the rest of their bodies." Those of their party whom the vampires preferred were usually nipped on the big toe; they would know nothing of the attack until they awoke in the morning "bathed in blood."

Violet liked to go riding at night; it was one of her few pleasures. Termite hills lay in the path and glowed in spots like coals in ashes. Fireflies and "other spots of electricity in decaying wood" made the jungle seem other-worldly. Henry went with her, carrying his gun, for jaguars still roamed the heights, and if you heard its cough, you were as good as dead. Some mornings she saw tracks around the house. The cats were drawn by her hens. There were half a dozen jaguar hunters living in the campos or in town. In Santarém, everyone told of the jaguar and the three men, an incident that occurred a few years before Richard Spruce arrived. One of the men was armed with a musket, another a *tercado* (saber), and the third, tall and powerful, went unarmed. They were hunting on the *mon-tanha* when the jaguar sprang at the third from the bush, but the man was strong and grabbed the beast's forepaws. They struggled until the cat wrenched one paw free and tore the man's scalp down over his eyes. The man with the *tercado* ran to his assistance, but the jaguar turned and wounded him severely. The cat sat between them, eyeing each disabled man as if wondering whom to devour first. At this point the man with the musket ran up and the battle was renewed. The "tiger" was killed, but not before wounding the third man, too. The man who'd been scalped still lived in Santarém, and people always pointed him out. He wore a black skullcap where his scalp remained tender. Violet had seen him and heard the stories; she didn't have to be reminded when Henry warned her not to stray.

Sometimes it seemed that the place was supernatural. Henry laughed when she said this, but she'd read his journals and knew there'd been times when odd feelings crept over him, too. It was as if she were being watched. More than once she disturbed a silk moth in its rest and its wings unfolded: staring back from the large pale wings were huge false eyes, deep blue spheres countershaded with ebony, laced with shiny silver flecks. After the initial jolt she realized that the "eyes" were there to startle predators, but it brought home to her that in the forest, eyes were hidden everywhere. Here on the Tapajós it was thought that the *curupira,* the spirit man of the forest, whom Henry's saviors on the Orinoco believed had cursed him, wandered beneath the trees with his wife. He sported one eye in the middle of his forehead, as well as blue teeth and huge ears; his wife was even uglier, with one eyebrow in the center of her head and her breasts beneath her arms. Violet wondered if she'd look like that after a year in the forest.

There were native witch/herbalists, too—*feiticeira,* who lived in solitary hovels at the edge of the forest, specializing primarily in love potions and forest cures. Henry Bates had come to know one named Cecilia, and although always very civil to his scientific party, she explained that some *feiticeira* were bitter women, familiar with various poisonous plants and not adverse to administering a debilitating dose to a rival in love.

Anything seemed possible in a place where the forest breathed in and out like a giant set of lungs. The canopy provided shade 65 to 165 feet beneath its top, acting as a cooling and humidifying system for all life below. In transpiration, the trees pumped water from the soil back to the atmosphere, making runoff more gradual. The forest was a great recycler. The trees sucked potassium, phosphorus and other nutrients into the root mat and attached fungi as they drew water up through their roots. Such luxurious greenery suggested abundance, but this was an illusion. The soil was so poor that the only option was to allow nothing to be wasted.

Then there were the storms. Violent thunderstorms were frequent, especially at night. Immense armies of clouds would swoop from the west. A suffocating calm preceded the onslaught, and then the sky disappeared, and gusts tore out rotten trees. Electrical explosions and fantastic rolls of thunder reduced everything to light and noise and water. As the sky merged with the land, it seemed to Violet that this must be what Noah experienced during the end of his world.

After the rains, there came a suffocating humidity so overpowering that physicians of the time believed it weakened the lungs, creating a "fertile field" for the tuberculosis bacilli. One thing was certain—the wet was everywhere. In the walls, in the floor, even in the bedding. Nothing ever dried, especially clothing, unless dried over a fire. A white, hairlike fungus covered books and papers; tools were attacked by inevitable rust. The wet seeped between each fold of skin and into one's private parts, creating a rash that itched unbearably and never went away.

Henry said it would pass: such discomforts were merely temporary. They'd build a house, plant their crops, acclimatize to the weather as thousands of colonials had before them wherever the British Empire planted the flag. Violet wanted to believe him so badly. Their first home was made of thatch, which sounded exotic but really was nothing but palm leaves. Violet learned from Judge Mendenhall's wife and daughter how to take a young, unripened palm frond, five to seven feet long, and with a shake and turn of the thumb and forefinger break the leaflets at right angles to the stalk "till it looks like a gigantic green fringe." These were tied to the rafters three or four together, one above the other. "While green it was pretty," she wrote, "but it soon looks shabby, although if well and thickly put on it lasts some years."

This first hut was a crude affair, fourteen feet long by nine feet wide, divided into two rooms by a partition and made of palm leaves tied to poles and rafters. Henry promised her a wooden house in a neighboring section of forest with a real plank floor and overlapping shingles to keep out the rain. To this end, he "started off early in the morning, and I saw him no more till evening." Since there were no sawmills within reasonable distance, he cut down the trees at the home site and split them into boards. Hours might be spent felling a single gigantic tree, and when it dropped, it brought others with it. Lianas strung from tree to tree like cables seemed to tie the whole forest together. They rained down in

twisted masses, dragging epiphytes and orchids down with them. Sarsa-
parilla was one such liana, and Henry could make some badly needed
money in town when that fell on his head. Other times, he wasn't so
lucky: The *juruparipindi,* or "devil's fishhook," festooned with broad sharp
thorns that could wound a man severely, would crash down. You never
knew what was up there. Termite nests might fall like clay pots, or sting-
ing ants drip like fiery rain. All a man could do was cover his head and
run. A tree's descent created pandemonium: Monkeys leapt from branch
to branch; parrots screeched; toucans, the most inquisitive of the observers, flew down and canted their heads as they investigated the
devastation.

Violet had her own problems at the *confederado* farm. Cooking was a
daily trial. Henry showed her how to make a fire—three logs, their ends
pushed together—and each day she squatted in the dust blowing on the
spark to get it going. "I get it burning, put on sauce pan or kettle, and leave
it to attend to something in the house and when I return the wood has
burned thru' and upset the sauce pan and its contents," she wrote. "This
was always happening, sometimes to the destruction of the kettle, always
to the fire."

Drawing water was an equally tiresome chore. Their farm was in the
heights to escape the clouds of insects and miasmic effects of "bad air,"
but the springs were in the valley, a half mile away on a steep incline.
There were no wells, no pumps; any attempt to dig for water resulted in
a black, brackish pool. Every day the men carried water uphill in what
seemed the labors of Sisyphus. The women walked to the springs for a
bath, but the humid return journey left them feeling "as tired, hot, and
unrefreshed as before." Trekking to the Tapajós took longer, and down
there one had to avoid the piranha. Nineteen years earlier, in 1852, Henry
Bates had found the Piquiá-tuba piranha ravenous: They would "attack
the legs of bathers near the shore, inflicting severe wounds with their
strong, triangular teeth." The things one took for granted at home—clean
water, a bath, no killer fish in the tub—were luxuries out here.

It soon became evident to Violet that Henry's promise of " 'temporary'
went on extending." She may have been the last to publicly admit the fact;
no doubt she heard grousing from the others about this hell on earth that

her husband had brought them to. She knew Henry heard too. Running sores erupted from her arms and legs, some so deep they would not heal, leaving scars that lasted to old age. The rainy season was suddenly upon them, and although there was no flooding in the heights, the rains and daily trips to the planting site got on everyone's nerves. They moved to the house Henry built long before it was finished "to save time with coming and going." The bedroom was raised two feet off the ground and floored over, except for a yawning hole, which they concealed with the bed. In the absence of a chimney, smoke drifted from the hearth through the doors and windows. Henry and his English laborers planted manioc, sugar, and tobacco in the Indian black earth, but the seedlings died after germination or refused to grow.

But this was child's play compared to the first real crisis. The English laborers had never dreamed of a life like this, "and soon left us one after another," Violet said. It is hard to tell if there was a precipitating event. One of the first deaths recorded in their group was a George Morley of Dorset and London, an older man, as Violet suggests, though his age is unrecorded. They left openly or in the night, singly or in pairs, back to Santarém, back down the river to Pará, back by monthly packet to Liverpool and home. Lack of labor could kill a farm as quickly as violence and disease. The "labor problem" was empire-wide, extending from small freeholds in the Queensland outback and Rhodesian veldt to rich plantations in India and Ceylon. When English workers deserted, farmers turned to native labor, but the indigenous idea of work was different from that of the British Isles. Henry tried hiring local Indians or the halfbreed peasants known as *caboclos,* but they were "independent and lazy," Violet complained, no doubt reflecting Henry's frustration. This was a worldwide clash of cultures as well as of classes, the struggle between owners and workers that too often turned into abuse, violence, or slavery. There is no evidence that Henry abused his workers, but he was always unsuccessful with them, a problem that hounded his every effort across the globe.

A hardness began to creep over Violet that she didn't recognize. In London she'd called herself working-class, but now she and Henry were owners, people with capital, no matter how meager, and in a moment they could lose it all. Possessions do something to people, even in the jungle: she shifted her sympathies, though she was never aware how much. Observations of settlers' wives and memsahibs across the empire proved time and again how crucial the woman's attitude was to a plantation's success: "If she handled the morning 'clinics' and other encounters with patience, sympathy and interest, a genuine *rapport* developed" between the two worlds of British owner and native worker, noted Empire historian Deborah Kirkwood in a study of Rhodesian farms. "[A] readiness to interest herself in the health" and problems of workers and their families "was undoubtedly appreciated." Violet had successfully negotiated the first hint of crisis—the complaint about the bread on the steamer—but her solution was to quietly embarrass the spokesman, and that proved to be a mistake. Now she had no way to win back her workers' hearts, and so she, Henry, and Henry's family were left to clear the fields alone.

But more than by labor, Henry was defeated by the soil. Despite the luxuriousness of the rain forest, the Amazon soils are more fragile than those in temperate zones. Unlike that found in temperate-zone forests, no thick humus layer exists on the Amazon forest floor. The ancient Indians tried to combat this by heaping generations of black earth, but they could not outwit ecology. The root system of the trees is shallow, usually concentrated in the top eight to sixteen inches of earth, and is three times as dense as that found in the temperate zone; it resembles the "root ball" one disentangles when buying a pot of mums at the local nursery. Nutrients are rarely stored in the organic matter littering the soil but are immediately reabsorbed; this rapid decomposition of forest litter is performed by fungi. Whereas in temperate zones recycling takes place in the soil, in the Amazon it occurs on top: research has shown that nearly all nitrogen and phosphorus are stored here. A 1978 study showed that 99.9 percent of all calcium 45 and phosphorus 32 sprinkled on the root mat near the surface was immediately absorbed, while only 0.1 percent leached through. The rapid growth of small roots and white fingers of mycorrhizal fungi lay behind this process: the forest did not depend upon the soil to grow from, merely to stand upon. The soil was not the principal source of nutrition, just the shallow engine for circulation and exchange.

When Henry cleared the land for crops, he exposed the soil to the full force of the climate. The daily rain compacted the surface, decreasing its permeability. When absorption declines, runoff increases, which means greater erosion. And there was the sun, raising soil temperature to the point where it cooked the bacteria that destroyed organic waste, meaning no humus could form. Ultraviolet rays, once blocked by the forest canopy, now produced chemical changes in the dirt, converting the nitrogen and carbon dioxide left over by decomposition into gas. Instead of staying in the soil, where they were needed as natural fertilizer, the two elements escaped into the air.

Soon afterward, the second, more terrible crisis materialized. The first wave of death began. Violet seems callous in her memoirs, only mentioning "burying two or three of the older members of the party" without listing names, but perhaps by then she'd decided that such hardness was the only way to survive. We know the dead from Henry's sketch of five crosses in the Confederate graveyard "made of some very hard Brazil wood," he wrote, and "erected in the Forest behind the Town of Santarem." The first cross stood over the worker George Morley; the second, over Anna Pedley, age fifty-five, the mother of John Wickham's fiancé, Christine; and the third was for Henry's own mother, Harriette, age fifty-eight, who'd survived so much to be killed by the place that had fascinated her. She'd filled her son with dreams of the tropics; he brought her with him to paradise, and nearly a year after arrival, on November 6, 1872, she died.

Henry never wrote of his mother's death. Violet stayed far away from the subject too. But Harriette's death marked a turning point, after which the others began to drift from Henry and his crazy hopes, and Henry sank further and further into himself. After 1872, he became a solitary ghost, a melancholy man. He must have tortured himself with guilt for bringing Harriette here to her death. From then on, he became more solicitous and protective of Violet at the same time that he dragged her deeper into his headlong schemes. Henceforth, until she could not take it any longer, she tried to match his toughness with her own.

What killed them so quickly, then would kill two others in 1875–76? No cause of death was ever mentioned, but the reason was certainly disease. Any number of parasites and microscopic killers live in the tropics, but in this region three culprits were more likely than the others: yellow fever, schistosomiasis, and malaria, Henry's old friend. Schistosomiasis, a liver disease attacking those who bathe in water where the plenopid snail is present, was least likely, since its main victims were children, but it should be mentioned because it has only two known provenances in the Amazon—back in Pará, and on the Tapajós. Yellow fever was more likely than that: The disease that in 1850 carried off many of Richard Spruce's friends in Pará recurred with a vengeance at ten-year intervals and came with the mosquitoes and rain. The virus struck quickly, beginning as depression and restlessness, then transforming into searing aches in the joints, back, and head. The victim's face became flushed, her pulse irregular, eyes glassy. Within five days she most likely would be dead. Sometimes a dark fluid would gush from a victim's mouth and this was the final horrific blow. A person could contract yellow fever and survive, but once the black vomit appeared, death was assured.

The most likely culprit, however, was malaria. Henry would have stood beside his mother as his own savior had, fanning away flies, bringing the cooling draught of sugarcane juice to her lips, remembering how even the simplest act of kindness was appreciated. But it wasn't enough this time. The egg-shaped plasmodium parasite killed her in five to seven days, rupturing the oxygen-rich red blood cells in huge quantities. Soon the hallucinations start. It is almost a blessing when the sufferer lapses into coma and dies.

By 1874, three years after arrival, Henry was ruined. A rift occurred within the original party. Those who'd survived so far—Henry's sister and brother and their spouses—set up house in Santarém. They dreamed of returning to London, but that took money. They established a school in town, saved for the passage, and counted the days.

Only Henry and Violet remained in the jungle. "We alone of the original party picked up and once more started afresh," Violet wrote. But starting afresh meant moving ever deeper into the forest. There'd be no backup if they made mistakes. They'd be alone.

CHAPTER 8

THE SEEDS

Henry might feel alone, but for the first time in his life he was about to be used by a higher power: Kew. He'd be turned into a tool, and he encouraged the conversion. If Henry and Violet had any hope, it came from halfway around the world.

The change began with publication. In early 1872, Henry received word that his book was ready to hit the London bookstands. His *Rough Notes of a Journey Through the Wilderness* would never be a hit like Livingstone's or Burton's journals, but the book did include Henry's sketches of jungle life, and one sketch in particular proved important: his drawing of the leaf, seed, and seed pod of *Hevea brasiliensis*. This drawing, more than anything else, would be the key to Henry's future.

Word that he was an author seems to have filled Henry's sails. This was still early, when he could dream that his small plot in Piquiá-tuba would someday be a huge plantation and he, a feudal lord. He apparently informed Drummond-Hay of his "success" on the Tapajós, but more important, in March 1872, he contacted Joseph Hooker at Kew. In a letter from Santarém, he told the director that his house was on "a spur just off from the forest covered table highlands S. of Santarem which occupy the triangle formed by the junction of the Tapajos with the Amazon." He always described his holdings in strategic terms: a triangle of land guarding all approaches; a view from the heights commanding his ancestral lands. "The waters of both rivers, islands and estuaries are taken into the view by new home," he wrote. He was, in essence, the lord of all he surveyed, and thus in position to send valuable specimens to Kew. Though

Hooker did not bother to reply, Henry was buoyed by his own confidence. He followed up the first letter with a package filled with tubers and palm seeds.

If Henry did one thing right during those disastrous years 1871–74, it was to maintain this correspondence with London. He apparently sensed that his best chance to succeed in the Amazon lay in some alliance with Kew. That he set his sights on Hooker suggests that the salesman in him sought ascendancy. He thought he could be a planter, but his real talent lay in self-promotion. In Hooker, he identified the one man who could salvage this tropical wreck he'd created and perhaps give it meaning. As he had with young Watkins and then James Drummond-Hay, Henry placed all his faith in one man.

And Hooker did take notice. Sometime in 1872–73, he read Henry's book and spotted his sketch of hevea. Although there was by now plenty of anecdotal evidence linking hevea to "Pará fine," not even Spruce had made the absolute botanical determination that *Hevea brasiliensis* was the sought-after Grail. Along came this badly organized book in which an obscure and cocky Englishman not only observed the collection of rubber but engaged in its tapping and curing. The drawing of the leaf, seed, and seed pod was a first. It convinced Hooker (and soon enough, Clements Markham) that Henry could identify hevea in the wild, not an easy thing to do in the chaos of the rain forest. Henry had always fancied himself a budding artist. The irony would be that the importance of his work lay far outside the art world.

Yet as important as this and related drawings would be to Henry's fate, they also contained a level of mystery that symbolizes so much of his career. Henry's simple, shaded drawing of hevea's "leaf and fruit" assured Hooker that he knew his business, and his descriptions of rubber tapping on the Orinoco established his reputation as a rubber expert, the latest in the fortuitous line of "men on the spot" on whom the empire's fortunes so often depended. But how well *did* he know his business? Although it could not have been known in London at the time, it has since been demonstrated that *Hevea brasiliensis* did not grow so far north. Henry and his assistants must have been tapping another variety on the Orinoco. An accompanying sketch depicted the removal of the entire bark of the trunk, a practice never reported in the Amazon among myriad travelers' accounts and one that would kill a tree. "His drawings of the leaf and seeds, if they did indeed represent *H. brasiliensis,* must have been of specimens found along the Amazon on his journey home," perhaps during his brief stop at Santarém, mused environmental historian Warren Dean. Hooker innocently put himself in the same position as Henry's family—placing his faith in one who was considerably less knowledgeable than he supposed.

But Hooker did know that another cinchona coup would help him politically. Three decades had passed since Charles Goodyear patented vulcanization, and each year industry discovered more uses for this nonconductive, waterproof, elastic material. The wars in Crimea, the American South, and the Franco-Prussian War proved its strategic and even geopolitical necessity. Shoes with vulcanized soles fused to canvas uppers and called "brothel creepers" were a big hit in the United States. In Great Britain, they became known as plimsolls. In 1870, B. F. Goodrich founded a rubber factory in Akron, Ohio; in 1871, Continental Kautschuk und Gutta Percha Co. started in Hanover, Germany; in 1872, Italy entered the race with G. B. Pirelli and Co. In 1849, when Spruce first came to Santarém and scoured the jungle for caoutchouc, the U.S. price for rubber was three cents a pound. In 1872–73, the price hovered around sixty cents and showed every indication of rising. Great Britain knew its power rested on ships, but ever since the first battle of ironclads in March 8–9, 1862, between the *Monitor* and *Merrimac,* the Age of Sail was inevitably transforming into the Age of Steam. With it came a change in raw materials, from timber and hemp to coal and steel—and rubber.

Clements Markham saw this more clearly than others. In 1870, he'd awakened to the need for rubber after James Collins's article compared its procurement to the theft of cinchona. In 1871–72, as Henry launched the plans that set in motion both personal and regional disaster, Markham began pushing his own designs. He appointed Collins to draft an even more focused report comparing the utility of the various known species of rubber-bearing tree. Collins's report favored hevea over castilla, guttapercha, or *Ficus elastica,* even though all the others were better understood scientifically and more easily acquired. Markham needed little further urging. Armed with the report, he enlisted the Duke of Argyll, Secretary of State for India, and Lord Granville, Secretary of State for Foreign Affairs, into his scheme. On May 10, 1873, Collins's report was forwarded to James Drummond-Hay in Pará with the request to find someone willing to collect the seeds. The letter mentioned "a Mr. Wickham, at Santarem, who may do the job."

A year after his first, unacknowledged letter to Hooker, Henry was getting attention. On May 7, 1873, three days before the official instructions were sent to Pará, Markham asked Hooker for advice. He informed Hooker of the Foreign Office orders to Drummond-Hay "to take steps to obtain a supply of seeds of the *Hevea*," then asked whether the seeds should be sent to Kew first "to be raised there, with the view of afterwards sending the young plants out to India." Eight days later, on May 15, Hooker said growing the seeds at Kew was a capital idea. He added:

> I have a correspondent at Santarem on the Amazon who is engaged in the business of rubber collection and I will write to him for particulars as to the mode of growth of the tree and the methods of collecting from it. I will also beg him to send to Kew a considerable quantity of the seeds.

A flurry of attention was being directed at Henry that May 1873, but he would not hear of it for another six months. In nearly simultaneous letters, Hooker wrote to Henry, while James Drummond-Hay was told by the Foreign Office to roust Wickham from the jungle and give him the news. But both letters were apparently misplaced, and the actors had changed. Drummond-Hay had left Pará for Valparaiso; although a promotion, the move was accelerated by his interference in local politics. The new consul was Thomas Shipton Green, manager of the Pará branch of Singlehurst & Brocklehirst, a London export firm. In the confusion, the letters would not surface for five months, in September.

While everyone waited, rubber seeds from an unexpected source materialized in London, thanks to the tireless James Collins. Collins had written many letters to rubber collectors around the globe, including Charles Farris of Cametá in Brazil, a town about sixty miles south of Pará. On June 2, 1873—while Henry's commissions from Kew and the Foreign

Office languished in some diplomatic limbo—Farris arrived in London to recuperate from fever, and he brought with him a packet of rubber seeds "quite fresh and in a state for planting," Collins said. Markham got busy when he heard. He set a price for the seeds—£2.10 per pound—and told Collins to buy every one. Since Farris had two thousand seeds, the Empire paid about twenty-seven dollars. "I thought it important to secure them at once," Markham told Hooker, "and deliver them to you with as little delay as possible." For once there were no bureaucratic quibbling, probably because, as Markham added, the U.S. and French consulates had already made a bid. It was an effective warning of the intentions of the other world powers, and two days later the seeds were sown at Kew.

Farris later told Lord Salisbury how he'd sneaked the seeds from Brazil. As he was leaving, customs officials entered his cabin and saw that he had a couple stuffed crocodiles with him. "Is that all you managed to shoot?" they asked. "It was a very disappointing trip," Farris replied. As soon as the ship left the harbor, Farris locked the door, slashed open the crocodiles, and revealed the two thousand seeds.

But they were not as fresh as Farris claimed. Only twelve germinated. Kew kept half for study, and on September 22, 1873, sent the others to the Royal Botanic Gardens in Calcutta. Despite the best efforts of the staff, every seedling died.

With this disappointment, Collins stepped from the picture. He'd be cursed by his pursuit of rubber, as were so many others. There would be no recognition of his work. The record of his contribution exists only because he was never paid. In 1878, he sent a second bill and a lengthy memo to the India Office asking for ten pounds for his services. He was clearly bitter about being pushed aside, and wrote: "I would like to take this opportunity to place on official record that if any honour be due for being the first person through whose instrumentality live plants of the Pará Indian rubber tree have been introduced into India, that honour is undoubtedly due to me." But he'd padded his ten pounds by a middling amount, and that sealed his fate. There was no greater sin in the parsimonious Victorian system than waste, and Collins had cheated by a couple pounds. Undersecretary Louis Mallet rejected the request. He accused Collins of "a gross attempt to impose on the Secretary of State," and added that he'd "already succeeded in obtaining £80 from this office for an utterly worthless report on Gutta Percha." There was no mention that this was the report rating hevea superior to gutta-percha and no mention that Collins's inquiries got the ball rolling on procuring the tree. By now, Collins was unemployed. In a few more years, he'd be broke and alcoholic. In 1900, he died in poverty.

But Collins's bad luck was Henry's good fortune. If Charles Farris's seeds had thrived, Kew could have abandoned Wickham to his own devices. When, then, did the dispatches from Hooker and the Foreign Office resurface and Henry finally receive his marching orders? It had be sometime in the late summer of 1873, a year after Henry's mother died. On July 23, 1873, Harriette Jane Wickham wed Frank Pilditch and John Wickham wed Christine Pedley at a double ceremony in the British Consulate in Pará, probably officiated by Consul Green. At that time, the consul failed to mention rubber, and so the letters were still lost. By September 1873, this was rectified.

Since Hooker's letter to Henry apparently no longer exists, we can only surmise its contents from Henry's giddy reply. The director seemed purposefully vague, and told Henry that an official commission from the Foreign Office awaited him at the consulate in Pará. Hooker's letter filled him with joy and hope. He assured the director that he was "glad to accept your offer to put me into communication with the partie you refer to in your letter," a conspiratorial tone that saturated his memories of this period. More important, Kew's commission became a way to justify his every wrong turn. It gave his life meaning and gave his imagination a boost. He was no longer Henry Wickham, deluded dreamer, but Henry Wickham, a spy in the enemy's camp, Agent of the Crown.

It also gave him the confidence to bargain for terms. By mid-September, Henry finally spoke with Consul Green, so that by September 23, Markham told Hooker that

The Consul at Para has written to say that Mr. Wickham proposes to establish a nursery of India rubber trees, and then to ship them direct to England . . . he says that in this way a large number of healthy young plants would be secured of uniform size and hardiness. The locality will be on the right bank of the Amazon . . . where he is now making a plantation of coffee. He of course asks for a remuneration for the time and care required. How would you advise us to answer?

Six days later, Consul Green told the Foreign Office that he'd asked Wickham for a cost estimate. Henry warned him that hevea seeds were highly perishable: The pods contained oil that quickly turned rancid. He thought it more practical to start a nursery on his farm and forward the young sprouts to Kew when they were hardier. Green apparently thought the idea expensive. He offered in his dispatch to "obtain any quantity that may be necessary at small expense," in effect proposing to undercut Henry.

But the Farris affair demonstrated hevea's fragility, and Henry's suggestion to let the seeds sprout, then ship them in portable greenhouses called "Wardian cases" had merit. He seemed to realize that his suggestion was opposed by the parsimonious Foreign Office, for on November 8, a month and a half later, he appealed directly to Hooker:

I have just received a letter from H. Majesty Consul at Para enquiring at what price I would supply government with the seeds (per hundredweight) of the Indian Rubber tree for introduction into India. Having considered the matter I submit the following suggestions. I am now making a plantation of coffee &c on the right coast of the Amazon, above the mouth of the River Curvá and below the town of Santarem. I would be willing with government assistance to establish a nursery for raising plants from the ciringa [sic] seed. I think the locality where I am now making my plantation would be admirably suited for the purpose. The navigable Amazon would enable the plant at once be placed aboard a vessel without need of further removal or transplanting. The plants could be grown from the seeds to a sufficiently hardy size in the native air before being removed and by this means a large number of healthy plants could be secured. . . . My experience of the ciringa tree gained by working them in the forests of the Upper Orinoco de Venezuela caused me think this the best plan for their successful introduction into India. On the other hand I doubt if the scheme of introducing them by the seed would succeed owing to the very oily nature of the beans which would be likely to become rancid in short time.

No one acknowledged Henry's proposal for another eight months, until July 1874. Eight months seemed a lifetime for a man like Henry. By then, his life had once again shifted. By then, he counted himself lucky to be alive.

Henry had two masters now: his craving for success and the "spymaster" at Kew. Sometime in late 1873–early 1874, Henry and Violet loaded their belongings in a curiara and struck out from Santarém. If anyone watched from the bank, it would never occur to them to see Henry as a spy. Violet didn't, and she knew him better than anyone.

Henry was transformed, not by success but by failure. Up to his mother's death, his dream had been for personal glory. Afterward, however, we hear a new note: "All for the Empire." He'd become a true believer in the British doctrine of world transformation, that nature's secrets could be secured and replanted—all for the improvement of man, the empire and her queen. But for all Henry's stated allegiance to England, he'd become Brazilian in the way he viewed life and the world. He'd absorbed that portion of the Portuguese character known as *saudade*. The term is untranslatable, a sadness of character that could only be known as longing—but a longing for what, Violet did not understand.

Longing was an intricate part of life on this river: the rivermen believed in miracles and in the stroke of luck, and the latter was Henry's personal god. "In a radiant land there lives a sad people," wrote Brazilian writer Paulo Prado, and this was indeed a melancholy land. Violet heard it in the dip of the paddle in the river at twilight, in the songs of the canoemen as they passed by. One of their most common songs was very wild and pretty: its refrain was "Mother, Mother," and told of the endless gloomy forests, the rivers and channels that echoed of monkeys and birds:

The moon is rising, Mother, Mother!
The moon is rising, Mother, Mother!
The seven stars are weeping, Mother, Mother!
To find themselves forsaken, Mother, Mother!

As the song faded in twilight, Violet imagined she and Henry were the last souls on earth.

There is some question today about their destination. Henry mentioned the River Curvá, but no such place exists. Some point to the Curuá, but there are two local Curuás from which to choose. The best known is the largest, a wide waterway about fifty miles west of Santarém that feeds into the Amazon from the north bank, but Henry and Violet would have been paddling against a strong current, and there were no substantial reserves of hevea on the north shore. By now Henry would have known this and focused his attentions on rubber lands.

The second choice is the Curuá du Sul, a small stream about twenty miles downriver from Santarém. The primary evidence is a note in Violet's diary citing their stay with a "Mr. And Mrs. R——," managers of a Brazilian plantation situated at the mouth of the river. At that time, the *confederado* Romulus J. Rhome and his wife managed the plantation of Cel. Miguel Antônio Pinto Guimarães, the Baron of Santarém—the same baron who'd hosted Wallace, Spruce, and Bates in 1849. Rhome was the only Confederate to be immediately successful. He arrived with lots of money, entered into partnership with the baron, and managed Taperinha as a sugarcane plantation. They built a still and distilled rum from sugarcane juice. Taperinha sits back from the Amazon on the short, scimitar-shaped Rio Maica, then called the Rio Ayaya; the Curuá du Sul feeds into this not far away.

Despite its tempting elegance, this answer has problems too. Violet said they were "several days" on the river after leaving Santarém, and Taperinha is no more than a day away by canoe. Henry was abysmal with directions and places, as shown by his disorientation in Nicaragua. Moreover, colonials commonly garbled Indian place names when transcribing them phonetically. The Curuá sometimes came out in travelers' accounts as the Cuvari, which mutated into the Cupari or Cupary. This confusion suggests a third destination, one that fits more neatly into Violet's descriptions and makes better sense when one considers Henry's goal.

Historians of science generally think that Wickham's hunt for rubber occurred along the Tapajós. By 1873–74, it was thought that the best grade of "Pará fine" came from this region, yet the source remained a mystery. Did it come from the east bank, near the town of Aveiro? Across the river on the west bank, from the trading town of Boim? Or deeper up the river past the future jungle empire of Henry Ford? Beyond Aveiro there is, in fact, a tributary called the Cupari River: set back from its mouth sat a huge plantation and cattle ranch owned by "Francisco Bros. and Co.," which employed, over the years, a succession of expatriate British and American managers. A twenty-four-hour paddle up the Cupari brought one to a village of Mundurucú Indians, as described in Violet's diary. Beyond this lay the cataract dividing the upper river from the lower (also mentioned in Violet's diary), and past *that,* rumors of virgin rubber land. Most convincing of all, the Cupari lay across the Tapajós from Boim, upon which so much of Henry's legend would turn.

Thus, a strong case can be made that this was the young couple's destination. If so, they paddled south along the east bank of the Tapajós. For twenty miles they followed the high rocky coast, the red sandstone cliffs rising 100–150 feet, waves bursting with a roar against the perpendicular stone walls. The summit was carpeted with luxuriant green forest. Palm-thatched cottages nestled in the hollows. The climate grew more humid—a heavy shower fell once or twice a week, with interludes of melting sunshine and gleaming clouds. By night the rains returned, followed by a chill.

After two or three days they'd come to Aveiro, the small straggling village at the head of the mouth-bay, where the Tapajós narrows to two and a half miles in width and the river is dotted with rocky islands. The town sat on a high bank and resembled a ghost town. The church was moldy and dilapidated; many of the houses seemed deserted. A primary school had been built by royal decree, but there were few children in town.

The people had been driven away. Aveiro was said to be "the very headquarters" of *formigas de fogo,* fire ants, and visitors landed at their peril. In 1852, Henry Bates visited Aveiro. A few years earlier the village had been completely abandoned due to the scourge, and people were only then moving back. The ground beneath town was perforated with their nests, the houses overrun. The ants disputed "every fragment of food" with the inhabitants and destroyed clothing due to their taste for starch.

Food was stored in baskets suspended from the rafters, the cords soaked in balsam, their only known deterrent. The legs of chairs and cords of hammocks were similarly smeared to give the residents some rest. The sting was like a red-hot needle: "They seem to attack persons out of sheer malice," Bates wrote. "If we stood for a few moments in the street, even at a distance from their nests, we were sure to be overrun and severely punished, for the moment an ant touched the flesh, he secured himself with his jaws, doubled in his tail, and stung with all his might." Even now, Aveiro had a melancholy air.

Eight miles south of Aveiro, Henry and Violet came to the Cupari River, and here Henry turned. He'd heard that wild hevea grew deep within the interior. The Cupari was no more than a hundred yards wide at the mouth, but very deep. The walls of forest rose a hundred feet on either side. Silence closed around them. Another ten miles and both sides of the river turned hilly. Another river joined from the east, and they came to a plantation owned by one of the old Portuguese lords. Houses occurred at such rare intervals that hospitality was freely extended to the passing stranger.

Henry left Violet with the American manager as he paddled on: another twenty-four hours up the river past the lower falls, on to where the channel was only forty yards wide, and up to a tribe of Mundurucú Indians who, it was said, still warred with their neighbors. The climate was more humid than on the Tapajós—the air was heavy with moisture, and showers were frequent here. He stopped and, wrote Violet, "once more made a hole in the primeval forest to put his house in."

If Henry and Violet were in the margins of civilization at Piquiá-tuba, here they were on the very margins of the margins, what sociobiologist E. O. Wilson called a "marginal environment," a nightmare habitat of limited food, water, or other resources, where no one willingly chose to live. Marginal environments were "nature's flophouses for the outcasts" but were important because the harsh conditions forced species to adapt quickly or die. Wilson studied tropical ants and noticed that when marginal colonies grew tough enough, they invaded the lush, comfortable world of the privileged ants and drove them into exile, turning the old guard into castaways. The baked sandy plains around Aveiro were a good example. There the fire ant had endured the wilderness until it conquered the town itself, and today it has successfully invaded Central and North America, driving farther north each year.

Henry and Violet were becoming marginal too. In fact, it was hard to escape how many human castaways lived out here. The *confederados* were a perfect example, most unable to fit in anywhere, though a few—the Rikers, Rhomes, Mendenhalls, Vaughans, and Jenningses—hung on by the skin of their teeth and would one day be among the richest families in Santarém. So, too, the escaped slaves from Barbados discovered a new life as fishermen and woodcutters on the north bank of the Amazon. The Indian tribes Henry encountered—the Woolwá and Miskitos in Nicaragua, the Mundurucú and Tupi-Guarani on the Tapajós, the embattled and mystical Maya on the Yucatan Peninsula—though decimated by disease and pushed back into the forest by settlement or conquest, also fought to survive. Jewish traders plied the rivers, and Henry heard of trading houses in Boim founded by families of Sephardic Jews from Morocco and Tangiers. On the Orinoco, he'd encountered Venezuelans in hiding and the children of Jesuit priests who'd died long ago.

For the next thirty years, strange tales of such people filtered from the headwaters of small rivers like the Cupari. Some subjects of legend were marooned by a price on their head; others were driven by an inner need to reject civilization forever. In Colombia, at the end of the Isamá tributary of the Rio Negro, it was said that a Corsican killer who'd escaped Devil's Island off the coast of French Guiana had enslaved a settlement of fugitives and created a rubber empire. Up the equally remote Uapés, explorer Gordon MacCreagh found a black strongman who ruled his water highway long after the Rubber Boom had died. His Indian slaves had fled up the inaccessible creeks; his palace of tile, lathe, and plaster crumbled with decay. His patent mechanical band had frozen with rust and no longer played. He lived in a former rubber shed surrounded by his remaining loyal subjects—the extended families of his several wives.

Henry simply was not ruthless enough to build this kind of empire. He took Indian workers with him up the Cupari, but he could manage them little better than his English laborers. Soon, as previously, he worked alone. He was becoming, in effect, a *caboclo*, a backwoods jungle peasant who owned his land until someone more powerful took it away. He acquired a taste for guarana and *acai* juice, lined his bureau drawers with the aromatic forest leaves and dried fruit called *xeros*, ate Amazon turtle, manioc, rice, beans, bananas, plantains, oranges, pineapples, fish, chicken, and eggs. His principal means of transportation was by dugout canoe. Twice a month he visited the *barracão*, or trading post, for supplies. Except for the company of Violet, Henry was a solitary man.

And it nearly killed him, the closest he'd come to death since his fever on the Orinoco. While clearing the forest, he lost his grip on his ax and he nearly chopped off his foot. "He stripped his shirt up and bound his foot and started off for the Indian village some miles away as fast as he could go," Violet wrote. By the time he reached the village, he was in shock from loss of blood: "Everything turns black and down he went," she said. A man working nearby saw him fall and ran for help. The Indians carried him to the village in a hammock, then manned a canoe and "brought him to the plantation where I was." But he survived, thanks to the ministrations of the American manager and his wife, and when strong enough to stand, he hobbled back to the jungle.

This time Violet went with him, first to an empty lodge in the Mundurucú village, then with her husband. Henry thought "he could make a pretty place of it in time." He burned the brush to clear it back and dammed the stream for water power. But Violet thought otherwise. The stream was a nice place for a bath, "and then I have said all there is to be said for it," she claimed. "I know few things less beautiful than a new plantation. The trunks and stems of trees lying about all black with fire."

There were new torments here she hadn't encountered before. Chiggers, ticks, centipedes, and scorpions abounded. She especially hated chiggers, the way they deposited their bags of eggs in her toes. The eggs had to be extracted before the digit became infected and grew as large and pulpy as an overripe plum. And there were snakes. On the Cupari, the most notable was the *sucuruju*, or anaconda. The serpent was abundant in this river. It lived to a great age and reached tremendous size. Bates saw one that was eighteen feet nine inches long and sixteen inches round at the widest part of its body—and this was considered small. He measured the skin of one that was twenty-one feet in length and two feet in girth, and heard of some "which measured forty-two feet in length." Natives in this region believed in a monster water serpent called *Mai d'agua*, the "mother of the water." Forty-two foot anacondas were monsters in their own right, and he repeated the tale of the *sucuruju* that "was once near making a meal" of a ten-year-old boy:

> The father and his son went . . . to gather wild fruit, landing on a sloping sandy shore, where the boy was left to mind the canoe while the man entered the forest. . . . While the boy was playing in the water . . . a huge reptile of this species stealthily wound its coils around him, unperceived until it was too late to escape. His cries brought the father quickly to the rescue, who rushed forward, and seizing the Anaconda boldly by the head, tore his jaws asunder.

Although Violet never came face to face with such a beast, the *sucuruju* were notorious for their depredations of livestock—and especially liked her chickens.

For all of Henry's hard work, he failed again. The house was never finished. The new attempt at planting coffee was abysmal. Although the hevea growing on the Cupari proved disappointing, in all likelihood he heard of the massive stands of rubber on the west bank of the Tapajós, across the river at Boim. His timing was bad: burning off forest and underbrush for new fields is best done in October, followed immediately by *coivara*, or clearing the largest trees not consumed by fire. It is important to do this before the rainy season. In tropical soils it seems there is a "nitrate pulse," which occurs in the first two weeks of the rains. Nitrate is the most useful form of nitrogen for plants, and if the pulse is ignored by farmers, the yields will be disappointing.

As much as Henry adopted the *caboclo*'s life, he never seemed to learn the central lesson: The *caboclo* was not a man of one vocation. To survive in the jungle, he was forced to be a little of everything, said anthropologist Emilio Morani: "a horticulturist, a rubber collector, a hired hand, a canoe-paddler, a cowboy, a collector of Brazil nuts, a fisherman, and he often earns a living from several of these pursuits simultaneously." Surviving in the Amazon meant tapping into every available resource, just like the white threads of fungus that drained everything organic that fell to the forest floor. Henry did one thing—planting—while searching for rubber. He never hired himself out, never collected or sold anything other than rubber. He was a specialist, not a generalist, and so, in this marginal environment, he failed.

In summer or fall of 1874, Henry and Violet returned to their starting point. The *confederados* sprinkled around Santarém were failing, too, and decided to solve their labor problems by forming a commune and working together in the fields. When Henry heard this, he joined again, but when they returned to Piquiá-tuba, their old house had been cannibalized. "My boarded floor had been taken up and used" for another house, Violet wailed. Half of Violet's baggage never returned from the Cupari; it seemed to disappear into the trees. They struggled on, clearing a nearby plot of land, building a new house, one "more comfortable than any" of their previous attempts, raised above the ground, with boarded floors and an adjoining kitchen. Maybe Henry was finally getting the hang of this pioneer life, and as if to christen this third attempt, they dubbed the house "Casa-Piririma"—the house of the Piririma palm.

Violet's father sent out another family of laborers to work for them but they, like the first group, soon grew "as disgusted as the others and went to those members of the party [now living in Santarém] who had made themselves comfortable at our expense," Violet complained. What did she mean? Had the party pooled their funds in London, then refused to divide the money equally when the majority moved back to Santarém, leaving Henry and Violet not only alone but penniless? Was it simply a statement of her feelings of anger and abandonment? She never says, nor does Henry, but the Wickham family became one more casualty of the jungle, and the estrangement was too wide to bridge. At least Violet's father still stood by them. He sent with the unidentified family a thirteen-year-old serving girl named Mercia Jane Ferrell from West Moors, Dorset, who stayed in the jungle with Violet as her only female companion for about a year.

During this period, young David Riker entered their lives. Riker's mother had come to Santarém with the first wave of *confederados*, but she could not stand the privation. In 1872, she took David and his younger sister, Virginia, back home to Charleston, South Carolina, but they all returned in 1874. "I was just delited and asked Mother to buy me a gun," he wrote in his memoirs: "She bought me a mussel loader and on our arrival I took to the woods and became a real good woodsman." By now his education was falling behind: "Father put Virginia and myself in school in town with the Wickham family who had opened a school to teach English." Henry's brother and sister lived in a house paved with red bricks on Quinze de Agusta in Santarém. While Harriette Jane "taught English at their house to a few children or rather young ladies," her husband, Frank Pilditch, taught English in peoples' houses around town. Henry's brother, John, was often sick. Though David Riker's father lived and farmed at Piquiá-tuba with the other *confederados*, David Riker and his sister boarded in the Wickham house. As they lay in bed, the young Rikers could hear them talking at night after supper on the veranda.

During this time, Riker first encountered Henry and Violet, who alternately lived in the new house at Piquá-tuba and with the rest of the family in Santarém. Henry was a mystery to the youth. He was "soft easy speaking, of a lonesome melancholy aspect." Sometimes he would get a wistful look on his face, mount his horse, and "ride out in to the country." Everyone assumed he was still looking for new plantation sites, still unable to abandon his visions of glory, even after they'd crashed so often. By then, however, he was either collecting rubber seeds or scouting out new stands of trees.

The negotiations for smuggling resumed when Henry returned to Santarém. In July 1874, Markham told Joseph Hooker that the India Office was willing to pay Wickham £10 for a thousand seeds, and Hooker passed this on to Henry in a letter dated July 29, 1874. Henry responded nearly three months later. He was gracious but pointed out that it would be too expensive for him to collect so few seeds. He still hadn't found the legendary trees of which he'd heard so much. Collecting seeds meant disappearing back down the Tapajós and into the jungle for an appreciable amount of time. Instead, he offered his own plan:

It is already very late to procure Indian Rubber seeds this season but if possible I will send you some with my best care. Although the sum offered by the government appears sufficiently liberal you will perceive that it will not pay me to go into the better districts to collect them in small quantities but if I may be guaranteed an order for a large number I am prepared to collect them fresh in the best localities and dispatch them to you direct during the ensuing season. In such case, would you wish me to make any observations as to localities of growth, soil, etc.?

He put the letter away for a couple of days, and on October 19, 1874, added a postscript:

I reopen this in order to add that, as the seed rapidly begins to shoot and spoil, I would be willing to take the parcel of fresh seed myself to Para and deliver it into the care of the Liverpool steamer about to sail that it might have the advantage of traveling (in) his cabin and under his care.

Henry may have thought that since he was the sole "man on the spot," he was in a favorable position to bargain. Perhaps he finally faced the hard realities of his position and realized how desperately he needed funds. In any case, Hooker and the India Office seemed to grow peeved. In October, Markham sent Hooker a note suggesting that the top ranks of the India Office were growing tired of Henry's demands:

With reference to Mr. Wickham's proposal to raise young India Rubber plants and send them here in Wardian cases; Lord Salisbury says—"What would be the cost of sending a gardener to superintend the packing and transmission of a single batch?" "We need not pledge ourselves for more." "If the experiment seemed likely to succeed, it might be repeated from time to time." It seems to me that if this was done, the sending of the plants would answer.

London elites had little tolerance for subordinates who demanded fair treatment—witness the fate of James Collins. Neither did they appreciate upstarts who presumed on their betters' good graces. Henry was an opportunist *and* an outsider. The first could be explained as ambition, but not when combined with the second. He was not part of the club, and his nursery suggestion, for all its merits, was turned down.

At the same time, however, they did relent on the issue of Henry's pay. By December, Markham proposed letting Henry collect "*any amount* of seeds at the same rate—£10 for 1000." The secretary of state soon authorized Wickham to collect 10,000 or more seeds at the £10 rate; Markham told Hooker that his office planned to give Wickham carte blanche and would pay him for all seeds he procured. Markham added: Pass the offer to Santarém.

So dawned the New Year of 1875. It should have started well. Henry and Violet had a new house, in a commune, where all were said to be working together. Henry had his commission from London locked in at a generous rate. His letters exude confidence, as if he knows where thousands of the best seeds can be found. The return to Piquiá-tuba seems to have brought some reconciliation among the Wickhams. At least they were on good enough terms to sit together for a group photograph sometime during the early part of that year.

Look closely at that photo, however, and hints of disaster filter through. By early 1875, the *confederado* commune was falling apart, and Violet would write, "once more each struggled on alone." Henry's crops were failing, and young David Riker remembered Henry disappearing into the bush without explanation: He was searching for seeds. Not just more, but the ones from *perfect* trees. His outward confidence when writing Kew was just a bluff: Yes, he could provide seeds in early 1875, but not those from the fabled source of "Pará fine." He could not mention his fears. He'd adopted the image of the self-assured colonial adventurer, an image he would maintain in one form or another for the rest of his days. The image was his capital, a force to which others responded. In the family photo, he stares brazenly at the camera, a smirk on his lips, arms akimbo, with a dark mustache and untrimmed beard, and a sheath knife hanging from his belt. He looked vaguely like Sir Richard Burton and was certainly more striking than his conventionally dressed brother and brother-in-law. Violet sits before Henry, looking pale, wan, and sick.

Frank Pilditch, the husband of Harriette Jane, seems vaguely amused. John Wickham stands proudly behind his wife Christine and one-year-old son Harry, born in Santarém in April 1874.

Henry's sister Harriette Jane is the sitting's focal point, and her eyes command the lens. She is dark like Henry and forcefully attractive, with a long face and strong chin. She sits in the center of the group, clad completely in white. One suspects she was the one who started the English school and kept it going, not Violet, as Riker would write in his memoirs. But Harriette also slumps in her chair. There is fatigue in her face; her cheeks are drawn. She is like Henry, never admitting defeat, but unlike Henry she will not succumb indefinitely to illusion. Her mother has died; a worker she knew; Christine's mother. She is tired, and there is a reason. Disease is on them again.

Sometime in 1875–76, a second wave of sickness hit the Wickham group, and this time it took the young. The first to die was Mercia Jane Ferrell, Violet's serving girl, age fourteen. Once again, Violet is laconic: "She stayed with me till her death" a year after her arrival, she reported; with that single statement one senses a deep loneliness. She is an exile like her husband, but without the attitude to shield her.

Then Harriette Jane died, at age twenty-eight. No comment is made. Hers is simply one of the five graves Henry sketched on the cemetery hillside. Little, in fact, was ever said about her. But if Henry was the family's imagination and heart, Harriette was the backbone. She was the "woman of independent means," according to the 1871 census; she ran her mother's shop on Sackville Street and ran the house in Marylebone. With her death, the family group began to splinter. Within a year, the band held together by her gravity would spin off to the farthest reaches of the globe, never to meet again.

With his sister's sickness and death, Henry finally seemed to realize a truth: *This jungle would kill him if he stayed.* On April 18, 1875, he wrote to Hooker from Piquiá-tuba, saying it was too late in the season to begin collecting rubber seeds. His tone was apologetic:

> I received a few other (seeds) from an up-river trader but when I got them
> they had already long shoots attached to them. There is now therefore

nothing to be done but wait for next season's fruit and I propose to go into a good locality early to collect and pack them myself for your order.

With that blunt statement, the empire's plans were delayed for yet another year. Then, uncharacteristically, Henry allowed a brief glimpse of his hopes:

Should you have opportunity of recommending me for an appointment in selecting, planting and tending the young "ciringa" in the East, may I ask you to favor me with your influence?

The seeds were not simply a way to make money—they were his means of escape. He'd seen too much death and failure on the Amazon. He was desperate to leave, and rubber was the key.

CHAPTER 9

THE VOYAGE OF THE *AMAZONAS*

Early in 1876, David Riker and his sister began to hear odd snatches of conversation as the Wickhams talked at night on the veranda. The two young *confederados* were supposed to be asleep; what they heard wasn't for their ears. "After supper the family would unite," he wrote, "and I could hear them talking and discussing about a trip up the Tapajos." The journey was a secret, never discussed before the two boarders. But if Wickham tried to keep his plans secret from the locals, he made them abundantly clear to Hooker in a letter dated January 29, 1876:

Dear Sir:

I am just about to start for the "ciringa" district in order to get you as large a supply of the fresh India Rubber seeds as possible. The season of the fall of the fresh (seeds) now commencing—I think it will be safest to avoid the European frost. I will therefore dispatch them with every care as soon as They can safely go.

An urgency surfaced in Wickham's dispatches that wasn't present previously. Perhaps he sensed a loss in faith by Hooker and Markham. The events of 1875–76 do, in fact, suggest that conclusion. Hooker was an easily irritated man, and Henry quibbled about everything: payment, method of shipment, when to collect seeds. The director was unused to impertinence. It would be easy to weary of this troublesome man in the Amazon. Henry Wickham was just another adventurer down on his luck;

he was not a professional trained at Kew. After the letter of April 1875, the suspicion arose that Henry might not be the man for the job.

A class structure existed in Victorian science just as it did in greater society. A botanist like Hooker was finely attuned to matters of professional status, sensitive to the hierarchy within the field. An ordinary collector, like Henry, existed at the bottom of the botanical world. As Joseph Banks, the patron saint of English botany, had written, "[T]he collectors must be directed by their instructions not to take upon themselves the character of gentlemen, but to establish themselves, in point of board and lodging, as servants ought to do."

In fact, London had explored other means for bagging hevea. Less than a month after Henry's disappointing letter of April 1875, Markham interviewed Bolivian trader Ricardo Chávez, probably through the aid of Consul Green. Soon afterward, Chávez moved up the Madeira River with two hundred Moxos Indians to the remote jungle town of Carapanatuba. The Bolivian was a *patrão* of the new corporate order, with a tight fist on costs and a mobile labor force that tapped, cured, and moved on. By locating on the Madeira, he'd entered a region that swallowed lives blithely, a road to the wild state of Rondonia and the disputed Acre territory, thought to be the greatest untapped rubber reserve in the world. Chávez collected rubber at the eastern tip of this wilderness, no doubt probing for some yet-undetected black lode. When he returned from Carapanatuba, he called upon Consul Green and said he had nearly five hundred pounds of rubber seed ready for shipment. On May 6, 1875, Green wrote that the four barrels of seed were on the way.

But when the barrels arrived in London on July 6, 1875, everything went wrong. Markham was not present to oversee the handoff. He'd left on May 29 with the Naval Arctic Expedition to Greenland and didn't return until August 29, exceeding his leave by a month. His star was falling in the India Office. The 1874 appointment of Sir Louis Mallet as Permanent Under Secretary of State in charge of the India Office put an end to that informal environment in which Markham dreamed up new schemes. Soon after his arrival, Mallet complained about the "laxity" that had been "allowed to grow up in this office"; a civil servant's personality "should be vigorously and systematically suppressed," he decreed. If

Markham had been present to shepherd Chávez's shipment, he might have had another cinchona coup. Instead, the clerk who received the barrels did not know what to do and cast around for advice. Ten days after their arrival, on July 16, 1875, he sent a few seeds to Hooker. The director immediately shot back a requisition, but by then the barrels were on their way to India. When they were unpacked, the seeds were useless—rancid or dead. The same India Office that grew incensed about paying £10 for Collins's services was now forced to pay £114 to Chávez, plus the price of freight to India, for four barrels of rotten seeds.

Henry did not know how close he'd come to dodging the bullet that would have ended his schemes—twice, in fact, first with Farris and now Chávez, and he was saved only because the seeds rotted when shipped overseas. He did not suspect the intra- and interdepartmental politics that would have such bearing on his future. But by January 1876, he sensed that any more delays on his part would be fatal to his prospects, and so, Violet wrote, "once again we started by boat" up the Tapajós mouth-bay. There was greater secrecy this time: "[H]e decided to collect himself though not in the neighborhood," she said. In addition to Henry and Violet, their party included the "little Indian boy who had been given to (Henry) to bring up, as is very common there." Taking the family was good camouflage. Those who knew him would assume Henry was off on another quixotic hunt for the perfect site of a plantation.

Why this need for secrecy? Although no restrictions existed in 1876 prohibiting hevea's export, South American officials still resented the cinchona theft. Something as valuable as rubber could be tied up in red tape if a canny official realized what was happening. Such a delay would kill Henry's hopes as the seeds turned rancid and died.

Brazilian officials were also sensitive to slights by the Great Powers, and Santarém was a site on the Amazon where such showdowns occurred. The famous case of Allie Stroop had played out just before Henry arrived. When a *confederado* died after arrival, he left his young daughter an orphan. On his deathbed he begged friends to send the girl back to family in the United States, but he was too poor to afford a ticket for her. The other Americans had cared for her for more than a year when a U.S. government ship appeared on the river on a survey mission. The captain offered to take her home for free, and everyone praised his charity. But when the ship reached Pará, the American consul was summoned by the government to give up "a minor who had been unlawfully taken from the jurisdiction of the Orphan's Court at Santarem." When the Captain refused to give up the girl, Brazilian authorities delayed his ship in port until extracting a promise that he'd deliver the girl personally to her relatives. The Americans had neglected through ignorance to take the proper legal steps; the judge in the case knew all the parties, and there was wide talk in town of what the *confederados* planned. But no officer of the Santarém court came forward to mention the necessary legal steps, and the judge would not back down. If an international incident could be made out of a charitable act, there was no telling the consequences of being caught in an act of smuggling.

So Henry decided to confide only in Violet and the boy. In February, they paddled upriver for two or three days to "the country house of an Englishman who lived there nearly all his life and was almost more Brazilian than English." The house was near Boim, a trading village on the west bank of the Tapajós, about midway up the mouth-bay, to the north of Aveiro. "We were in great danger once from a squall that came up suddenly," Violet recalled, "but I scarcely realized the danger and so got more 'kudos' for pluck than I perhaps deserved."

February to March was the period during which hevea's trilobed seeds began to ripen on the tree. By early March, the three were ensconced in the Englishman's *sitio* on the Tapajós; they apparently used it as a base for collecting. Henry "left me there," said Violet, "while he and the boy went off into the woods to collect the seeds, returning on Saturdays to start again on Monday, also buying all that were brought to him." On March 6, he wrote Hooker that "I am now collecting Indian Rubber seeds in the 'ciringals' of the river being careful to select only the best quality. I am carefull [*sic*] packing them. I hope soon to leave with a large supply for England."

Less than a month later, on April 1, 1876—apparently before anyone saw Henry's note—Clements Markham penned a note to Hooker informing him that the secretary of state was sending Robert Cross to the Amazon to collect hevea.

Cross was a veteran of the cinchona expedition and always eager to return to South America. On one of his later trips, he forwarded to Markham a small bag of *caucho* seeds, and this helped convince Markham that Cross was the man for the job. Markham's note to Hooker said the gardener was granted four hundred pounds "to cover all expenses and include remuneration." Before he left, Cross was to call upon the old and ailing Richard Spruce for advice. Hooker was also to provide the gardener with a letter of introduction to Wickham. If the two should meet, Wickham would essentially relinquish control of the hevea project. Cross's orders were the closest thing in writing to an admission that London no longer trusted Henry.

For two and a half months, from early March to mid-May, Henry collected alone with the boy and bought seeds directly from Indian and *caboclo* collectors, as Markham and Spruce had done with cinchona. This would not be enough for the huge quantity he'd need to make a profit and insure germination. What has never been emphasized is that by now Henry knew the Tapajós region well enough that he could more narrowly target his search, and he maximized his take by buying from middlemen.

The heart of his search lay in the highlands behind Boim. The source of the seeds would become a matter of economic consequence fifty years later when Henry Ford entered the picture, but Henry Wickham was never circumspect. As early as 1902, he wrote that "their exact place of origin was in 3 degrees of south latitude . . . in the forest covering the broad plateaux dividing the Tapajos from the Madeira rivers." Henry headed due west to the highlands behind Boim, which stretch west for fifty miles and rise 250–300 feet above the river plain. Throughout his account, Henry reiterated one point: The best seeds were found in the highlands, not at the water's edge.

However, 3° S latitude covers a lot of territory and does not begin to explain how he gathered an abundance of seeds. The answer lay in the village of Boim and its history, which is forgotten by all but a few. Boim was the oldest village in this part of the Amazon Valley, founded on March 9, 1690, a few days before Santarém. Like Santarém, it had been a Jesuit mission established at an Indian village, and in 1841 a new church dedicated to St. Ignatius of Loyola, founder of the Jesuit order and patron saint of soldiers and spiritual retreats, was built on the sandy bluff overlooking the Tapajós' wide mouth-bay.

Boim's location beneath the high plateau was important, but more important for Henry was its infrastructure. Four families of Sephardic Jews had come to the village from Morocco in the mid-1800s and set up trading houses. Indians and *caboclos* from up and down the river and deep within the interior came to their posts with all manner of goods, but the region's specialties were rubber and Brazil nuts, then called Pará nuts. So successful were the trading houses of Cohen, Serique, Azulay, and one other, whose name is now lost, that for its time, though Santarém might have been better known to the outside world due to its more obvious location at the joining of the Amazon and Tapajós, Boim was more important commercially. Oceangoing freighters heading back to Europe from Manaus would anchor in the mouth-bay, where they were met by nimble Portuguese sailboats loaded with jungle goods.

Like the *confederados, seringuieros,* and Britons, the Sephardic Jews in the Amazon were one more instance of the nineteenth-century reshuffling of "marginal men" to the most forgotten places of the world. Boim's four trading families had come from Tangiers in Morocco: they stopped briefly in French Guiana, then dropped south to Pará, where a synagogue was founded in 1824. In Pará, Jewish merchants called themselves *klappers,* or "door knockers," who peddled their goods door to door. Soon the market grew tight, and the best chance to make a living lay in filling a canoe with a stock of goods and setting off into the endless Amazon. "One must take great care in the jungle on entering, for one gets lost easily," wrote veteran trader Abraham Pinto. "Some travel with a compass, others are guided by the sun, for at times one cannot see anything because of the great height of the trees. It is best to mark the trees with an ax, or by breaking the branches, indicating the path so you can return."

At first, their role in the rubber trade was primarily that of small-time middleman, exchanging knives, cloth, and cooking pots for balls of smoked latex, which they then exchanged for more barter goods with the Pará *aviadors.* Some became a mixture of *aviador* and *patrão* themselves, a niche filled by the trading houses in Boim. As the young men grew wealthy, they'd return home—one of the owners of the "Franco & Sons" cattle ranch at the mouth of the Cupari graced the front of his house with two ornamental palms he'd taken as seeds from Tangiers. They traded with anyone who stopped. Henry's insistence on seeds instead of balls of rubber might seem eccentric, but numerous transplants bought seed to start plantations; the *confederados* on the opposite shore did the same. Neither they nor their Indian suppliers realized that by selling to Henry, they ordained their own ruin.

Henry collected fast, in every way available—from the Boim houses, by purchase, by his own hand. A number of commentators have argued that the speed with which he worked eliminated quality control and that he could not insure the provenance of his seeds, yet all the seeds and latex gathered in the Sephardic houses came from the highlands behind Boim. A mule trail led directly west from town up a gradual sandy slope; after two miles this turned into a plateau covered with jungle and some rubber, but not of the same quality as on the higher plateau farther in. After six miles, Henry and the boy started up a long incline called the Serra de Humayta, which finally opened into the high plateau three hundred feet above the river. Here the jungle was heavy but open; it had a light degree of undergrowth and very few palms, plus a climate so unusually dry "that the people who annually penetrate into these forests for the season's working of the rubber have to utilize certain lianas [water-bearing vines] for their water supply since none is to be obtained by surface-well sinking," Henry wrote. The plateau covered 1,600 square miles, or over a million acres, but Henry apparently headed for the ancient sites covered in a deep, stiff Indian black earth similar to the soil found atop the pyramidal hills near Santarém. This soil was so fertile that farmers from Boim would hike the long distance to grow their crops in it, and it was here that Henry sought his trees.

The center of their collecting was a village named Agumaita, or "Highlands," about seven to nine miles into the plateau. The hevea trees there were large and straight, attaining, Henry said, "a circumference of 10 ft. to 12 ft. in the bole." Working with as many Tapuyo Indians as he could hire on short notice, "I daily ranged the forest, and packed on our backs in Indian pannier baskets as heavy loads of seeds as we could march down under." Like the majority of tropical trees, hevea's bark was gray on the surface. To check whether or not a tree was really hevea, he'd scrape it clean to reveal a bark that resembled the "colour of a light bay horse's coat." Such cleaning took time, but was essential, for in humid regions like this "the bark is thickly coated with growths of moss, ferns, and orchids." The tree's own flowers were small and green or creamy yellow; the young green leaves secreted a nectar around which honeybees buzzed.

Whether by design or luck, Henry collected the best seed available. These were perfect trees. Their silvery trunks, much like a poplar's, soared aloft for one hundred feet. On the upper branches grew small, three-lobed leaves with undersides of silvery white, giving way at the tip to green, sweet-smelling flowers. A later study revealed that "out of seventeen varieties, (Henry) chose seeds from the black, or best grade of tree," said William C. Geer, former vice president of the B. F. Goodrich Company. Tappers distinguished between types of tree by color—the black-, red-, and white-bark type, distinguished by the color of the hard bark beneath the periderm. The black-bark version was said to yield more latex and a better-quality rubber than the red or white, but more important than that for Henry was the fact that Agumaita was said to be the home of the "mother tree," a gargantuan seed producer growing straight from the black earth and surrounded by her progeny—giants in their own right if not standing by this leviathan.

It sounds too good to be true, a fantastic tale harking back to myths of a Mother Tree believed by many Amazon tribes. But there may have been at least one tree that epitomized the one in the story. There is a witness, at least of its remains. Elisio Eden Cohen, author, historian, and Boim's postmaster, is also the scion of one of the old Sephardic houses. Now in his sixties, he'd had the stump of the Mother Tree pointed out to him when still a child. It was in the heights above Boim, surrounded by its descendants, growing from the black earth as the stories said. Cohen watched as seven men barely stretched their arms around it, standing fingertip to fingertip. "The mother tree had died," he said, "because of the *caboclo* practice of using kerosene when cutting into the bark to get latex. The kerosene made the latex flow more freely and in greater amounts," but by rubbing it into the wound, the tappers "killed the tree over time."

Such a cathedral of trees took Henry's breath away. "[D]uring times of rest, I would sit down and look into the leafy arches above," he said, "and as I gazed, became lost in the wonderful beauty of the upper system overhead." It was the same transcendent experience he'd experienced in the rubber forest of the Orinoco, but it was also an extremely dangerous time to gather seeds. The fruit of the rubber tree was a three-chambered nut like a horse chestnut. Each segment contained a speckled seed resembling a slightly flattened nutmeg. As the seeds ripened, the outer envelope dried and tensed until it burst with the sound of a pistol shot. The rich, oily seeds exploded from the pods, flying outward sixty to one hundred feet before dropping to the ground.

The sound of the shot signaled a race of life and death on the forest floor. The base of the massive trunk would be littered with seeds, and like the tree in Genesis, a serpent lay coiled nearby. The pop of the seeds drew the rodent agouti to the feast, and the venomous *jararaca,* or fer-de-lance, the most common venomous snake in Amazonia, awaited a feast of its own. Seed collectors must be watchful at such times. Each bite of the snake's fangs packs an extraordinary amount of potent yellow venom, and the snake is quick to strike. Every female has sixty to eighty young, all with their deadly machinery in good working order. At that time, there was no known antidote for the venom—one bite, and death was assured.

So they worked, crawling up the heights like a line of carpenter ants, scrambling through the brush at the sound of each shot, then stumbling back down the escarpment, weighted down like stone-bearing slaves. Back at Henry's base, the village women wove large, open baskets of the same design as those found along the Amazon today:

I got the Tapuyo village maids to make up open-work baskets and crates of split *Calamus* canes for receiving the seed, first, however, being careful to have them slowly dried on mats in the shade, before they were put away with layers of wild dried banana leaf betwixt each layer of seed; knowing how easily a seed so rich in a drying-oil becomes rancid or too dry, and so losing all power of germination. Also I had the crates slung up to the beams of the Indian lodges to ensure ventilation.

This was where Henry's jungle experience came in handy. Everyone knew hevea's propensity for going rancid, but no one had figured out a way to prevent that from happening. If Kew and the India Office had sent along the portable greenhouses as Henry had requested, the problem would have been solved. But Henry's masters were parsimonious to a fault, and the fact that they would send such Wardian cases with Robert Cross bespeaks their distrust of an outsider.

Thus, Henry was left to his own devices against unfavorable odds. Rubber seeds contain linamarin, a glucose derivative, which provides the energy boost needed for germination. Linamarin is a toxic compound also found in cassava, the principal root crop along the Amazon. Unless dried, soaked in water, then rinsed or boiled, cassava is a certain last meal—of cyanide. As the linamarin decomposes by hydrolysis during storage, it produces hydrocyanic acid (HCN), a colorless, poisonous solution of hydrogen cyanide in water that smells like bitter almonds and is better known as prussic acid. Hydrocyanic acid is also explosive and is used in many industrial processes. Although there are no reports of exploding rubber seeds, it is easy to see why they decompose so quickly once water takes hold. A heavy deluge would have destroyed Henry's cache, either causing his seeds to germinate early or begin the unstoppable rot. It was still the rainy season, so he had every reason to worry. But for once he got lucky and somehow avoided the storms, probably because he collected in these drier highlands. When Henry got the seeds down to the river, he dried them gently in the air, packing them between banana leaves to soak up excess oil, and left them swinging from the rafters in the cooling river breeze—the only expedient possible to stave off a build-up of moisture that would lead to early germination or mold. His precautions were exacting and ingenious, exhibiting an understanding of rubber for which he was never credited.

By mid-May he'd collected seventy thousand seeds, an incredible number, considering all the odds against him. It seemed to assure success—if he could keep them alive. But the very quantity created a new problem.

The historian John Loadman calculated in his book *Tears of the Tree* that the seeds weighed three quarters of a ton. Add to that the weight of banana leaves and pannier baskets, and the gross weight was nearly one and a half tons. Based on volume, Loadman estimates as many as fifty hemispherical baskets with a diameter of twenty inches suspended from the rafters. Henry must have gazed at them, hanging up like bells, and realized he had a problem.

How was he going to get this huge load home?

The jungle had a way of coming up with solutions when least expected, and in this case salvation materialized in the form of an ocean liner. It moored in the middle of the river off Santarém, and the captain invited the local planters aboard.

The ship was the SS *Amazonas*, "first of the new line of Inman line steamships—Liverpool to the Alto-Amazon direct." This was her second voyage, both times under the command of George Murray, an Inman captain in his midthirties. The previous run suggested that she put in at Pará in mid-April 1876 and returned there in the second week of May. Crew records suggest a complement of thirty-two men: two deserted or did not show at port during the first few days of sailing, but they were replaced in Lisbon; no grog was allowed onboard, but other than that records did not indicate dissension or trouble. Release documents showed that the crew was paid in full when they returned from the Amazon on June 2, 1876: no crew member seemed missing, but officers were neither mentioned nor named.

Henry's version was more dramatic than anything suggested by the crew manifest and ship records, since it hinged on shipboard intrigue. At first, the ship's appearance on the river had all the hallmarks of Henry's Orinoco fever dreams. Here was the apex of oceanic technology, right outside tiny Santarém. The liner sounded its great whistle, and the ship's boats came off with their uniformed officers—including two gracious "supercargos," or cargo-masters—inviting Santarém's ragged elite aboard. "The thing was well-done," Henry recalled. They rowed out at night under the shadow of the massive ship, dressed out in blue lights.

They were served a sumptuous supper in the wood and brass saloon, the congenial Captain Murray presiding. Violet no doubt thought she'd died and gone to heaven, as if she'd returned home to London to one of the fancy restaurants her father had taken her to in Covent Garden or Regent Square.

The next morning the ship lifted anchor and headed upriver for Manaus, scheduled to load a hold full of rubber for transport back to Liverpool. "I then thought no more about the episode in rumination on any conceivable means of effecting my purpose with regard to getting out a stock of the Pará rubber tree," Henry continued. It was an "unlooked-for" and pleasant diversion, nothing more.

But then, he said, "occurred one of those chances, such as a man has to take at top-tide or lose for ever":

The startling news came down the river, that our fine ship, the "*Amazonas*," had been abandoned, and left on the captain's hands, after having been stripped by the two gentlemen supercargoes (our late hospitable entertainers!), and that without so much as a stick of cargo for return voyage to Liverpool. I determined to plunge for it. It seemed to present an occasion either "to make my spoon or to spoil the horn." It was true that I had no cash on hand out there, and to realize on an incipient plantation, in such a place and situation, quite out of question. The seed was even then beginning to ripen on the trees in the *Monte alto*—the high forest. I knew that Captain Murray must be in a fix, so I wrote to him, boldly chartering the ship on behalf of the Government of India; and I appointed to meet him at the junction of the Tapajos and Amazon rivers by a certain date.

Henry redoubled his efforts, again crossing the Tapajós, climbing the path behind Boim into the high forest, trudging the seeds back down. "There was no time to lose," he repeated like a mantra. He must have been a pain to live with, though Violet keeps mum.

Captain Murray was also a pain, or as Henry said, "crabbed." In Manaus, his supercargoes had stripped the vessel, absconding with the incoming cargo. Instead of selling the cargo to buy a full hold of fresh-season rubber, they sold it off and vanished, and Murray waited on the Rio Negro until he finally grasped what had happened. He sent his ship's officers into Manaus to search for the men, but they'd disappeared into the jungle. As he lifted anchor, he realized how completely he'd been duped; the Inman Lines might fire him for the debacle.

According to environmental historian Warren Dean, it was unnecessary for Henry to wait for an infrequent steamer like the *Amazonas,* since by the mid-1870s Santarém was visited every ten days by steamers from an English company and almost daily by steamboats owned by importers and local shippers. There was also a steam launch in Santarém owned by a Swiss resident who hired it out for the Pará run. But this is based on a misreading of Wickham's motives and fears. If Henry was so worried about word of the seeds leaking out, Santarém—with its state officials and businessmen—would be the last place from which to ship them. Oceangoing ships already anchored in the mouth-bay of the Tapajós to trade at Boim. Henry, though vague in his account of the details, planned to meet the *Amazonas* within sight of the two rivers, not in sight of so public a place as Santarém.

There may have been real reason for such subterfuge, other than Henry's paranoia and love of intrigue. Two articles written by International Society of Planters' associate W. A. Wilken in 1940 and 1967 claimed that Wickham had been warned by Brazilian officials in Santarém not to export rubber seeds. Wilken met Henry in 1925, three years before he died. Henry related that in early 1876, when he'd started collecting the seeds, he'd been given permission by Brazilian authorities to collect and export hevea. But then, when the consignment was packed and ready for shipment—or at least nearing that state—the Brazilians reneged, telling him now that he "would not" or "might not" be allowed to ship them after all. "This suggests that the rapid charter [of the *Amazonas*] was to beat a possible change of mind among Brazilian officials," Wilken posited.

It is easy enough to see what went through their minds. Santarém was a small place, and the Brazilian officials would have learned of Henry's plans despite his best efforts at secrecy. At first, they probably thought nothing about the hapless Wickham sending a few seeds to London, but as they heard tales of his persistence, of his plans to start a nursery, or his requests for portable greenhouses, they no doubt remembered the cinchona fiasco in Peru. It was then they began to renege. Loading the seeds on the Tapajós, out of sight of Santarém and its resident officials, was the only chance he had to ship the seeds undetected.

Not only is there no record of the *Amazonas* stopping in Santarém, there is no mention of rubber seeds in the cargo manifest, a point discovered by historian John Loadman. In fact, the bill of entry signed at the Liverpool Customs Office on June 12, 1876, has intriguing details. The *Amazonas* was not, in fact, empty when she got back home. She'd picked up a load of timber, "nuts," "capini," a resin used in varnish and perfume, and 171 cases of India rubber while still in port at Manaus. Either the tale of the supercargoes was false, or the *Amazonas* picked up the cargo on credit based on the strength of the Inman name. This was not unknown in the Amazon. Most of the rubber trade was handled on credit, and this would be the underlying cause of economic collapse in 1913. On the return trip at Obidos, she picked up "819 bags of Pará nuts"—Brazil nuts. Perhaps the line entry was convenient camouflage by Captain Murray once he understood what he was getting into; perhaps it was thought wiser not to mention the seeds at all.

Running counter to these doubts are two new sources: the village lore of Boim, and a casual line in Violet's newly uncovered diary. According to Boim historian Elisio Cohen, the *Amazonas* anchored within sight of Boim and Wickham paddled out to meet her. This is not inconceivable: a trading vessel like the *Amazonas* would have been very aware of Boim's trading houses, and Murray would have known how far up the mouthbay he could safely steam at this time of year. Such an arrangement had the added advantage of secrecy—there was no possibility that the *Amazonas* could be seen in that location from Santarém.

Violet also addresses the mystery in a most casual way: "When [Henry] had collected and packed about 70,000 [seeds], we started for England on board the first steamer here." No mystery at all to her: It was simply a matter of seizing the first opportunity. Henry himself in his March 6 letter to Hooker said, "I hope to leave with a large supply for England," which suggests steamers stopped frequently below Boim. In any case,

Henry, Violet, the adopted boy, and the seeds met the *Amazonas* at a given place and time, and they did so precipitately. They appeared as a speck on the water and Henry convinced a stranger to trust him in a desperate adventure, as he had done so many other times. "What seems most likely," speculated Warren Dean, "is that Wickham managed to persuade the captain to accept himself, his wife, and their baggage on credit and that he later reimbursed the line with money the Indian Office paid for his seeds."

What's more shocking than Captain Murray's decision to participate in Wickham's risky scheme was the sudden manner in which Henry left behind his remaining family. Only Violet would have known what was up, and perhaps even she was not fully informed until the last minute that they were accompanying the seeds to London. She would have been overjoyed: She was finally leaving this hellish place and returning to her family. For the past five years, they'd gambled everything on a new life, and it had nearly killed them. Now they gambled again, betting everything on the seeds.

One can only imagine the shock to Henry's surviving family when word drifted back that he and Violet had abandoned them. Henry, who'd led them into this tropical deathtrap, apparently did not offer them the chance to go home. The rift was complete, and the once close and hopeful family completely disintegrated, never to meet again. John Joseph Wickham, his wife Christine, and son Harry left the Amazon two years later. In 1878, he settled in Texas and became a cattle rancher. By 1881, Frank Pilditch, widower of Henry's sister Harriette Jane, had settled with his parents in London, where he practiced again as a solicitor. In 1882, he married Alice Molson Symon, a woman about twenty years his junior, by whom he had at least five children. All that remains of the graves of the others are faded images in Henry's sketch of the cemetery. The date is 1876. He was engaged in his desperate adventure when he made this last stop. The sketch was his only way to ask forgiveness and say good-bye.

Yet Henry's problems did not end when he and the seeds boarded the *Amazonas*. The seeds were "slung up fore and aft in their crates in the roomy, empty forehold," but Henry had little time to feel relieved. Captain

Murray was "crabbed and sore from the experiences with his two rascally supercargoes," but as Pará drew near, "I became more and more exercised and concerned with a new anxiety, so as not much to heed Murray's grumpiness." They were obligated to call at Pará to obtain official clearance before the *Amazonas* could lawfully put to sea. "It was perfectly certain in my mind that if the authorities guessed the purpose of what I had on board we should be detained under plea for instructions from the Central Government at Rio, if not interdicted altogether." Any delay could increase the odds of his seeds beginning to germinate or decomposing to a rancid cyanide mush; once either process started, he was ruined. Good intentions were not enough in this venture: as he said himself, his understanding with Kew and the India Office was "a straight offer to do it; pay to follow result."

As the *Amazonas* neared Pará, he shut and secured the hatches. Once in port, the ship showed no sign of loading any cargo. According to one historian, she was known to be without cargo; although the bill of entry shows otherwise, "a number of Brazilians had been much amused by the discomfiture afforded the Inman Line," which suggests that *something* happened in Manaus, even if the details are still unclear. Papers were cleared; no inspection was required.

It was in Pará that the most fanciful accretions to Wickham's tale occurred. Part of this can be laid on the imperial enthusiasms of later storytellers, but in a large measure Henry, his flawed memory, and his habit of bombast must take the blame. Although Pará was "an obstacle of appalling magnitude," as one commentator described it, Henry did have "a friend in court"—Consul Thomas Shipton Green. The Honorary Consul had worked against Henry's interests when setting up the failed shipment of seeds from the Bolivian *patrão* Ricardo Chávez and was probably involved in the disappointing attempt of Charles Farris to smuggle seeds in his crocodiles. But he'd also handled the correspondence between Wickham and London, and his ultimate duty lay not in his personal feelings but in advancing the interests of Great Britain. Nothing within Green's sphere of influence was more important than obtaining a separate source of rubber for the empire, and he jumped at the chance when

Wickham magically appeared. Green "quite [entered] into the spirit of the thing," wrote Henry:

[He] went himself with me on a special call on the Barão do S———, chief of the "Alfandiga," and backed me up as I represented "to his Excellency my difficulty and anxiety, being in charge of, and having on board a ship anchored out in the stream, exceedingly delicate botanical specimens specially designated for delivery to Her Brittanic Majesty's own Royal Gardens of Kew." Even while doing myself the honour of thus calling on his Excellency, I had given orders to the captain of the ship to keep up steam, having ventured to trust that his Excellency would see his way to furnish me with immediate dispatch.

In other words, Wickham and Green bluffed their way through, not exactly lying about the ship's cargo, but not exactly telling the whole truth, either. To add weight to the request, they appealed in the name of Her Brittanic Majesty, as if the plants were to be set right before the throne. Queen Victoria's name had clout in a country where Britain was the leading foreign investor.

There seems little question that the meeting with Pará's officials took place, though again the details are misty. The only baron with the initial "S" who made his home at the time on the Amazon was the Baron of Santarém, a venerable old gentleman who'd been one of Henry's neighbors and who, aware of recent history, would have been suspicious of the ruse. It was known, after all, that Wickham had wanted to be a rubber planter. Pará's port director was a commoner named Ulrich. Green would have been there, since ships, customs, and ports were well within his scope of duties, and he soon would help Robert Cross in his own quest for seeds.

In the early 1900s, a story had credence that Henry also called on the governor while in Pará. Before leaving, he allegedly told Murray to raise anchor and move slowly downstream. It was a quiet social evening in the governor's palace. Other guests were present, and Henry was well received. He was a known Amazon "character," so he would have had entertainment value at such an affair. There was interest in town about the *Amazonas*'s closed hatches, and again Henry was said to have repeated his concern about the "delicate botanical specimens" destined for Her Majesty's personal garden at Kew. If there was one thing to which Henry was always attuned, it was his audience and his effect upon them. He was a showman, and this touch of panache no doubt delighted the small group of Brazilian and Portuguese nobility. The evening was pleasant and cordial, and Henry took his leave in a spirit of goodwill. He boarded a cutter, made all speed to the *Amazonas*, joined her downriver, and put out to sea. "I could breathe easy," he said. The seeds, and the Wickhams, were free.

The scene at the Customs House raises the question of whether or not Henry broke the law. His request was very specific: He was shipping "delicate botanical specimens" to Kew, an appeal based on Article 643 of the Brazilian Customs Regulations, which stated:

> Products destined for Cabinets of Natural History, collected and arranged in the Empire by professors for this purpose expressly commissioned by foreign Governments or Academies, or duly accredited by the respective Diplomatic or Consular Agents, national or foreign, will be dispatched without opening the volumes in which they are encased, a sworn statement by the naturalist sufficing, and duties will be charged according to the value which he gives them, in accordance with a list in duplicate which he must present.

This is an extremely liberal regulation, one based on scientific trust, a quality that has disappeared amid current fears of biopiracy. The most generous interpretation that can be given is that, at best, Wickham and Green bent the law. True, the seeds were destined for Kew, but merely as a waypoint. And these seeds weren't for study but were destined for Britain's plantations in India and the Far East for purely commercial purposes. Their delivery to Kew and other Royal Gardens around the world was to sow the seeds and raise them to maturity, thus producing stock to sell to planters and commercial nurseries.

With Wickham a new idea was added: biopiracy. Minerals and metals can be guarded and sold by the countries in which they are mined, but crops can be grown elsewhere. Even if a plant's original habitat occurs completely within a nation's boundaries, it can still be sown in more favorable conditions. Markham's cinchona coup was a prime example, but there have been hundreds of others. Pineapples, found in South America, did well in Hawaii. The failure in the 1840s of potatoes from Peru spelled disaster for Ireland in the Great Famine. The early Virginia colony, Zimbabwe, and scores of other countries would all depend upon tobacco revenue. Brazil itself has been both victim and victimizer. In 1747, the Brazilian adventurer Francisco de Melo Palheta charmed from the wife of a French governor in Arabia a sample of coveted coffee seeds, which remains today one of the nation's most profitable exports. Although it was not expressly written that "whatever Victoria wants, Victoria gets," the essence of colonialism for every Great Power was assured markets and expanded riches and resources. Henry knew he could be stopped by Brazil, since such tactical moves were part of the Great Game. But it never entered his mind that he might be doing anything wrong. When it came to plants, the whole world was open ground.

In fact, it's doubtful that Henry would have backed off from his plan for any reason. He was not particularly ruthless, but neither did he overly concern himself with ethical issues. As he later said, he was a practical man. He was trying to survive the jungle, and this was his last way out of what had become a damnable place. His ethical guide, like his empire's, was the Protestant ethic: "God helps those who help themselves."

Henry's theft was no different than that by scores of others before him—and yet, in a fundamental way, it was. He did not steal one seed, or even a hundred; he stole seventy thousand. Like the anaconda, the sheer size of the subject made one pay attention. Thirty-four years after Henry's theft, the British rubber grown in the Far East from Henry's seeds would flood the world market, collapsing the Amazon economy in a single year and placing in the hands of a single power a major world resource. In 1884, the state of Amazonas levied a heavy export tax on rubber seeds, and in 1918, Brazil banned their export entirely. By 1920, when Henry was being knighted and called the "father of the rubber industry" in Great Britain, Brazilians dubbed him the "executioner of Amazonas," "the prince of thieves," and called his theft "hardly defensible in international law."

Biopiracy in its modern sense refers to the appropriation, without payment and usually by patent, of indigenous biomedical knowledge and genes by foreign corporations, institutions, or governments. The rosy periwinkle of Madagascar is often cited as a classic case in modern international law. Research into the plant was prompted in the 1950s by the periwinkle's use in native medicine, and it resulted in the discovery of several biologically active chemicals—most notably, vincristine—that were instrumental in the fight against various childhood cancers. When pharmaceutical giant Eli Lilly patented and marketed vincristine, it made billions, but Madagascar never made a dime. There was a complication, however. The locally known medical properties were not the same as those discovered by Eli Lilly, and when the company filed its patent, the flower had been replanted in numerous tropical countries. Since the researchers did not necessarily obtain local knowledge and plant samples from Madagascar, this muddied the legal claims.

In a looser sense, though, biopiracy is about power and its imbalance— the historical fact that poorer countries have been high in resources, while richer nations want—and can take—what they have. In that sense, Henry's theft became a symbol for every act of exploitation visited on the Third World. The issues revolve around ownership rights. Who owns the earth's riches? Current international law holds that nations own their resources, yet the counterargument is as old as Clement Markham's for taking cinchona: Nature, and her "improvement," belongs to mankind.

Was Henry a smuggler, aided by his government, a modern freebooter or privateer? The British Empire had a history of this, especially when the target was a Latin regime. This point of law, both in its letter and wider spirit, would generate plenty of heat in the ensuing years, especially in the first third of the twentieth century, when the quest by governments to control rubber was as frenzied as today's similar frenzy for oil. Interpretations of Henry's act shifted with the times and political, economic, and environmental winds. In 1913, the year Brazil lost its world monopoly in rubber to Great Britain, O. Labroz and V. Cayla of Brazil claimed that authorities were aware of Wickham's plans, but this seems an exercise in saving face for their nation. In 1939, the U.S. Department of Commerce sought from Brazil a report on the circumstances surrounding Wickham's exploit, and the reply from Pará pointed out that, since no one foresaw the possibility of establishing hevea plantations elsewhere, no law then existed that specifically prohibited the export of seed. Henry's theft could be seen as a triumph of the imagination. The Brazilian report came as world war loomed on the horizon. Ford's empire on the Tapajós was already established, and Brazil hoped for vast forest plantations—and a second boom in rubber due to the world's strategic needs. As the century progressed and Brazil finally understood that the richness of the Amazon Valley required federal management, Roberto Santos, Brazil's premier historian of the Amazon, asserted that even in the absence of specific regulations, no one had the right "to appropriate the goods of others when there is a sure owner or a defined jurisdiction." Critics scoffed that Santos seemed to possess "some higher vision of property, of nature constituting a national patrimony," yet this was exactly the consensus that came out of the 1992 Earth Summit in Rio de Janeiro, which produced a convention entitling nations to a share of the profits from substances yielded by their flora and fauna and which led directly to today's stringent regulations against biopiracy. It all started with Henry, at the Customs House in Pará. According to contemporary definitions, Henry Wickham and his wife were smugglers who acted at the behest of more highly placed smugglers—who believed they acted in the name of empire.

And so Henry, Violet, and their precious seeds sailed from the Amazon, never to return. Although a cliché, this is truly a case of sailing into history. They steamed past the huge floating lighthouse on the shallows of Bragança. They vanished into an Atlantic that, on this voyage, would be calm and blue. Henry took the hatches off the hold storing his open-air crates of seeds. He made sure they were secure on lines fore and aft, swinging in the breeze, safe from the ship's rats in the hold. He dealt with changes in climate as they crossed the equator. He was something of a mother hen. But Violet saw him more peaceful than he'd been in a long, long time. Henry reasonably thought he'd earned his stripes and the empire would be grateful. Henceforth, he said, life would be easy. Little did he know that his trials—and those of patient, practical Violet—had only just begun.

PART III

THE WORLD

I will work harder.

—Upton Sinclair, *The Jungle*

CHAPTER 10

THE EDGE OF THE WORLD

Henry delivered his seeds nearly three months to the day after Alexander Graham Bell made his first successful telephone call. Ten days before the seeds arrived, a train called the Transcontinental Express arrived in San Francisco, eighty-three hours and thirty-nine minutes after leaving New York. Henry delivered his seeds in the year that Dr. Nikolaus Otto and his assistant, Gottfried Daimler, built the first successful internal combustion engine to run on a four-stroke cycle. Each was a harbinger of the forces that would make rubber, from 1880 to 1910, the world's most market-sensitive and sought-after "new" commodity. Nevertheless, delivery of the seeds received scant notice in the press, overshadowed in Britain by the "Eastern question" and in the United States by the massacre at the Little Big Horn.

He certainly arrived like a conquering hero. The *Amazonas* docked at Lisbon, Le Havre and, finally, Liverpool on June 10, 1876, but Henry couldn't wait. He and Violet left Captain Murray at Le Havre with a promise of repayment, then crossed the channel and hitched a ride to London carrying a small bag of seeds. Henry arrived at Kew by hansom cab at 3 A.M. on June 14.

Hooker was an insomniac, and each night he would recite poetry to himself as he lay in his upper-story bedroom in the redbrick Georgian mansion overlooking Kew. He'd nearly fallen asleep when suddenly there was a rattle at his window. He jumped up, pulled apart the white curtains and spotted a lone figure beneath the chestnut trees fringing his yard.

He couldn't believe his eyes: The rude fellow seemed to be tossing pebbles at the glass.

When Hooker opened the window and snapped for the impertinent man to stop, the stranger stepped from the shadows and shouted something about seeds. Hooker rushed downstairs in his nightshirt and threw open the door. The nighttime skulker, who wore a wide tropical hat and clutched a Gladstone bag to his chest, introduced himself as Henry Alexander Wickham. The name rocked Hooker back on his heels. He led him to a study littered with rare floral prints and Wedgewood medallions and asked him what was in the bag.

"A sample of the rubber seeds you requested. Seventy thousand, altogether." Henry launched into his tale and the amazed director had to find a chair.

One witness was William Turner Thiselton-Dyer, the assistant director at Kew. In time, he would be Hooker's son-in-law and his successor as director. By then, he'd gained a reputation for pomposity, but during his first years as Hooker's assistant, he served as a buffer between the prickly director and the rest of the world.

The jungle adventurer descended on Kew like a bolt: the curved lane before Kew was paved in brick, and the hansom cab could be heard clattering from a block away. The last time they'd heard from Wickham was four months earlier. They assumed he'd messed up or missed the fruiting season again, and so no longer took him seriously. Robert Cross had already started on his collecting mission to Pará. He was somewhere in transit between a final visit to Richard Spruce in Scotland and his ship, docked in Liverpool, soon bound for Brazil. In the excitement over Henry's arrival, Cross was forgotten entirely. He'd leave Liverpool on June 19, and arrive in Pará on July 15. He literally crossed paths with the *Amazonas* in port and probably watched it unload.

"Not even the wildest imagination could have contemplated" the results of Henry's arrival, Thistelton-Dyer recalled. Hooker ordered a special night-goods train down to Liverpool, and by June 15, all seventy thousand seeds were sown in a vast greenhouse called the "seed-pit" and placed in the care of R. Irwin Lynch, a veteran foreman of the tropical department. Given Kew's history with hevea, Lynch was worried, as was

Thiselton-Dyer, who checked the greenhouse several times. "We knew it was touch and go," the latter recalled, "because it was likely the seeds wouldn't germinate. I remember well on the third day going into the propagating house . . . and seeing that by good luck the seed was germinating." An unsigned internal memo still found in Kew's archives and dated July 7, 1876 spelled out Henry's triumph: "70,000 seeds of *Hevea brasiliensis* were received from Mr. H. A. Wickham on June 14th. They were all sown the following day, and a few germinated on the fourth day after."

Those seeds that sprouted did so quickly. Hevea's germination usually occurs three to twenty-five days after planting, but the first sprouts poked above the soil on June 19, on the fourth day. Henry must have hovered over the planting trays like a worried mother; this was not just his future at stake, but a vindication of every disastrous decision he'd made. After planting, the seeds absorb moisture, going from less than 13 percent liquid to about 50 percent in a matter of hours. The process, called *imbibition*, is rapid once it starts. The rubber seed swells and ruptures its nutmeg-colored skin. The sudden infusion of water dissolves gibberellic acid, a plant hormone in the endosperm very similar to steroids. It turns on genes in nuclear DNA that trigger hydrolysis of the seed's large starch reserves. The starch turns into sugar, and, with the help of this built-in fuel, the first radicle soon emerges from the shell. The first pair of leaves usually appears within eight days.

By July 7, more than 2,700 of the seeds had germinated and been potted. According to the unsigned memo, "Many hundreds are now 15 inches long and all are in vigorous health." The fact that 2,700 germinating seeds represented a mere 3.6 percent of the total shipment did not seem cause for criticism—it was, after all, 2,700 more rubber plants than Kew had ever grown in its greenhouses before.

By the second month, however, little signs began to show that Henry rubbed Joseph Hooker the wrong way. In addition to rubber, Henry brought with him other plants he thought might have value, including seeds of the piquiá tree. He described these to Hooker in a note, suggesting their possible use. He tried to cast himself as an amateur botanist in hopes of accompanying the rubber seeds to Ceylon, as he'd mentioned in the April 1875 letter, but amateur botanists were a breed of human that Hooker despised. Henry wrote that he'd "made some experiments in planting" rubber that could be valuable. Time would prove him right, but Hooker was not sold. The director coolly noted the Latin names of Wickham's specimens in the margins "as if to underscore his ignorance of botany and Kew's already perfect familiarity with them."

On August 20, 1876, the *Evening Herald* ran a short story on the seeds. The young sprouts were said to cover "a space about 300 square feet, closely packed together. A number began to grow almost immediately, and many within a few days reached a height of 18 inches." Special cases had been built for their shipment to Ceylon, Burma, and Singapore. The program, described as a success, could only have been reported with Hooker's approval, but nowhere was there mention of Henry's role—not even his name.

About the same time that he was being stonewalled by Hooker, Henry approached Clements Markham. The hero of cinchona generally tried to play fair by those who'd aided him in his imperial intrigues, and he now tried to do the same for Henry. Twice in July 1876, while the young sprouts were growing, he mentioned to Hooker by memo that "Mr. Wickham seems to have taken very great pains with the seeds," a reproachful hint that Henry deserved a fair measure of gratitude. Hooker was not swayed; he replied that he doubted Henry's abilities. Perhaps Hooker counted on Cross's efforts, although no word had come back from the veteran Kew gardener. In a July 19 letter, Markham was especially adamant:

> I have had a long conversation with Mr. Wickham, who is willing to accept employment in connection with the introduction of caoutchouc cultivation into India, either by making further collection of seeds in the Amazon Valley or by taking out plants to India (Ceylon and Singapore), and giving advice about sites, etc., and the establishment of plantations. . . . I am inclined to recommend that Mr. Wickham should be employed to take the caoutchouc plants out. . . . He might be instructed to see them established at Peradenia and at Singapore, to put himself in communication with the Madras Government and the Chief Commis-

sioner of British Burma, and to give the benefit of his knowledge and experience as regards the selection of sites in Tenasserim and Malobar, etc. Will you kindly tell me what you think about this?

Markham's letter is interesting for several reasons. First, Henry had made an incredible concession: He'd return to the Amazon to hunt for more seeds. The tropics had nearly killed him twice. It killed his sister and mother, and he could only avoid its grasp so long. But he saw himself completely wed to hevea, wrapped in its embrace as though by the strangler fig, *matapalo*. He'd do whatever was needed to maintain that claim.

Secondly, Markham had observed Wickham at length and thought he knew his strengths. Rather than a botanist, he saw Henry as rubber's advance man. Planters were a conservative lot. At that moment they were doing well with coffee and tea. More necessary than any gardener was the presence of a loud and brash promoter, a shoe that fit Wickham perfectly. British bureaucracy was a lumbering beast. Markham hated it, dreamed of ways to outflank it, and worried that government intractability would strangle all that was creative in reams of orders and obfuscation. His letter laid out Henry's proposed marching orders: See to the seeds' safe establishment in Ceylon and Singapore, then hound the colonial governors until, in exasperation, they agreed to give rubber a try. He so believed in his insight that he was willing to push Hooker for an answer: "Will you kindly tell me what you think of this?"—a remarkably direct request, considering the dry and uncommitted flavor of Victorian bureaucracy.

But Hooker would not be moved. One gets the idea that he rarely changed his mind. There was something intractable in his makeup that wasn't part of his father's nature—an arrogance, as if convinced of his own botanical infallibility. In his July 20 reply to Markham, he explained that "we have no knowledge of his horticultural competence, in taking charge of cases or in selecting sites," but in private he seemed enraged that Henry would appeal to another quarter. Considering the sudden defensive tone that appeared in Wickham's letters of July 20 and August 1 to Hooker, there is good reason to believe the director lit into Henry with barely controlled rancor. He backpedaled. "I did not mean to suggest my taking entire charge of the plants. I made experiments in planting [rubber] on the Amazon. Now it appears to me that all the skill of the most trained gardener will not supply this information. I saw Mr. Markham today with regard to the Hevea; he merely said that you thought it expedient to send a gardener."

Markham advised Henry to "await developments" while writing a memo to the India Office recommending that Wickham be posted to India under its own authority, not Kew's. But Markham's days were numbered. Under the irascible Louis Mallet, he no longer had the clout he once enjoyed. Even though Markham was instrumental in obtaining cinchona, even though he'd repeated the feat with rubber, Mallet saw him as a holdover from the days when the East India Company was filled with adventurers. Clements Markham was in no way reconciled with Mallet's new vision of bureaucracy, where a prompt, daily appearance at the desk was a civil servant's chief virtue. A year earlier, in September 1875, a public confrontation between Markham and Mallet got to the point where Markham demanded an apology from his superior for describing him as "dishonourable." Mallet complained to Lord Salisbury, *his* superior, that "the time has come when he must be told very distinctly that he must . . . comply with official rules, or go." Markham's 1877 resignation was already inevitable in July and August 1876. Martinets would block the plans of both Markham and Wickham.

Henry had one last line of attack—to write a report describing everything he knew about hevea and prove that he was the man for the job. He advised that "the Malay Peninsula is most likely to combine the climactic conditions required for the Indian rubber tree of the Great Valley of South America," a suggestion that would have saved the empire three decades of lost time and money if he'd been heeded. He cautioned that hevea grew more slowly under forest shade than in open plantation conditions: "I have known trees, grown in the open, seed abundantly after the third year." Most important, he warned that just because rubber trees were first discovered by explorers on floodplains or along riverbanks did not mean that these were the best growing conditions. The best specimens were found on higher ground, back from the river. Later tests proved him right and showed that trees planted in soggy ground failed to develop an adequate root system.

Yet even that final bit of prophecy was doomed. Wickham's report was buried in the files of the India Office, uncritiqued, unused. "What is more," said Edward Lane, who unearthed the report in the 1950s, "its coffin bore the wrong name: An examination of the document proves that it is Wickham's in handwriting, in style, and in content. Yet someone— presumably in the India Office, but possibly at Kew—has written underneath the title, 'by Robert Cross?'"

The report would be as buried as the man. Henry was paid for his services—£700 according to most accounts, though £740 according to a memo by Hooker—but he'd outlived his purpose and become inconvenient. When the first shipment of seeds left for India in August, he was not with them. Henry continued his petition to accompany the next shipment, but Kew washed its hands of him with a lie. He was told that "though Kew authorities advocated it, the depreciation in the value of silver" undercut his posting, Violet wrote. A sympathetic Kew gardener, most probably Irwin Lynch—described in Kew biographies as a decent sort who might have felt ashamed at the way his nation treated Henry— added to that final payment a Wardian greenhouse filled with 175 Liberian coffee seedlings.

Henry bore no grudge against Hooker, convinced for the rest of his life that the director and he were twin visionaries. He never realized that the director opposed his appointment to India or that he'd disparaged him as an uneducated opportunist, a prejudice that Kew factotums would echo for the rest of Henry's days. Henry remained ever grateful to the irritable and duplicitous director, blaming his rejection on the India Office, never imagining that the permission to return with an unlimited number of seeds came from that very source. He never figured out that his best friend in the empire was not Joseph Hooker, but Clements Markham.

In mid-September 1876, unwilling to wait any longer for what he knew would be an official rejection, Henry, Violet, and the ever-anonymous Indian boy took a slow boat to Queensland, the case of Liberian coffee beside him. The official jungles of London had defeated him; he would return to the kind of wilderness that he understood.

Two long and futile journeys began in August and September 1876, both dogged by disaster, both forgotten on the far side of the world. The first one involved Henry's seeds. A month before the Wickhams sailed for Queensland, 1,919 of Henry's seedlings were packed in 38 Wardian cases, placed in the care of Kew gardener William Chapman, and stowed aboard the P&O steamship *Duke of Devonshire,* bound for Colombo, Ceylon (now Sri Lanka). The P&O Company had caused a revolution in sea transport: Their ships arrived in port as scheduled, something previously unknown. Yet they also insisted on prompt payment of freight charges; otherwise, the cargo was not released from the hold. When the *Duke of Devonshire* arrived safely at Colombo on September 13, the captain would not release the seedlings to the director of Kew's botanic garden at Peradeniya, because the India Office had failed to pay the fees. Word was telegraphed to London, and three days later the fees were paid. Of the original shipment of 1,919 seedlings, approximately 1,700 survived, thanks primarily to gardener William Chapman's ministrations and the fact that an enraged Clements Markham expedited the payment back in London.

The next batch was not so lucky. On August 11, 1876, the India Office sent one hundred seeds to Singapore. These were not accompanied by a gardener and also got held up for nonpayment of freight. This time, every seed died. Kew and the India Office kept trying, and by the end of 1876, Kew had distributed 2,900 plants to its branch gardens in the Far East, to collectors in the British Isles, and small lots to other British colonies.

On November 22, 1876, Robert Cross returned from the Amazon, bearing with him 1,080 sickly hevea seedlings from the swamps around Pará and sixty Ceará rubber plants, known as *Manihot glaziovii,* which produced an inferior grade of rubber. Kew only kept four hundred of Cross's struggling hevea and gave the rest to commercial nurseryman William Bull. By spring 1877, Bull reported that only fourteen of these still lived, while Kew admitted that twelve hevea plants from Cross's mission had managed to survive.

Although Cross's rubber seeds were sickly, he did write a March 1877 report that would have enormous consequence for the British rubber industry. Because he collected hevea on the islands and in the swamps around Pará, he assumed that the best place for its growth would be in the hottest parts of India, a conclusion in direct contradiction to Wickham's. "The flat, low lying, moist tracts, subject to inundation, shallow lagoons, water holes, and all descriptions of mud accumulations, miry swamps and banks of sluggish streams and rivers, will be found best" for hevea, he wrote. Although Wickham would contest this opinion for decades, no one listened. Cross was a trained expert; Henry was not. Since Henry had been dismissed by Kew, everyone thought him wrong. Cross's mistaken observations ensured that those who tried growing hevea in the 1870s and 1880s did so in some of the most miserable, disease-ridden landscapes in the world.

Robert Cross's opinions, the result of his training and judgment, helped delay the beginning of the British rubber monopoly by decades. In 1881, when Cross was hired by the India Office to supervise the growth of cinchona and rubber in the Nilgiri Hills, hevea was not catching on with planters. It grew more slowly than *Castilla* and Ceará, an important point for those hoping to turn a quick profit. The easy availability of Asian varieties like *Ficus elastica* also worked against the choice of hevea. Due to his experience with the sickly plants of Pará, Cross tended to agree and paid more attention to the development of Ceará and *Castilla*. Hevea would grow untapped in Kew's branch gardens and untried by Eastern planters for years.

For the next two decades, the British cultivation of hevea stalled. Wickham's trees and their offspring were sent to Ceylon, Singapore, and Malaya, but also to more inhospitable environments like the Dutch East Indies, Indochina, and other foreign regions. Because the trees were slow-growing, planters were not interested in tending them, even when Kew gave them away for free. Yet these were the decades, 1880–1910, when the three great developments dependent on rubber—electricity, bicycles, and automobiles—increased its worldwide demand at a rate that nearly doubled production every five, then every three years.

Early electrical systems were terrifying. On Manhattan's Pearl Street, in the world's first domestic lighting system, exposed, uninsulated wires were the norm, with horrible consequences for linesmen and children. Thanks to the insulating properties of rubber, commercial lighting was in wide use by the 1880s, though domestic use would not catch up for another four decades. By 1880, "submarine telegraph wires" spread all over the globe from London, the most famous being the Channel Cable from Dunkirk to Dover and the Transatlantic Cable from Cornwall to New Brunswick. Electrical traction made possible subways in London, Paris, and New York, while in scores of other cities, cable cars became the norm.

The 1890s would be the decade of the bicycle. The seven million bicycles found worldwide in 1895 used most of the world's rubber, a boom that would not have occurred if not for the invention of the "pneumatic rubber tyre." Although there had been bicycles previously, they rode on solid rubber tires. These were puncture-resistant, a boon on roads where nails were frequently shed from horseshoes, but they lacked suspension, were hard to steer, and were an unpleasant ride. This changed by the late 1890s. The market was flooded with steel tubes, ball bearings, variable speed gears, and high-quality chains. Above all else, it was flooded with replaceable rubber tires and inner tubes, mass-produced in the factories of Dunlop in Birmingham, England; Michelin in Clermont-Ferrand, France; and Pirelli in Milan, Italy. The bicycle was cheap and popular. People suddenly had a means of freedom that had been unknown.

The first decade of the new century was the first decade of the automobile, and all the components of the bicycle were put to use. The United States made few cars in 1895, but fifteen years later it produced two hundred thousand—more than the rest of the world. By 1920, there would be 12 million cars registered in the United States, and approximately 2 million built that year. More than half were Henry Ford's Model Ts.

Henry Wickham was not part of this. Kew's rejection cut deep. He'd defined his hopes through rubber; his success gave meaning to his jungle ordeal. Except for Violet, hevea—and the triumph it represented—was his one great love. Now it had been stripped from him, and like a jilted lover, he grew bitter.

For the next twenty years, Henry wandered in self-imposed exile at the far edges of the British Empire. He dragged Violet with him to some of the most inhospitable and dangerous environs in the world. He never wrote of his motives, but his actions spoke for themselves. While the rest of the world grew rich off rubber, he embarked on a quest to succeed as a pioneer planter who discovered the next miracle crop. When he returned to the center of empire, he would do so in triumph. He'd be important again.

The quest began in Queensland. Violet never said in her memoirs how they chose such a place, but in 1872, agents promoting emigration to Queensland began to spread throughout the British Isles. They pitched it as a land of riches. Magic lantern shows of beautiful scenery accompanied a barrage of success stories: the discovery of gold in Gympie in October 1867, in Cape York Peninsula in May 1869, and in Charters Towers in 1872. Tin mines in the Stanthorpe district seemed inexhaustible. Sugarcane was farmed in the north and the south. Railways spread west from the seaports. There was no shortage of jobs for the workman, while land for the squatter, or "pastoralist," sold at ten pounds per 80 acres. The small towns springing up in the bush were just like those back home, with schools, newspapers, musical unions, and cricket clubs. The Immigration Act of 1872 offered free passage to laborers, but Henry and Violet paid their own way, since they traveled first-class to their new home.

As with most advertisements for paradise, much was left unsaid. The agents did not mention the floods and crop failures or the fact that whites and "aboriginals" were still engaged in a vicious racial war. It has been estimated that from 1840 to 1901, the violence in the Australian frontier claimed the lives of 2,000–2,500 whites and 20,000 aborigines; estimates vary on the decline of the native population, but historians believe it may have dropped from a high of one million in 1788 to 50,000 in 1890. In the year of Custer's Last Stand, Henry and Violet entered a 678,000-square-mile battleground that was more like the American West than the tropical jungle from which they'd just hailed.

Henry entered a stage filled with racial hatreds unlike anything he'd encountered in Latin America. There he'd been part of a sheltered class. As an Englishman, he symbolized such ideals as progress and liberty. In

Australia, he was one of the colonizers. The cycle of hostility and reprisal started as settlers invaded aboriginal hunting grounds; then it swept across the continent as pastoralists destroyed aboriginal communities and drafted survivors into the alien roles of stockmen, native police, or maids. For each attack there was retribution, often during the yearly corroborees in which young warriors were initiated into the rites of manhood.

North Queensland, where Henry and Violet were headed, was still home to various hatreds. Carl Lumholtz, a Norwegian anthropologist who spent thirteen months on the Herbert River in 1882–83, met a farmer who boasted that when he killed a black, he cremated the body to destroy the evidence. Kidnapping was more common than slaughter—so common, according to the 1884 *Queensland Figaro*, that "There is nothing extraordinary in it." In 1892, a former prospector wrote that many of the aboriginal women and children found on sheep and cattle ranches between Normanton and Camoweal were kidnapped in slave raids: "These children are brought in and tied up . . . and if they manage to get away and are caught, God help them." As late as 1901, there was a "mutual understanding" throughout the area that "a runaway black child could be hunted and brought back," one observer said.

The Wickhams left on September 20, 1876, aboard the barque *Scottish Knight* for the three-month, sixteen-thousand-mile emigrant run. These trips could be nightmarish, plagued by shipwreck, lengthy becalming, starvation, and epidemics of contagious disease, but by the standards of the day, their voyage was uneventful and pleasant enough, it seemed. They traveled first-class with the Indian boy and a nine-year-old English serving girl whom Violet's father had no doubt once again secured. Henry tended to his coffee plants packed in their greenhouses, while Violet observed the small society of hopeful strangers thrown together in cramped quarters. The single men were berthed forward, single girls aft, and married couples amidships as a buffer against wanton immorality. The ship's doctor assumed responsibility and "supreme authority" over the emigrants. He delivered babies, performed one or two successful operations, enforced cleanliness, intercepted love letters between fore and aft, and acted as chaperone. The latter duty was doomed to failure.

One newly married couple was so ardent they were dubbed Romeo and Juliet. Several weddings took place, Violet wrote, "in spite of the restrictions on board."

They arrived in mid-December 1876 at Townsville in North Queensland. In addition to Kew's gift of Liberian coffee, Henry brought from the Amazon some Brazilian tobacco, and his real hope was to cultivate the leaf on a huge scale. It seemed that his timing was perfect. An outbreak of rust had attacked the sugarcane crop during 1874–75; then in early 1876, the corn crop was wiped out by an unknown scourge that rotted the cobs. On July 6, 1877, he bought 160 acres for £20 in the Herbert River country near Cardwell, then four months later added another 596 acres for £149. In 1881, he bought 300 acres for £225, for a total holding of 1,056 acres.

By so doing, he overextended himself. It cost about £110 to set up a 160-acre farm, while fencing, required by law, cost £40 per 160 acres. It is safe to say that within four years he used up his £700–£740 payment from Kew for the rubber seeds, and was in debt for at least £100. He put himself in the hands of the storekeeper-banker, or "gombeen man." There were so many farm failures in the 1880s and 1890s that the gombeen man raised his commissions. The average interest rate paid by a farmer like Henry was 35 percent.

"Dear Land," as they called it, cursed farmers in Australia and proved extremely expensive. One of the nation's great scandals in the second half of the nineteenth century was the failure of tens of thousands of farmers to make a living from their holdings after years of slaving away. By 1880, Henry had ten acres of land planted with various crops and a garden, but his dreams lay in five acres planted with tobacco. He dried and cured the leaf himself. He invited local storekeepers to inspect his crop, and they "assured him if he could produce that quality, he could find a ready sale" anywhere in Queensland, Violet wrote. It seemed a vindication of his earlier failures, but by then his costs had edged past his income and he'd never enter the black again. He was a "cockie," or "cockatoo farmer," a term of derision for the small, struggling landholder. The Herbert River was sugarcane country, and there were very few cockies around like him.

But Henry was also a novelty, and by 1877 word had filtered out of his adventures on the Amazon. A neighboring sugar planter invited him to stay until he got settled, and Henry reciprocated by planting most of Kew's Liberian coffee—possibly the first such coffee in that region—on his host's land. Since his holding was well-timbered, "once more there was the old work of cutting down the site for the house," Violet wrote. He built a rustic log cabin "which rather amused his neighbors, as here in Australia things were far more civilized" and people lived in light-frame houses. Once more, they lurched from catastrophe to catastrophe.

Their Australian disasters sprang from the elements and began immediately. On their land stood two empty thatch cottages. As Henry was raising the cabin, they moved into one of these. One morning Henry asked Violet to collect some grass. "When the dew was off I set fire to it so as to clean up around the house," she wrote. She looked away for a moment and then, "to my horror, I saw that the fire had extended to the cottage and completely destroyed it."

They moved into the remaining cottage while Henry finished the house. Two years later, he repeated Violet's error on a much grander scale. One Sunday in 1879, as he wandered around the farm, he noticed that the wind was blowing away from the house and decided to burn off a new clearing. No sooner was the stubble alight than a sudden gust blew a shower of sparks back on the roof of their new house; in a few minutes, the house was ablaze "end to end." They lost everything—family letters, photos, even the sword that Henry's warrior grandfather had carried with him on the Nile. "Saddles, flour, etc., might have been saved, being in an underground room," Violet lamented, "but we forgot and simply" watched in shock until everything was reduced to a smoldering pile of ashes.

They built anew, this time on the side of a hill. Instead of another log cabin, Henry built a shingle frame house two stories high in front, but backed into the hill face so that in the rear the roof stood at ground level. They used the lower floor as a kitchen. Henry dug an elaborate flue to carry smoke to the chimney, but it "leaked the whole way and spoiled everything in the bedroom . . . and the smoke circled in clouds around my head." Henry built a new kitchen away from the house and added a new roof of overlapping sheets of corrugated iron. Although this eliminated the fire danger, it turned the house into an oven, while the drumming of rain on the metal roof during the frequent thunderstorms drove Violet crazy. "Rain does not express it," she wrote. "Sheets of water falling on an iron box and you inside it is the most trying thing to the nerves I know."

Then the third disaster struck. In May 1881, the sky opened for a week. The rainfall was twelve inches, and the floods were the greatest since 1870. "After having burnt out, it seemed necessary for me to try the water cure," Violet said. Henry was away on business for some weeks, and Violet was alone. She'd been watching a distant storm roll nearer when the gale perked up. The thick foliage on the plain whistled until the heavens opened and the torrent descended. This storm seemed more personal than most, more malevolent, for the wind crept under the iron roof and lifted and flapped the corrugated metal sheets with each new gust. She knew her roof would go soon: There was no stopping fate and the wind. There was something either unflappable or fatalistic about her response. After all she'd been through with Henry, maybe it was a combination. The wind was blowing too hard to seek help, so "I fell asleep, dreaming that the house had capsized and was rolling down the hill in to the creek." By morning the entire roof had blown off. Once again, everything they owned was ruined.

To aborigines, this land was created in the Dreamtime, the time of Creation, made sacred by the lingering presence of spirits. To Violet, it was turning into a land of violent dreams. She'd gaze across the landscape and spot the whirlwinds marching steadily through the dust. There were small ones, their funnels measuring a hundred yards or less in diameter, and there were the giants that drew trees, henhouses, the roof of one's house up in the vortex until it stalked too far and faded. There was a sameness to the land, just as on the Amazon, but there everything was saturated and soaked, while here the enemy of comfort was aridity. The Amazon was a massive beast that hissed as it passed. In Queensland, a river was a wide sandy bed with a shallow stream trickling through the center—or not even that, just dry, scorching sand with a water hole every two miles.

There were as many dangers in Queensland, it seemed to Violet, as on the Amazon. There were scorpions and tarantulas, and alligators that hid in the salt marsh until you waded too close, then carried you off with a massive clack of the jaws. There were nine-inch green centipedes whose sting was so agonizing that you wished you were dead. Horses would walk to a fire and thrust their heads into the smoke for relief from flies and mosquitoes; two would stand head to foot and flip their tails, mutually keeping the pests away. Leaf-cutter ants destroyed crops, returning night after night and "cutting off the leaves and young shoots and carrying them off" to the nest. Nothing seemed to stop them, not even when Henry borrowed a blacksmith's bellows and blew smoke and sulfur into the nests for hours. "You could see it issuing at different exits all over the plantation, even on the other side of the creek," Violet wrote, but that night, the ants returned.

The worst, by far, were snakes. The place was a paradise for snakes: big black snakes with rose-colored bellies and tiny black snakes no thicker than a pipestem whose poison could kill a chicken after one step and three or four drunken pirouettes. There were poisonous whip snakes that ran the spectrum from dirty brown to emerald green. Their poison seemed to put a man to sleep. One walked him back and forth to keep him awake, and farmers discovered that if they could get the victim drunk on rum, he would survive. There were snakes in the house, under the house, underfoot, in the corners, crawling into bed. Once Violet went to the creek for a bath and saw the tail of a snake disappear beneath her dressing gown. "You may imagine I slipped out of it as quickly and quietly as I could."

Violet's greatest war was with constrictors. They fought an anaconda in a salt marsh that was "as big round as [Henry's] body." They shot it and tied it with a rope that snapped several times like twine, until Henry finally pinned it to the ground with a pitchfork. It measured twenty to twenty-four feet in length. They used its tanned skin in place of canvas for a deck chair. Constrictors liked to glide into the henhouse at night. Once she killed one that was fifteen feet long and that had swallowed a chicken up to its legs. One night she awoke on hearing the hens cry out. "Henry, get up!" she hissed, shaking him from sleep, but he wouldn't budge. "The fowls are just having bad dreams," he groaned. She rushed out alone and found three chickens wrapped in a constrictor's coils.

During this time, Henry had become northern Australia's sole promoter of tobacco. Though he might not grow much at home, he was growing more than anyone else, which meant he could consider himself the region's authority. In 1884, he published a four-page pamphlet under the authority of the Queensland government entitled "Directions for Tobacco Growing and Curing in North Queensland," a forerunner of his more widely-read 1908 treatise on rubber. He was turning himself into a crop promoter, if such a thing existed at the time.

During this time we also see the first sign of a growing distance between Violet and Henry. She was often alone, as Henry was frequently absent, physically and emotionally. No word of childlessness was ever mentioned in the writings of Henry or Violet, nor in family tales. Violet never seemed to regret the absence of children in her life, and Henry was too driven to care. In compensation, Violet grew tough and independent. She hopped on her mare, Fairy, for a ride through the bush. "I did not need anyone with me," she claimed proudly, "though once or twice I had a gentleman apologize for not escorting me." She visited the neighboring ranch and brought home some beef in a saddlebag, "though I expect I considerably scandalized those neighbors." She rode twenty miles to town to shop—town in the bush was two stores, a hotel, and a courthouse—sometimes taking along the Indian boy. She did her best to manage a social life during these trips, but during her decade in Queensland she went to the races once, went to one sporting event, and paid one formal visit to a neighbor to welcome a visitor. Violet did not whine, did not complain, and showed back in the Amazon that she could be as tough as her husband. But sometimes the loneliness of the bush peeks through her memoir, such as when she writes that those three outings were "the only recreations I had during our . . . years in Queensland."

Most of her waking hours were spent managing the farm, since Henry was usually away. The cultivation and harvesting of tobacco and sugarcane demanded a good deal of labor—cheap labor, if possible. During this time, the labor force in Queensland consisted almost entirely of South Sea islanders brought from their villages to work for a period of three years. As a rule, these workers were tough Melanesians who went by the general term "Kanakas," a Melanesian word for "man." The abuses connected with the "recruitment" of islanders made Kanaka labor one of the burning issues of Queensland politics. Slavery had been abolished from the American South, yet a society of huge sugarcane plantations resembling the antebellum South had grown up in Queensland, and residents were sensitive to accusations that a new slave system existed in the far reaches of the British Empire. To counter this, a number of government commissions investigated charges of abuse, and they almost always concluded that the work force was not mistreated. It certainly was a paternalistic system, as illustrated by an editorial in *The Queenslander* on May 14, 1881:

> The Kanaka is at best a savage, often tractable and biddable, but still undeniably a savage whose short contact with civilisation works little change in him. He is a child to be protected from the ill-use or deception of cruel or designing men, and is to too great an extent incapable of guarding his own rights.

Although an islander's service was regulated by the government, the real abuses took place during "recruitment," far from the public eye. The "blackbirders," as those employed in the labor trade were called, ranged through the islands recruiting shiploads of workers, and they lied to the natives without conscience. Few carried interpreters, but signed islanders up for their three-year stints based on a little pidgin English, trade goods, and pantomime. "A favorite device," one contemporary newspaper reported, "was to hold up two or three fingers and to imitate the cutting of cane and grass or the digging of yams. One gentleman with a sense of humour took a yam and bit it three times." Others impersonated missionaries and promised the gifts the clergymen usually brought. The tribesmen were left with the impression that they were only going for a short cruise to see the wonders of the white man's world. They were astounded by what they'd bought into and often required what was euphemistically called "breaking in." Kidnapping was rampant, often at the end of a gun.

The consequences were inevitable. Several genuine missionaries, including an Anglican bishop, were murdered by natives. The biggest story for 1880 in Queensland was the May 30 massacre aboard the trading schooner *Esperanza*. When the ship was seized by a party of natives, the captain, two white seamen, and four native crewmen were killed and the vessel plundered and burned. Natives were killed during a punitive expedition. The blackbirders were excoriated and moved to virgin territory in the islands east of New Guinea. From 1883 to 1885, nearly seven thousand people were kidnapped or duped and sent to the farms of Queensland.

Since she was the farm's de facto boss, Violet managed the Kanakas. Apparently she and her neighbors did a good job:

The chief Magistrate of the district is the Polynesian inspector and whenever he visits a district he has them all assembled for inspection and to hear complaints. They are well treated as a rule, shed tears in many cases at leaving when their term is expired, often returning for another term after a short time at home. They are provided with clothes, food and a fixed wage which they draw on very sparingly till their time expires, when they spend it all. The two first important things are a gun and trunk, after that knives, tools, shirts, bright blankets.

On the whole, she liked her Melanesian workers, though she tended to look on them as children needing careful supervision:

[W]e found them trustworthy, working as well without their master's eye as with it. . . . They are difficult to manage when sick, losing heart instantly. However strictly cautioned as to diet, they eat all manner of indigestible food. The wife of the owner of the neighboring sugar plantation was in the habit of making delicacies for the sick Kanakas but from our own experience I am sure it was useless.

Not even cheap labor could save their farm, however, and then Henry accidentally sold it out from under them. From 1880 to 1885, he'd partnered with a man named Hammick, but in 1885, Hammick wanted out.

The chronic lack of capital, the labor costs, the fluctuating prices, the weather, and bad luck were just too much for him. The son of a wealthy planter offered to buy Hammick out, and Henry felt he might get a break, since the new buyer seemed wealthy. His star seemed to be rising. The market for tobacco was finally opening up, and the new partner, though young, seemed willing to pump funds into his plans.

Then Henry made a stupid mistake. Hammick was so eager to pull out that Henry agreed for him to be compensated by a bank loan that would be repaid by the new partner. That's when it happened: Henry acted as the loan's guarantor. Without warning, the new partner backed out. He sent Henry a letter stating that his father disapproved of the venture, and their verbal agreement could not be enforced or ratified. Henry had depended on trust and never signed a contract. Now he had to repay the bank, and that ruined him. He held on another eighteen months, but in early 1886, a depression hit Queensland and his hard-earned farm was virtually worthless.

"I have often wondered," Violet later wrote, "whether it was not a plot between the new and old partner; but I suppose not, as they could neither of them have supposed Henry could be so unbusinesslike."

For the second time, he'd been too trusting. In spring 1886, Henry and Violet packed the few belongings they had left and took the three-month voyage back to England. He repeated history, but not as he'd wanted: he returned penniless, as he had from the Orinoco, not in triumph as he'd dreamed. The sum for the sale of his farm was, according to family stories, "probably little more than sufficient for the fares home."

CHAPTER 11

THE TALKING CROSS

Then, when all seemed bleak, another savior appeared, like Captain Hill in Nicaragua, young *confederado* Watkins on the Orinoco, James Drummond-Hay in Pará, Violet's father, and Clements Markham. At each point in Henry's life when his future seemed ruined, a stranger responded to some quality in him that evoked sympathy or trust and bucked him up again. *On to the next wilderness,* the savior seemed to say. *On to better things.*

This time, however, we do not know the savior's identity, only Violet's note that one existed. In summer 1886, shortly after his defeated return to London, Henry "agreed to join a friend in journeying to British Honduras," she said. On November 12, 1886, he sailed on the *Godalming,* and on December 18, he disembarked in Belize, the capital city. It was the first time in a decade that Violet had been home to visit her family, and now her wayward husband was heading back to South America. "I let him go back some six months in advance," she wrote, resigned.

During those six months, it must have seemed to her that old friends and family lived in a different galaxy than the one she'd loved as a girl. It was a faster, brighter world than any she'd known in the tropics, and she couldn't help but feel left behind. In 1878, two years after Henry and she took the *Scottish Knight* to Queensland, electric lights had just been introduced in London. Now they seemed to be everywhere. The underground trains, reaching far out of the city, were powered by electricity, their cables insulated in rubber from the Amazon. In 1881, the papers said, London's population topped 3.3 million, making her birthplace the largest city on earth, larger by far than New York, with only 1.2 million. The year before, the Englishman John Kemp Starley had introduced the "safety" bicycle, and soon, people said, every Englishman would own one. She tried to imagine 3.3 million Londoners all on their bicycles at once, riding down Piccadilly, ringing their bells for the right-of-way.

London was the world center for finance and transport, and all England was in motion. In 1881, the clipper ship *James Stafford* crossed the Pacific Ocean in twenty-one and a half days, a world record. Almost anywhere on earth was accessible in a matter of months. Soon there would be no place untamed by civilization, no place to run or hide. Even the vast American desert was no longer intractable: on September 4, the elusive Apache chief Geronimo had surrendered in a place called Skeleton Canyon, Arizona, thus ending the last major U.S.-Indian war. The papers said forty-eight thousand new homes were built in London last year. She'd stroll through old neighborhoods and they'd transformed into colonies of French, Italians, Russians, or Greeks, every space filled with settlers far from home.

Violet knew she'd eventually have to leave this modern world to rejoin her husband, and in April or May of 1887, she did.

British Honduras was a tiny place, an 8,867-square-mile strip of beach and jungle. The low coastline was swampy, with thick mangroves blocking the passage inland, punctuated by rivers. The colony was tucked like an armpit beneath the shoulder of Mexico's violent Yucatán Peninsula and bordered on the west and south by Guatemala. The latter claimed British Honduras as part of her territory but never pressed the issue, and the British refused to leave. In that way, it was like Nicaragua's Mosquito Coast: The Spanish claimed but never settled the unhealthy strip of mangrove swamp, and only wanted it back when the British crept in under their noses. Most historical sources agree that the origin of Belize City, and thus British Honduras—for the two were synonymous during their early history—occurred in the 1600s. The earliest settlers at the mouth of the Belize River may well have been British privateers hiding from the Spaniards, for no coastline was better suited for guerilla warfare. The maze of islands and concealed channels was perfect for staging sudden attacks and retreats. The channels led into the trees to become a web of lagoons. The shore was protected ten miles out by a wall of coral that fringed the coast from the Yucatán down to Guatemala, a distance of about two hundred miles. Inside this strip water was smooth even when rollers pounded on the reef, but to pierce that reef one needed knowledge of its breaks and channels, something the Spanish never learned.

There was an economic reason for settlement, however, and the buccaneers found it first. Very soon after landing they turned the minute colony into an important source of logwood, a dyewood that grew along the coast and was one of the great prizes on pirate raids. It was a short step from plundering logwood to cutting it in the interior, and by 1670 logwood sold for about £100 a ton, a good profit in those days. By 1705, the British shipped most of their logwood from the Belize River area.

By the end of the eighteenth century, demand for logwood declined as new technology and better natural dyes were adopted by dyemakers, but as so often happened, the forest provided a new moneymaker. Mahogany, or *Swietenia mahogani*, began replacing logwood as the colony's principal export as early as 1771. By the early nineteenth century, mahogany exports climbed to twelve thousand tons, providing a yearly revenue of nearly £20,000. Mahogany was a handsome red wood that had been popular with eighteenth century cabinetmakers and now was used in shipbuilding, construction, and later in railway carriages. Although the best trees grew in the limestone sands of the north, they could be found throughout the jungles of British Honduras. The trees were scattered, like rubber, and scouts pierced the forest in search of new trees. Crews disappeared into the wilderness in August, cut crude logging roads from the trees to the nearest river, and had to be out before the summer rains in May, which turned the roads into quagmires. Once dragged out, the giant trees were floated down rivers swollen by the storms, then halted by booms in the river mouth. They were formed into huge rafts and floated to the wharves of the Belize City timber companies for shipment abroad.

In 1863, when British Honduras became a Crown Colony, overcutting had taken its toll and the business was in decline. New companies, squeezed from better logging areas, roamed far afield to find new trees. As early as 1790, they began raiding Spanish territory. Such "foreign wood" was forbidden, but as competition stiffened, the companies sent cutters into Guatemala and Mexico to cut mahogany then quietly ship it to Belize as "local wood." By the 1820s, cutters operated north of the Hondo River, which formed the border with the Yucatán, and south of the Sarstoon River on the Guatemala side.

When Henry and Violet arrived, the "plantocracy," as locals called it, was the most powerful force in British Honduras. It held onto power even as supply ran short and profits fell. Its stranglehold on land stifled farming, which was prohibited on logging property. This meant that the entire population depended on imported food. Rather than change, the plantocracy entrenched, consolidating their capital, resisting all attempts at reform. By 1859, the British Honduras Company, which originated as a partnership between old settler families and a London merchant, emerged as the colony's predominant landowner. It spread like an amoeba in the 1860s, usually at the expense of competitors, who were forced to sell their land. In 1871, the firm became the Belize Estate and Produce Company, a London-based business that owned about half the privately held land in British Honduras and acted as the chief force in the colony's political economy for over a century. There were other companies that operated on the borders and in the deeper forests, but they did so by arrangement with Belize Estate and Produce.

Thus, Henry and Violet came to British Honduras during one of the most openly corrupt periods of its history. The population was changing. As more black immigrants moved in from the Caribbean, and particularly Jamaica, the white population dropped from 4 percent in 1845 to 1 percent in 1881. The timber houses still controlled the colony, but as the white settlers moved out, the houses came under absolute foreign control, usually from London. The Crown did not control British Honduras; the timber companies did, a fact that set British governors at odds with the "monied cutters." The mahogany houses maintained British Honduras as a private timber reserve, and to do so they controlled the press, the government, and the courts. When the Wickhams arrived, British Honduras was in danger of becoming a colonial dead end.

The main opponent of the mahogany houses was the colony's new governor, Sir Roger Tuckfield Goldsworthy, a hero of the Indian Mutiny.

Appointed by the Foreign Office in 1884, he'd alienated the plantocracy by summer 1885. Belize City had a history of yellow fever and malaria. Built on a muddy flood plain, surrounded on three sides by water, and rising a mere eighteen inches above sea level, the city was a sink of stagnant canals, a breeding ground for mosquitoes. When Goldsworthy took office, he sought land reforms and awarded all public works contracts to improve sanitary conditions to a local man who was no friend of the plantocracy. Within two years, the colony's treasury, which had a £90,000 surplus when Goldsworthy took office, was in the red. The improvements and the governor were blamed. Goldsworthy also seemed bent on improving conditions for the nonwhite immigrants. Belize's *Colonial Guardian* railed that he "never for one moment ceased to be a friend of the least reputable portion" of the population. The plantocracy-controlled press delighted in calling the governor "the most hated man in the colony."

Henry and Violet sailed into this storm like blind mariners. The politics would determine their lives, because Henry and Goldsworthy had become close friends. The two had met in 1877–80, when Goldsworthy was colonial secretary of Western Australia and Henry was Australia's sole promoter of Brazilian tobacco. They took to each other immediately, and Goldsworthy's governorship may have been the deciding factor in setting Henry's course toward Belize. The governor was fond of the rubber thief. Both were brash and reckless, and both believed without doubt in the superiority of the British Empire. When Henry sailed on the *Godalming* on November 12, 1886, Goldsworthy was going in the other direction, having boarded his steamer for England on November 2. He'd been called home by the Foreign Office to answer questions about the furor surrounding his administration. The two old friends passed each other on the high seas.

Goldsworthy may have prayed that he'd never have to return. Yet he was sent back in spring 1877—around the same time Violet arrived—and he would stay until 1891, the longest tenure during the nineteenth century of any colonial administrator in British Honduras. The Foreign Office never explained its reasons for sending him back, and his return threw the plantocracy from elation into despair. "Whether it was simply a matter of completing a routine tour of duty or whether it was to teach the townspeople some proper respect," moaned the *Colonial Guardian,* "it was surely a crushing blow."

Despite the controversy, Violet put Goldsworthy's affection for her husband to good use when she arrived. The fear of God grabbed her by the throat when she saw Henry's ramshackle homestead nine miles outside Belize. It was a nightmare flashback of all she'd endured in Brazil and Queensland, and she acted with alacrity. "A friend and I persuaded him to take a Government post," she wrote. When Henry presented himself to the governor, his employment was immediate. For the next two years he was diverted from his self-immolating dream of becoming a planter, sent instead across the colony as a *locum tenens,* a substitute magistrate when the regular office-holder went on leave.

According to the *Honduras Gazette,* the official colonial organ, Henry did a little of everything. From May 1887 to December 1888, he served as acting district magistrate, acting sub-inspector of the constabulary, and foreman of the works in the Toledo District in the extreme south of the colony. From December 1888 to May 1889, he was justice of the peace and acting district magistrate in the Orange Walk District along the troubled border with the Yucatán. He was then appointed fruit inspector, which took him up and down the coast checking deceptive practices by banana exporters. In 1890, he became inspector of forests, checking trespasses on crown lands by the mahogany companies, a position that earned him their enmity.

"This period was the easiest and most pleasurable of my life," Violet wrote. While Henry was away, she stayed in Belize. It didn't matter to her that the homes were flimsy and unpainted, the streets narrow, the open canals choked with sewage. For once, she was around friends, not stuck in the middle of nowhere. Belize had its beauties: the open verandas were shaded by red-blooming poincianas; red, pink, and cream oleanders glowed behind white picket fences. The races and tongues were a virtual Babel—Creoles of African descent, black Caribs, mestizos of Spanish and Indian blood. The market was a tropical cornucopia: loggerhead turtles, bananas, papaya, custard apples, red chili peppers, breadfruit, yams. "Being in contact with the Governor I was invited to such social functions as went on there, such as balls, tennis parties." Gone were the days of fighting constrictors, watching her house burn to ash, or nearly drowning in the Amazon.

And Henry was sent to some of the wildest places imaginable, which meant that, once again, he became known as a local character. Years later, in an official memo dated September 16, 1892, this assessment would be made:

> Mr. Wickham is a large-framed idealist, dreamy, sympathetic, artistic, a great wanderer and naturalist in tropical America, but not well quali-fied for official or commercial business. He was appointed by Sir R. Goldsworthy, who had a great liking for him, to be Fruit Inspector, in which capacity he had to check sharp practice by the contract purchasers of bananas for the weekly Mail steamer at the villages along the coast, and afterwards to be Inspector of Forests, to check trespasses on Crown Lands by Logwood and Mahogany Cutters. In this capacity he did some useful work by fits and starts, but also made some mistakes.

Above all else, it was always remembered that he was Roger Goldswor-thy's man.

❧

Henry's government service may not sound exciting, but tucked away in colonial archives is evidence that he spent his tenure hunting for pirate gold, evading sharks, conquering a supposedly inaccessible mountain, and negotiating with Mayan revolutionaries who believed God spoke from an enchanted crucifix and told them to kill any white interloper found trespassing on their land.

The treasure hunt occurred first, in the form of a yacht filled with Americans. According to a dispatch by Goldsworthy to the Foreign Office, the *Maria* arrived in port on January 28, 1888, amid a shroud of rumors. Soon afterward, Mr. John Benjamin Peck presented himself to Goldswor-thy. Peck was a retired special agent with the U.S. Treasury. On January 1, he'd cast off from New York in the *Maria* with some friends. In the hold was a "boring machine" designed for digging through sand and coral. Peck had an old map marked with the location of treasure worth half a million dollars "to be had at or near the island of Turneffe to the north of Belize."

Goldsworthy was highly amused. "I believe [Peck's] errand to be somewhat fanciful," he informed the Foreign Office, but just in case it wasn't, Peck agreed to place the riches in the Colonial Treasury. The Crown would keep 10 percent and relinquish the remainder to Peck and his friends. There was an additional provision—that Wickham tag along as watchdog, observing the excavations and inspecting the treasure, if found.

From January to March 1888, Henry camped with the Americans on Turneffe, a barrier island at the very edge of the sea. On the lee side, coral reefs glimmered through the water in shades of opal, emerald, and turquoise; on the ocean side, the air throbbed with the boom of the crashing surf. There were several wrecks on Turneffe, including the English merchantman *Mary Oxford*, lost in 1764, and the HMS *Advice*, wrecked in 1793. The most exciting rumor concerned a Spanish galleon said to be carrying eight hundred thousand dollars in gold specie, lost on Turneffe's northeastern tip in 1785. This seems to be the treasure sought by Peck and his adventurers.

At first, Henry was as skeptical as Goldsworthy, but by March the search had focused on Half Moon Cay, a speck forty-seven miles out from the coast in a line of tiny coral islands known as Lighthouse Reef. Half Moon Cay was a hatchery for red-footed boobies, and the birds wheeled around them as they worked. Spiny-tailed iguanas called "wish-willies" blinked at them from the underbrush. A seventy-foot lighthouse stood at the cay's southern point, a fixed white light that could be seen in clear weather twelve miles out to sea. An iron human skeleton, painted white, was fixed to the lighthouse. Its keeper, "A. Martin," earned the annual equivalent of $480 to stay in this supremely lonely spot. Peck and his partners probed through twelve feet of coral quicksand with iron rods and grew convinced that crates and chests lay underneath, but every time they dug a hole, sand and water rushed in.

"[Henry] believes they really were on the spot, as they brought up such things as might be expected," Violet wrote later. "But the inrush of water was too much for them." In March, Peck and his partners began their return to the United States for a coffer dam to sink in the spot. But on the voyage home, the *Maria* was "wrecked on the way and never returned," the yacht sinking somewhere in the Caribbean with all hands during a gale.

Thus ended another El Dorado dream. After Peck sailed off, Henry pulled a stunt that entered local legend. Violet and Henry lived on an island in sight of Belize—probably Haulover Island—and one night Henry was delayed late in town. He could not secure a boat, and the quarter-mile channel separating him from Violet was notorious for sharks. Sir Eric Swayne, governor from 1906 to 1913, gave a sense of the danger:

[W]ild tales are told of men who have missed their footing, have fallen into the sea, and have been shot up into the air again minus a leg, to fall back again into a seething mass of sharks. There is a well-known shark known to the fishermen as Sapodilla Tom, who is popularly supposed to be of an enormous size. One of the pilots who is not, I think, deficient in imagination, gravely informed me that Sapodilla Tom had on several occasions swum alongside his 45-foot sloop, and when the nose of the shark was level with the bow of the boat the tail was level with the stern.

Despite such tales, Henry feared that Violet would worry about his safety, and so jumped in the channel and swam home. Violet's reaction is not recorded.

In April 1888, Henry was once again offered the chance to wander. In the southwest lay the Cockscomb Mountains, then called the Corkscrews, an undulating granite and quartz massif whose 3,680–foot Victoria Peak was the highest point in the colony. Gold was rumored there, the district clouded in mystery. The natives believed the jungle to be home to Sisimitos and Sisimitas, hairy people of the forest who wore their feet backward and liked to eat humans. Any approach to Victoria Peak was believed impossible. "Strange as it may seem in a colony so old, and only eighteen days from England," wrote geographer J. Bellamy in an expedition account for the Royal Geographic Society, "the interior is less known than Central Africa." Authorities hoped that opening up such virgin territory might relieve the "congested state of the mother country," Bellamy wrote, but the expedition's true purpose was to scout out undiscovered reserves of rubber and gold.

They started on April 4, 1888, proceeding up the forested South Stann River in five dugout canoes paddled by Carib porters. Exploring this wilderness was unnerving and strange. A settler cutting through the dense tropical growth would sometimes come upon a hewn block of stone covered with hieroglyphics or carved into a monolithic head, all that remained of the Mayan civilization that vanished from this part of the world. Some said it disappeared when the climate changed and rains became torrential, others that epidemics killed them all. The forest was always man's enemy, with its fevers, its extraordinary vitality, creeping over man's works like an insistent sentient being. Once it got a head start, said Sir Eric Swayne, it was "impossible . . . to recover lost ground."

On April 11, the party struck the spur of the main peak, and they rested awhile. A series of escarpments rose from the forest like volcanic upthrusts. The perpendicular rocks were covered in a thick, beautiful moss that seemed to make ascent impossible. Everything was crowned with incredible growth: vines crawled across the smallest twigs, orchids with sweet-smelling purple blossoms grew on ledges and crests. While Goldsworthy waited for Henry, who'd been delayed, Bellamy pushed ahead to look for gold. Its signs were abundant in the washing of the sand and clay and specimens of quartz, which also showed signs of lead and silver. The party began its ascent up the 1,250-foot Bellamy Peak, then to the 1,800-foot saddle, where they made camp. On April 15, they prepared to climb Victoria.

Henry went ahead, alone. Trailblazing seemed to be his role. "Mr. Wickham continued the ascent, which he managed by climbing round the heads of the spurs, over many difficult and dangerous places," Bellamy wrote. "[F]inally, after a precipitous and arduous climb, especially up the last 500 feet, he succeeded in reaching within a short distance of the summit." Yet he couldn't make the last few feet. The moss was so thick that "the final ascent became in sensation very like crawling over the edge of a great sponge," said a guidebook of the time. "One could thrust in an arm up to the shoulder before reaching the perpendicular face of the rock with the tips of the fingers." Henry repeatedly attacked the final ascent, but each time he slipped back. In the afternoon, Bellamy and Goldsworthy were walking along the saddle when they met him "returning with the good news of his success in finding the peak accessible, but he was terribly exhausted with his exertions and want of food and water."

The next day the entire party retraced Henry's route, and five climbers made it to the top with ropes attached to overhangs. Henry stayed below, too spent to continue. The five mountaineers, "having recovered sufficient breath, celebrated the ascent by giving three cheers for the Queen and Governor." That night, the exhausted party turned in early: "During the night one of the Carib porters shrieked in his sleep, and this so alarmed his companions . . . that they rushed through our camp, upsetting mosquito nets, tents, themselves, and everything else in the darkness, imagining that Tapir Peccary or some other evil genii of the place were among them."

※※※

Of the many gods in this wilderness, the Christian ones were most deadly.

Since its beginning in 1848, the Caste War of the Yucatán made it impossible for a light-skinned person to travel in the eastern Yucatán and come out alive. Only the indigenous Maya were safe; Caucasians and light-skinned mestizos were killed on sight. The Spanish had battled nineteen years to conquer the Yucatán Maya, who, unlike the Aztecs in central Mexico, were never permanently subdued. The Caste War began as a political, rather than racial, uprising when three Mayan revolutionaries defending communal land rights against Spanish owners were executed in Valladolid. But the ancient hatreds of the Santa Cruz and other Indians were so great that the goal quickly became the extermination and expulsion of all Caucasians.

It was a particularly bloody war, with years of racial violence on both sides. From 1847 to 1855 alone, approximately three hundred thousand people died. The massacre of Spanish settlers and townspeople received the most publicity, partly because two British peace commissions sent from Belize watched as forty Spanish women and fourteen men were executed despite efforts to pay off the executioners. Only little girls were spared—and one small boy who later told his story. When Spanish refugees flowed into British Honduras, the Indians crossed the Hondo River and defeated a contingent of troops sent to the refugees' aid. The breakaway state inspired other Mayan communities to revolt. In 1870, the Icaiche Maya attacked Corosal Town in the colony's far north and then Orange Walk Town ten years later.

In 1850, the Mayan insurgents were on the brink of defeat, when the war took a religious turn and the Talking Cross appeared. Mexican folk-Catholicism had for centuries produced a string of prophets and miracles, each claiming to be a messiah or appearing at critical moments of struggle. In the 1700s, the Indian leader Tzantzen emerged in the north. In 1810, during the Mexican War of Independence, the miracle-working nun Sor Ercarnación arose. The Talking Cross was not God Himself, but Santo Jesucristo, God's intermediary, able to speak to the Maya, His Chosen People. It appeared beside a *cenate,* or natural well of drinking water, and promised the desperate Mayan fighters that, if they continued the war against the whites, they would be invulnerable to bullets. The place where the Cross was found was transformed into the town of Chan Santa Cruz, or Small Holy Cross. A church was built around the Talking Cross, from which it continued to talk to its followers, the *Cruzob.* Eventually, the Cruzob and the British reached an uneasy peace: If there were no more attacks, the British would unofficially supply arms to the insurgents to fight their old foes, the Spanish. But there was still violence. Sometimes the mahogany cutters penetrated too far into Cruzob territory; sometimes individual Indians and whites were killed. By 1887, responding to the rumors of arms sales, the Mexican government filed a formal complaint and asked that the practice end.

In January 1888, while Henry hunted buried treasure, events unfolded along the colony's border with Yucatán that would send him on his last great adventure for Goldsworthy. On January 8, the Santa Cruz chiefs sent a letter to Goldsworthy in response to the Mexican charges. "We are . . . a people living under our own laws and are peacefully governed by men of our own race," it said. The Cruzob needed firearms for hunting and "for our own protection," the chiefs entreated.

Soon after this letter arrived, William Miller, the colony's assistant surveyor-general, rode into the Yucatán as far as Chan Santa Cruz. Miller's was not an official mission. Although obviously mapping an area unseen by whites for decades, he said, in an account written for the Royal Geographical Society, that he simply went out of curiosity. If this *was* a secret mission, it was only partly successful. He traveled twenty-five miles from Corosal to Bacalar, site of the famous massacre, where he spotted a number of human bones in an old church. He traveled another eighty-five miles along a flat, straight road to Chan Santa Cruz, only to find the town deserted and the Talking Cross moved another forty-seven miles north to the town of Tulum. He met the governor, Don Anis, who lived four leagues outside Santa Cruz: "When I arrived there he had just lost the sight of one eye, and believing he was bewitched, he had killed the man and his wife whom he suspected of doing it, the day before my arrival." Don Anis was still in a bad mood. When Miller asked questions about the Empire of the Cross, Don Anis replied, "Why do you want to know?" When Miller suggested riding to Tulum to see the Cross, his men refused. He returned home unharmed but unenlightened.

On December 15, 1888, Henry was appointed Justice of the Peace of Orange Walk, across the River Hondo from Yucatán. Sometime after this, Goldsworthy asked his friend to contact the Santa Cruz Indians again. We know very little about this mission; no official correspondence has been found. Due to the Mexican complaint, any contact between British Honduras and the Cruzob demanded secrecy. All we know is a passage in Edward Lane's biography that apparently came from family tales:

> [T]he governor, fearing a raid by the Santa Cruz Indians, invited Wickham to make diplomatic approaches to the tribe. Forcing his way on horseback through dense bamboo country, Wickham persuaded the

tribal chief to maintain the peace. . . . Veneration for the [Talking Cross] is so profound that no stranger may look at it. Nevertheless, Wickham caught a glimpse through a convenient peep-hole; he may have been the first European admitted to the tribal territory since the massacre of the Spaniards.

It makes sense that Henry was chosen to go. As justice of the peace, it was his duty to maintain peace along the border. He had a history in Nicaragua, Venezuela, and Brazil of working well with Indians; it was just with whites that he could not get along. The timing was right, and he was reckless enough to do it, plunging into forbidden territory in the same way that he launched himself at Victoria's Peak, without training, alone.

Henry's best protection was his respect for Indians. He believed them capable of anything. He rode to Santa Cruz on a road that was eight feet wide and kept clear by Indian work crews. Every few miles a cross was propped up with stones and covered by a little shelter of palm leaves. A garrison of 150 armed men was stationed at the Chan Santa Cruz fort, but the village itself was a ghost town.

The road to Tulum was rougher than the one he'd just ridden, a four-day journey on a narrow jungle path fraught with danger. A few years earlier, when a Catholic priest arrived by sea, he was taken to the Cross and interrogated. Displeased with the priest's unannounced incursion, the Cross demanded the priest's execution. Since then, few outsiders had attempted to enter the Yucatán.

Deep down, Henry believed himself invincible, so he went. The church housing the Cross in Tulum was shaped like a crucifix, with the sanctuary itself forming the upright and the guards' quarters forming the perpendicular arms. The Cross sat in the center in profound darkness, in a separate room called the *gloria*. Two sentries guarded the door to the Cross. Only four people—the high priest, the Cruzob's commanding general, and their wives—were allowed inside. The priest talked for the Cross, but how this was accomplished, whether by ventriloquism or through a concealed speaking tube, has never been learned. The rest of the sanctuary was filled with people muttering in prayer. A strange hollow whistling issued from the Cross before it spoke. As the holy words poured out, the worshippers pounded on their chests or blew ardent kisses in the air.

Since Henry's visit was sanctioned, he was not interrogated by the Cross, and he lived. He said he saw the Cross through a peephole, but with all the guards and worshippers present, this is probably braggadocio. Nevertheless, his was a great honor, and as Edward Lane asserted, few if any other Europeans had been admitted to the presence of the Cross and survived. The border calmed down after Henry's mission, and from then until the Caste War ended in 1901, there was peace between the British and the Cruzob.

The Maya had a direct relationship with the Almighty. When Henry left, his hosts would not have thanked him for coming. Instead, they'd say, "Dios botik," God thanks you.

❋

Sometime after this, Henry began hearing his own voices, commanding him to wrest status from the soil. "Alas," wrote Violet, "back came [Henry's] old longing for plantation life, being his own master, and in spite of all I would do, he saw something that took his fancy and got what he called a valuable concession . . . and away we went, 60 miles or so away from everyone and everywhere, to plant India rubber, cocoa and bananas." During his travels as inspector of forests he came upon what he described as "the finest block of land in the Colony," a 2,500–acre plot on the south bank of the Temash River, the most remote river in the most remote southern district of British Honduras, only six and a half miles from the Guatemalan frontier. Five houses were located on the Temash River, and Henry's would be the sixth. His concession fronted the deep river. He paddled into the forest past old abandoned plantations to large, horizontal blocks of limestone hewn by the ancient Maya. The land across the river, to the north, was owned by the mahogany cutters Messrs. Cramer and Company, but according to surveys the south bank was all Crown Land, open for agriculture. Based on his reputation "as the man who brought the rubber seeds from the Amazon" and a promise to grow rubber, Henry signed Lease No. 22 on January 1, 1890, at an annual rate of $500, payable for ten years. If, after that time, "value to the extent of $10,000 to consist of India Rubber trees can be proved to the satisfaction of the Government to have been planted on the land . . . [an additional] grant for 5000 acres will be issued."

It seemed his dream had come true, the chance for which he'd struggled so long. But there were bad omens. In 1889–90, fever swept through Belize, taking several acquaintances—the Rev. Mr. Nicholson, the prominent barrister-at-law W. M. Storach, the merchant Robert Niven. There was enough death that correspondent G. S. Banham for the *New York Herald* portrayed the city as a charnel house. On June 23, 1889, Banham was prosecuted for "maliciously fabricating false reports to the detriment of the colony by representing it as ravaged by pestilence," for which, on August 5, he apologized. During that time, death nearly caught up to Henry, too. He "had an attack of the fever and as nearly as possible died," Violet wrote. "The Roman Catholic priest came over and administered a very heavy dose of quinine which checked the fever till the doctor returned." When he recuperated, his obsession with his plantation intensified. His latest brush with death made him think about what he wanted most from life—and how, at age forty-two or forty-three, half his life was over.

When he felt able, he set to work like a dog. He thought the fertile soil well-suited for coffee, cocoa, tropical fruit, and rubber. He planned to pay his way by cultivating bananas, then plant the fast-growing *Castilloa* rubber to ensure a good return before the ten-year lease expired. It may seem strange that he did not plant hevea, his "blessed tree," but *Castilloa elastica* was native to British Honduras, and in the 1880s many still believed it to be more profitable. He built another log house, with an iron roof and veranda. "He lived contentedly enough," wrote Violet, "working early and late through all sorts of difficulties."

But there were rumblings that could make anyone uneasy whose land sat close to the plantocracy's. On May 5, 1890, a special hearing of the Supreme Court considered the complaint of Don Filipe Yberra Ortol against Messrs. Cramer and Company. Ortol wished to restrain the firm from cutting logwood on his land, which sat on the border with Yucatán. He had sole right to work his land, he said, but the suit was dismissed. Then, on July 8, C. L. Gardrich, editor of the *Independent,* was found guilty of contempt for publishing Ortol's open letters to Mexican woodcutters about his treatment. In essence, a gag order restricting unfavorable press against the plantocracy was imposed on the colony, and the editor was ordered to pay court costs and a two-hundred-dollar bond.

It simply was not wise to stray too close to the mahogany men. They controlled the law, the courts, and public opinion; they were a law in themselves. Some of their holdings stretched over a million acres. With the high rents and taxes, the sale or lease of land to small settlers like Henry was blocked as much as possible. In order to ensure a constant supply of timber, owners would only cut one-twentieth of their land each year, just selecting trees that had grown more than seventeen inches in diameter; in the course of twenty years, they'd come back when smaller trees had grown. This rotation allowed them to stay solvent by fixing the levels of capital and labor in advance, but it also led to confusion and court suits regarding ownership of plats. When the plantocracy returned to a site, their surveyors checked to see if there was anything they'd missed earlier or if any squatters had moved in.

By December 1891, Henry had built his house, cultivated forty acres, planted ten thousand banana trees, four acres of cacao, a few dozen oranges, lemons, and mangoes, and a small number of *Castilloa* rubber trees. He seemed to deliberately avoid hevea while building his reputation on its theft, but one could never call him lazy. His improvements amounted to "exceedingly good work," wrote the government, which estimated the house value at $1,500 and his concession's at $4,000. Yet his profits were meager—his account books showed a monthly balance between $13 and $47.96—and he'd been unable to pay anything on the two years' rent then due. That month, in a letter to the Colonial Secretary, he pointed out the hardship of paying $500 in advance and asked to be allowed to invest as much as possible in developments in the early years, then pay the equivalent increase on improvements at the end of his ten-year term.

He was asking for a favor, but the time for favors had passed. He was Roger Goldsworthy's man, and when Sir Roger was governor, his request might have been granted. But Goldsworthy had been transferred earlier in 1891 and was now the governor of the Falkland Islands, those cold, lonely specks in the South Atlantic that would be repeatedly claimed and contested with Argentina. His replacement, Sir C. Alfred Maloney, made twelve thousand pounds, the highest salary for any colonial governor at that time. The plantocracy and its press rejoiced. The *Colonial Guardian* of October 4, 1890 said that "honest men, as a rule, [kept] aloof" from Goldsworthy. A new wind was blowing in British Honduras, but it wasn't favorable for Henry.

In early 1892, the inevitable occurred. The colonial secretary rejected Henry's compromise solution. If he did not pay his rent, the lease would be forfeit. And there was an addendum—most of his concession belonged to Messrs. Cramer and Company.

This clause seems openly malicious and corrupt. Of Henry's 2,500 acres, Messrs. Cramer and Company said that 1,470 belonged to them, and the government agreed. The surveyor who "discovered" the flaw in Wickham's title was the same who originally drew the map safely placing Henry's property on Crown land. The surveyor general said he had no other map of the Temash, pleading that his department was too busy to conduct an accurate survey. Everything came back to Henry: The mahogany company could sue him for trespass. The government could escape liability by canceling the lease for nonpayment of rent. As former inspector of forests, Henry should have known which lands were Crown and not blame the government for his own error. Messrs. Cramer offered to sell "their" land back to the government at two dollars an acre, on the condition that they had right of way on the only path from the river to the back lands, where the mahogany grew. But Henry's house was on that path, along with his bananas and rubber. To buy back his land from the government, he'd have to tear down his house and uproot every crop he'd planted.

He went to court, the only option left to him. On May 11, 1892, his solicitor submitted a lengthy statement to the secretary of state citing Wickham's experience as a tropical planter, his official posts in British Honduras, and his value to the British Empire for bringing the seventy thousand hevea seeds to Kew. He was a man of "great and rare experience," an asset to any colony. Wickham's position was critical; he could neither work on the property, nor invite partners to invest, and creditors were circling. An investor from Guatemala had been ready to sink £2,000 into Wickham's plantation when he heard about his legal troubles. Two others—"Mr. D. Wells" and "Mr. Strange"—had already arrived in Belize when they learned that all that Wickham owned was a lawsuit. All three backed away. The dispute dragged on and Henry was going broke.

So in the spring of 1892, Henry appealed to the queen.

In 1892, Victoria had been on the throne for fifty-five years. Her Jubilee Celebration five years earlier had been one of the triumphs of her career. She was compared to Elizabeth, but Elizabeth had ruled a little island of merely 5 million people, while Victoria ruled nearly half the world. Her name and face were stamped on coins and documents in cities that had not even been established when she came to the throne as a girl. When British subjects acted out their separate dramas for the empire—when General Gordon died at Khartoum, or Henry fretted over his seeds on the Amazon—at some point, they thought of Victoria and ransomed their lives and hopes to her.

Victoria understood perfectly the importance of her colonies. They had to be protected. Everything rested on their backs—the riches, the dominance, the vital British interests. The most fundamental interest of all was the ability to trade and invest throughout the world, an ability that lay at the heart of every major foreign decision, and it had been so for decades. The word "imperialism" in the 1880s and 1890s was usually associated with a desire for territorial expansion, but at the core of this expansion lay the guaranteed markets, limitless resources, and free market capitalism the Victorians enjoyed and revered. In such a world, corporate interests were of greater importance than individual interests. After all, corporate profits benefited the greater whole. Imperialism was a faith as well as a business, and as it spread, so spread the mystique of empire. Prefiguring Calvin Coolidge, she knew that the business of England was business.

In due course, Henry's statement of his value to the empire was returned to the colonial government, denied by the queen in her handwriting: "Let Justice be done. Victoria R. & I."

It was an oracular condemnation as final as any by the Talking Cross. With that royal snub, Henry—once again—was ruined.

CHAPTER 12

RUBBER MADNESS

Few Victorians who'd done as much for their empire could claim that they'd been personally betrayed by their queen. He'd been ruined before—Santarém, Queensland—but this third time worked its negative charm.

There was a brief denouement to Victoria's note: On May 4, 1893, Henry's lease was cancelled. Four months later, on September 7, the courts awarded him $14,500 in damages, plus legal costs, but this was not enough to cover his land and the amount he'd spent on the house, crops, and improvements. A hurricane hammered the colony that summer, ripping the roof off Belize's Catholic church, dragging five ships ashore, and leveling banana plantations up and down the coast to such an extent that no fruit would be shipped for six months. What did he care if this corrupt colony blew away and went to hell? He might have felt as if his bottled rage had found expression in the howling wind. To make matters worse, a "skin complaint" that Violet never detailed became unbearable, and she needed treatment in London. Shortly after the verdict in his case, Henry left, never to return.

He was deeply embittered, scarred as visibly as the soft bark of hevea slashed by *seringuieros*. It was during this period when the world began to know him as an opinionated, silver-maned imperialist blustering from the edge of the world. "His keen analytical mind and authoritarian manner made him a difficult partner," Edward Lane said of this time of his life. "[T]here was much of the intolerant dictator in his makeup, although he was unswervingly loyal and . . . generous towards his friends."

But that was the way planters were expected to act in the backwaters of the empire, and that's where he moved next—to the antipodes. In the British Empire, the antipodes referred to New Guinea, Australia, and New Zealand, despite the fact that none overlapped the antipodal points of the British Isles. Henry must have felt at times that his strength was at an end. His funds certainly were, as was his patience. He might not abandon his desire to create a plantation, but he would do so miles from contact with his fellow countrymen, in an isolated empire of his own design.

He'd always been driven and obsessed; now, it seems, he went a little crazy. We can trace the psychic route today. He started at Papua New Guinea, an island of "appalling roughness and disrupted character," where mountains soared up as inaccessible cliffs, then fell away into "rock-walled gorges, through which roil rivers, their courses blocked by boulders and ever-rolling stones." Henry O. Forbes, a professional Victorian adventurer, said that "during many years of travel in rough countries, I have encountered nowhere such difficulties as in New Guinea." Nevertheless, settlers in England and Australia saw in New Guinea a "great and salubrious 'Treasure Island'" with gold trickling down the rivers.

This was too crowded for Henry's tastes, so on the island of Samarai, at the eastern tip of the New Guinea peninsula, he bought a ten-ton lugger and named it *The Carib* in honor of his time in the Caribbean. Remembering the tales of Captain Hill, his Cornish savior in Nicaragua, Henry sailed east toward the neighboring archipelagoes—the D'Entrecasteaux Group and the Louisiades, those exotic-sounding pinpoints raided by blackbirders a decade earlier when Henry farmed in Queensland. The former consisted of three or four high, rocky islands that thrust from the depths to great heights. The gradients were obscenely steep, ascending from four hundred to nine hundred feet in a mile, with scant flat land. The Louisiades were the polar opposite, a sprinkling of hundreds of small, low islands reaching east like a tentacle from the toe of New Guinea to Rossell and Sudest (now Taguta) islands deep in the South Pacific. Smack in their middle he found a lonely coral atoll called the Conflict Islands, named after an unlucky British warship that wrecked on a reef in its western extremity. In 1888, the year New Guinea was declared a British possession, a survey of the Louisiades noted that the Conflicts, "like the Cocos Islands, in the Indian Ocean . . . may, someday perhaps, be planted with cocoa nuts, and bring in a fine revenue as the Cocos Islands did." This was Henry's plan.

The Conflict Group was made up of 23 coral islands looped around a central lagoon that was five to seven miles wide by twelve miles long. There were two excellent channels into the lagoon, one on the east and the west, if anyone decided to visit—which no one really had. The island group was truly the middle of nowhere: eighty miles from the eastern tip of New Guinea, six hundred miles from Cairns in the northern tip of Australia. No regular shipping lanes crossed close. It even lay outside the path of tropical cyclones. The islands varied in size from one to twenty-four acres across; if they all were lumped together, they'd comprise a land mass one and a half times the size of Monaco. Henry settled on Itamarina, a six-acre island in the lagoon's center encircled by a smaller inner reef. On a map, Henry's new home resembled a castle wall, with a wide moat and inner keep. He'd curtained himself off from the world.

Like many wanderers, Henry dreamed of a South Seas paradise, especially after hearing Captain Hill's stories of white beaches, carefree people, and cool trade winds. But the reality was antipodal, and Henry would have caught a glimmer of this during his disastrous years in Queensland. On most of these islands, explorers found that tribal warfare and cannibalism were not only part of life but a central part of some tribes' religions. In the Louisiades, cannibalism was a reality, and the hatred left over from the raids of the blackbirders assured that Europeans were sometimes the meals.

Since Henry employed Kanakas on his farm on the Herbert River, he would have heard the bloody tale in 1878 of W. B. Ingham, a popular but unsuccessful Herbert River cockatoo farmer much like him. Ingham had turned beachcomber and bêche-de-mer fisher after his farm went bankrupt. Bêche-de-mer, better known today as sea slug or sea cucumber, was a profitable delicacy in China. As a police official pointed out, the "business is a dirty one but profitable, and seems to possess attractions for the lowest class of whites and Manilla [sic] men, who have no scruples whatever in dealing with their black employees." One dove over the side of a boat into shark-infested lagoons to pluck the mollusks from the coral, and employers cared little about their divers' exhaustion or fear. The Queensland government hired Ingham to cruise around the islands and gather information about the fisheries, and in December 1878, a group of Ingham's "boys" threw him overboard as they floated off an island in the Louisiades. Ingham laughed at their high spirits and swam back to the boat. As he grasped the thwarts to pull himself up, his "boys" cut off his hands, then pulled him aboard to finish the job.

Ingham was so popular around the Herbert River that the main township was named after him. His death gave natives a demonic cast, reinforced two years later when parties of Chinese and European bêche-demer fishers were found similarly murdered. Several of these victims were brought to Queensland for burial, and their condition was so gruesome that Australians demanded that Great Britain annex New Guinea and the surrounding islands to protect peaceful fishermen.

Now Henry came to a place that no European had settled. On March 5, 1895, he signed a twenty-five-year lease at the rate of one pound per annum under British New Guinea's Crown Lands Ordinance. In return, he'd grow sponge, cultivate pearl oysters, and harvest coconuts for copra. A clause gave him the right to purchase the Conflicts during the term of the lease. The terms were his most favorable yet, and he arranged it as he had in British Honduras—by making friends with the colonial governor, Sir William Macgregor, whose *Handbook of Information for Intending Settlers in British New Guinea* he'd read during his two years in London following the failure in Belize. Henry recruited Kanaka help from the surrounding islands, built a trading store and crude barracks on the central island of Itamarina, and cleared land for a plantation on Panasesa, one of the outside ring of islands looking west toward New Guinea.

Sometime between March 1895 and April 1896, he almost lost his land again over contractual details. According to a tale told to distant relatives, he'd overlooked a clause in his lease that stated that any unplanted island could be designated a reserve for islanders traveling in the area. A friendly trader told him that the government yacht *Merrie England* was on its way to his central island with just that purpose in mind. Henry and his Kanaka workers hopped in canoes and catamarans, paddled across the lagoon to his coconut plantation on Panasesa, loaded up on coconut seed, and spent the night by torchlight planting Itamarina. Shortly after dawn the *Merrie England* docked and officials confronted Henry. He showed them that Itamarina was, indeed, planted with coconut and was therefore exempt from the law.

The affair apparently stuck in the craw of the Port Moresby officials, for in April 1896, Governor Macgregor visited the Conflicts during his annual island tour to check on Wickham again. He found him ensconced on Itamarina, living in an open shed. He'd apparently seen worse on the islands, for he did not seem surprised. Henry had already started changing his plans, Macgregor wrote:

> This gentleman has been making trial of the sponges found in the vicinity. He turns out a promising looking article, but its market value has not yet been ascertained. Recently he has been giving his attention to the planting of coconut trees—of which he has already put in several thousand.

He did not mention Violet, so in all likelihood she had not yet arrived. She lingered in London for a year, then came in late spring or early summer 1896. She found her husband in the Itamarina shed, surrounded by piles of black sponges. Henry's initial response was one of shock; nothing was prepared. Three quarters of the shed was "roughly ceiled to make a loft or sleeping place reached by a rough ladder, about half the lower story floored in native fashion, a few inches from the ground, with split cane. Here I remained some months while a more permanent house was built on a neighboring island," she said.

Violet was forty-six; her tanned and increasingly grizzled husband, fifty. Did Henry think they could go on like this indefinitely? Twenty-five years ago he strode into her father's shop on Regent Street, and she'd been swept away. Boundless energy and confidence leaked from every pore. He'd been to the jungle, *twice,* and nearly died, but he always bounced back. *Like a rubber ball.* Her friends laughed, and all she could do was shake her head and laugh along. It was true. You simply couldn't keep Henry Wickham down.

There'd never been a question that he loved her: He'd swim a shark-infested lagoon to be with her, live unconcerned in squalor until her arrival, then gaze around ashamed of his surroundings, and spend back-breaking hours just to make her comfortable. She loved him for that: loved him for his hopeless optimism despite the string of failures, for his naïve confidence that someday, if he just worked harder, he would prevail. His life wouldn't make sense otherwise. But as much as he loved her, he loved his craving more. With each new failure, he'd become more intractable, convinced that his was the only way. Decoyed by the mirage, he'd gone too far, and she'd gone with him. Henry's strength might be unending, but hers had limitations. They'd tilted together at Henry's windmills for a quarter of a century, and with sadness she began to realize, as she gazed at Henry standing abashed among his stinking, blackened sponges, that she'd been defeated by this man.

She held on for another two and a half years, watching Henry bounce from scheme to scheme. First there were sponges, but their farming was more than Henry bargained for. They were gathered by two men in a dinghy diving from the side or using iron hooks attached to ten-to-twelve-foot poles. When the dinghy was full, they returned to the island and spread the sponges on a table in the sun for several days to allow the black gelatinous membrane covering the sponge to die and dry. Their house and stores on Itamarina always had a dead-fish smell, which never went away. After a few days in the sun, the sponges were placed in an enclosed "sponge kraal," where they were washed by lapping water for another six days. They were beaten with sticks until the decayed outer cover dropped off. Then and only then did they resemble the amber-colored sponges that Violet knew from London's stores. Greek sponge brokers sailed among the islands. They arranged his sponges into lots, keeping some and rejecting many. The automatic costs kicked in—$\frac{1}{2}$ percent for "wharfage," five percent for "brokerage," two percent for "drayage"—a total of $7\frac{1}{2}$ percent for taking a sponge, and carting it away, and selling it. Henry decided sponging was unprofitable, so he went to the next best thing.

This was planting coconut palms, from which he sold copra, dried coconut meat used as a source of coconut oil. Ripe coconuts were split with a machete and laid out in the sun to dry, then the meat was scraped out and dried again on raised platforms to protect it from land crabs. This was pulverized with rollers, steamed and pressed at about 6,500 pounds per square inch. High-quality copra yielded about 60–65 percent coconut oil. What remained was called coconut oil cake and used as livestock feed. Since Henry could hire Kanaka farmers for five to ten shillings a month, he thought he could turn a profit, but the copra brokers took their percentages just like the sponge merchants, and Henry moved on to bêche-de-mer.

The sea slug was abundant in his lagoon and fetching high prices in China as an aphrodisiac, but Henry's divers were very slow. He tried breeding oysters for mother-of-pearl shell, growing papaya trees and cardamom plants, and harvesting hawksbill turtles, but the only time his workers caught them was when they came ashore to lay eggs. "Then they turn it on its back and a half a dozen or so haul it home with songs of triumph such as I suppose they have used for their human captives," Violet wrote, showing her lack of patience for the Louisiades and their people. "They make a fire on its breast plate and kill and cook it at the same time."

All Henry needed, he told Violet, was *more capital!*—the mantra of his age. He'd steam back to London to sniff out investors, leaving Violet to manage affairs. He wanted a total of £22,000, added to his own personal investment of £2,000. With that he'd create his own trading empire in the South Seas. He'd build a steamer for £6,000 to haul cargo and passengers, plus a smaller one to collect exotic produce from the islands. He'd hire a hundred native workers, sailors for his ships, clerical workers, European overseers, scientific experts, and so on. His estimated profits on copra and pearl-shell would be £9,400 the first year, increasing to £26,100 by year three. The scheme *would* work, he promised, because nowhere else in the world did such conditions exist for raising mother-of-pearl. A South Sea investor named J. G. Munt heard Henry's pitch and later described it as "nonsense." The Conflicts were "not a locality where anyone would, or could, work Mother-of-Pearl for the reason that the tides or currents are far too strong," he said. "What few shells are to be found are very inferior."

The one thing Henry did produce in abundance was loneliness, and Violet sampled that in full. "During the whole time of my sojourn there I never saw another white woman or left these two islands" of Itamarina and Panasesa, she wrote. Except for Henry, she had no more than perfunctory contact with anyone. Expatriate society was ultraconservative. Cut off from the exhilarating change of the city, the residents developed a siege mentality, which included more stringent and unbreakable barriers between whites and "blacks" than any she'd ever seen. Planters took for granted their "right" to exploit cheap labor while bestowing on them the joys of civilization. "We expected to be respected, have privileges, be superior," one colonial woman in New Guinea later said.

> In return, we were the Rock on which such fragile structures as honesty, fair play, protectiveness, obligation were to be erected. Generally, we lived a life of natural *apartheid*, in which the native inhabitants went their way and we ours.

This meant that Violet rarely saw Europeans with whom she could relate. Those she saw were sponge brokers and copra buyers. Her contact with islanders was superficial or distant. In calm weather, the natives would canoe between islands or, on larger trading missions, lash eight or nine canoes athwart, plant two sails "shaped like the claws of a crab," and put in at island after island, collecting tons of sago and other goods. They'd pull into Itamarina when they wanted water or tobacco, or when becalmed.

> I did not come in contact with their family life or seen [sic] anything of the women, beyond the one or two who came ashore for water. They wore their hair in apparent ringlets, but when you get closer you see the ringlets are just matted locks. Their dress is composed of petticoats of cocoanut leaves bound round the waist. When they put on two or three of these skirts they look something like the old ballet girls. This is all except tightly fitting bracelets [woven] around the upper arm, in which they tuck anything they wish to carry, and perhaps a bead necklace. The men take pride in their appearance, they paint their faces and rub themselves with cocoa-

nut oil and comb out their wool into most becoming, mop-like head dresses. Occasionally they shave it into eccentric shapes. They use brace- lets like the women and put them to the same purpose, tie bracelets under their knees with hanging tossels of shells which rattle as they walk. They have a hole through the nose and when fully dressed wear a large crescent shaped piece of shell there. Both men and women have the ears pierced and stick all manner of things in them for ornaments.

She did not participate in any society; she only observed. In places like Port Moresby and the Solomon Islands, a strange tension evolved, amount- ing to sexual hysteria. The 1926 White Women's Protection Ordinance carried the death penalty for rape and attempted rape, but its passage seemed based more on rumors of interracial lust than objective reality. By the 1930s the fear mutated to the absurd—an islander could receive 100–150 lashes for an imagined lewd glance, while "boy-proof" sleeping rooms, enclosed in heavy chicken wire, were installed at government expense in all houses where white women lived. But here on the Conflicts there wasn't even eye contact that could be misinterpreted, and Violet wasn't the sort to indulge in that kind of lunacy. But she was susceptible to loneliness, and the hermit's life did strange things to people.

All she could do was stare to the west, where Port Moresby lay beyond the horizon. Maybe it was just as well she didn't go. Although Port Moresby was dubbed "a quite civilized town" by promoters—a tropical London, with hotels, stores, reading rooms, and running water—in truth, said world traveler Henry O. Forbes, the town consisted of the original native village,

> a few Government weatherboard buildings, a residence or two for the
> officials, the mission station, one store, a three-celled jail, and the Gov-
> ernment printing office—which *was* the hotel—dotted anyhow along a
> couple of miles of shore. The "water supply" is a dribbling discharge
> from a . . . pipe, conducted for some hundred yards. . . .

Finally, she could take it no longer. One morning, "we woke to find our boys had gone off with one of the boats leaving only [Henry] and one

Malay man to manage the large boat." They took off after the thieves in the undermanned vessel, "and I was left alone for nineteen days not knowing but that they were wrecked, and wondering how many weeks or months before I would be rescued."

Violet kept her fears to herself in her memoirs, but they are not hard to surmise. She knew that Henry might have fallen prey to shipwreck, drowning, sharks, or the tide, and she already had admitted her awareness of local cannibalism. In most cases, cannibal tales served as an "agenda" for the Great Powers to annex new lands: Those engaged in cannibalism were less than human and needed saving from themselves. In the Louisiades and New Guinea, however, cannibalism was not an unreasonable fear. She'd have known about poor beachcombing Ingham during her ordeal in Queensland. As late as 1901, the missionary James Chalmers and eleven others were last seen rowing up a mangrove creek on a densely wooded island in the Gulf of Papua when they vanished from sight forever. Seven weeks later an expedition learned their fate: When the missionaries had entered a Goacribari village, they had been stabbed with cassowary daggers, beheaded, cut up, mixed with sago, and eaten the same day.

Violet's reaction when Henry finally returned was not recorded. But soon afterward, she gave her husband an ultimatum. *It's this place or me. I can't live like this,* she told him. He had to choose.

✺

Back at home, the newspapers called this the Rubber Age. The world was mad about rubber, and Henry was half mad that he'd never be involved. It was better to hide away.

The bicycle craze had gripped the West during their time in British Honduras. Now, as they quarantined themselves in the Conflicts, the world entered the Age of the Automobile, the second great development to fuel the madness for rubber. Pneumatic tire design accompanied that of autos every step of the way, and soon the tires would be purchased by the millions. Rubber exports from the Amazon jumped from a yearly average of 9,386 metric tons in 1886–90 to 14,939 tons in 1891–95.

By the time Violet was ready to call it quits in the Conflicts, the United

States was on its way to being the world's largest consumer of rubber. Ransom E. Olds cranked out thousands of cars each year. Henry Ford was nine years away from producing the first inexpensive Model T. By the turn of the century, the auto industry was becoming one of the world's most complex and interlinked industries, with hundreds of interdependent parts, and rubber was a great reason for that success. Tire and tube manufacturers would consume 60–70 percent of the rubber sent to the United States. Five major tire and rubber companies emerged in the three decades after 1870. North American rubber imports jumped from 8,109 tons in 1880 to 15,336 in 1890. From 1875 to 1900, in keeping with its sudden, surprising ascendance as a world power, the United States consumed half of all the rubber produced in the world.

And where were Henry's seeds? Not only was he excluded from the profits, it was as if his seed theft had never occurred. It didn't seem to matter. It was a failed enterprise that took the lives of his mother, sister, and others and seemed to illustrate the foolhardiness of dreams. By now, the story of Wickham's hevea theft had filtered back to the Amazon, but not even the smallest ripple of worry spoiled the calm assurance of the Brazilian rubber men. They laughed at the early British attempts at domestication. The growers in Ceylon and Malaya who received the seeds soon after 1876 made a complete mess of their attempt to grow and tap rubber trees. "God has planted for us," Brazilians boasted. If rubber trees were meant to grow in rows, God would have planted them that way. Statistics seemed to support their lack of concern. In the first year of the new century, four tons of plantation rubber from the East trickled into the market. That same year, exporters shipped 26,750 tons from wild trees.

This was the belle époque of the Amazon, the decades from 1880 to 1910 that are still nostalgically called the Boom. By 1907, as many as five thousand new men a week flooded into the valley past Santarém. It was the biggest boom since the Klondike, a black gold bonanza that ignited the economies of Bolivia, Colombia, Venezuela, Peru, Ecuador, and Brazil. In 1906 alone, the £14 million in rubber that came down the Rio Negro paid off 40 percent of Brazil's annual debt. It was said that 100–300 million virgin rubber trees still existed in the unexplored forest, scattered across an area of two million square miles. Although that sounds like another dream of El Dorado, it was a fact that in each of those three decades, Brazilian output alone rose by ten thousand tons. The Amazon held the world's only viable supply of hevea, no matter what the pathetic Wickham had done—and hevea was the gold standard now. Those who didn't participate in the Boom were foolish. Even the great steel magnate Andrew Carnegie had lamented, "I ought to have chosen rubber" instead of steel.

There were no labor problems either. Although conditions for *seringuieros* had grown ever more degrading, new tappers arrived. One effect of the Boom was a settlement of the valley like nothing seen since the original Portuguese conquest. Thousands of small expeditions ascended hundreds of small tributaries, searching for undiscovered trees.

But each new surge up the river made *seringuieros* more dependent on the *patrão*. The cost of living in Pará and Manaus was two to four times greater than in London and New York, but the *aviadors* compensated by selling ever higher down the line. At each link of the chain, practices were structured to drive up costs. Brokers on the Amazon charged higher prices than their counterparts in London and New York. The steamship companies charged whatever they wanted for shipment. The tax on exported rubber was exorbitant and drove up costs by as much as a third. The manager sold supplies to the *seringuiero* for a final profit of 50–200 percent, and the isolated *seringuiero* had no choice but to buy. His economic helplessness made him the victim of every swindle imaginable. Beans and rice were sold at seven times their price in Rio. The "trade gun" became notorious—a muzzle-loader with a wire-wound barrel, which would unwind and fall apart after fifty shots. The gun sold so well that it was made in Europe specifically for the Amazon. In some distant *estradas,* food was rotten and swarmed with maggots by the time it reached the workers, but the *seringuiero* either bought the spoiled meat or starved.

Legally, there was no obligation for a *seringuiero* to remain; he could pick up and go at any time. But managers had ways to ensure that their labor force stayed. They built their base camps in strategic locations so no tapper could slip past unnoticed. They hired gunmen to check on the ranchos. In Peru and Bolivia a fugitive could be returned to his master until he liquidated the debt; he could also be sold to another manager for the price of the debt, and this "truck system" of debt peonage was widespread. Because the rubber houses drew labor from the parched wastelands of the northeast, unhappy *seringuieros* were far from home with nowhere to run. They came to the Amazon by the shipload, thousands of hard, desperate men willing to bet everything on "Pará fine." By the peak of the Boom in 1910, an estimated 131,000–149,000 men were tapping from 21.4 million hevea trees on 24,000–27,000 *estradas* deep in the Amazon.

It was a changed world from when Henry tapped *goma*. He'd been an independent owner-operator, a *posseiro*, as had generations of *caboclos* before him. By law, the tapper was supposed to receive 60 percent of the value of the rubber he produced, and the average tapper produced about 1,750 pounds a year. But there were a number of ways to separate a tapper from his money, and the *seringuiero* who eked out enough to cover his living expenses without going into debt was a wise, or lucky, man. By the turn of the century, the *aviador* was a big-time capitalist, hiring anywhere from two hundred to five hundred men and shipping them into the field at his expense, then advancing each tapper £40–£70 in provisions, arms, medicine, clothing, and other basics, which he priced at 30–40 percent over what he paid. He was in turn exploited by the wholesale merchants in Pará and Manaus, who gave him credit up to £40,000. These merchants were funded by investors and speculators in London and New York, who received their payment in rubber at season's end.

For the men at the pyramid's base, the attrition was frightful. A January 1899 report by the U.S. Consul in Pará said that for every hundred new recruits, seventy-five would die, desert, or leave because of disease. A *patrão* in the Upper Amazon could plan on losing a minimum of five out of twenty-one workers. Death rates as high as 50 percent were recorded among tappers in Bolivia.

Even Henry would not have been blind enough to tap rubber now. And he never had the capital necessary to employ so many men. The closest he might have come was to let the Indians gather for him, as they had for the traders in Boim. The western side of the Tapajós was sewn up by the Boim trading houses, but on the eastern shore, tributaries like the Cupari were open for business, stretching back to the forested table-land as far as the Cuara du Sol and beyond. This was the home of the Mundurucú Indians who'd saved Henry's life when he nearly chopped off his foot. A 1912 survey of land claims on the Tapajós listed "400 tame Mundurucu Indians" who'd been coaxed into gathering rubber. Though they were often cheated, they were not forced at gunpoint to collect rubber as in some of the horror stories filtering from the Upper Amazon. Instead, they wanted the goods rubber could buy—the steel machetes, iron pots, and dyed cloth they could not produce on their own. Their conquest was by seduction, the same strategy employed by the legendary Crisóstavo Hernández. Although the rubber trade destroyed their culture, it did so quietly, by paying tribesmen to spend their time gathering rubber rather than in traditional farming or hunting. The Boom was relentless as it crept up the most remote tributaries, sniffed out rubber, promised a life of ease.

What if Violet and Henry had stayed? They might have become rich, like some of the *confederados* who'd held on. Judge Mendenhall built his lonely plot of land at Piquiá-tuba into a "model, prosperous plantation." The Jenningses and Vaughans intermarried, raised rubber and sugar, and were on the way to owning a two-thousand-acre cattle ranch outside Santarém. A "Dr. Pitts" would write to the *Mobile Daily Register* that he raised sugarcane, cotton, papaya, squash, five kinds of sweet potato, Irish potatoes, and a variety of beans. "I have made enough to live well on and am better pleased than ever," he said. The *confederados* who stayed introduced the plow, harrow, spade, and rake to the area. Some Brazilians copied their techniques, and they fared well too.

None did better than Harriette Jane's student David Riker. In 1888, Brazil decreed unconditional emancipation, ending slavery—the reason Riker's father had settled there. But David was of a different generation and considered himself Brazilian. The old conflicts and prejudices meant nothing to him. When his father had railed about the mixing of races, David had listened respectfully, but he didn't really care. He'd married a mixed-blood Brazilian woman from the wastelands of Ceará and eventually had fourteen children with her. He expanded his father's cattle ranch at Diamantino into a huge landholding; in 1884 he planted rubber, and by 1910 this had grown significantly. That year he sold out to a consortium of English investors, who incorporated it as the Diamantino Rubber Co., Ltd. Riker sold it for six thousand dollars, a fortune at the time. He retired, at age forty-nine, a rich man.

Hundreds of similar plantation companies sprang up across the world. They were promoted haphazardly, managed by men who knew little or nothing about rubber, and described to buyers as if they'd existed for years. As the price of rubber continued up, any investment seemed a sure thing. In the 1860s, rubber in the United States sold for six to ten cents a pound. Over the next thirty years it gradually rose to about sixty cents. In 1903–04 the average price was still about 68.2 cents per pound. A plantation company like Tapajós Pará Rubber Forests Ltd. was typical: Incorporated in 1898, its directors included a London merchant, a coffee planter, a printer, and an accountant. Since it never appeared to do any business, it probably never existed anywhere except on paper, like many such companies. As quietly as it started, it folded in 1901.

In the summer of 1905, the madness escalated. The price of rubber hit $1.50 a pound and stuck, something no one had ever seen. In the Stock Exchanges of New York and London the effect was electrifying. Capital bottled up in England during the Boer War sought a release; in the United States, the Rubber Boom coincided with an economic expansion so rapid and comprehensive that by 1900 the market was glutted with cash for investment and speculation. To attract money from smaller investors, plantation companies offered "low denominational shares," or shares going for under a pound or a dollar, an unheard-of offering. Lured by the gimmick, the public became unhinged. Everyone wanted a share in rubber, but so many companies had formed that they could only make a profit if rubber's selling price stayed at this unnatural high.

The "bubble" covered the world. Anywhere on the planet that a plant wept latex, stock companies followed. They formed to harvest *Landolphia* vine in the Congo, *Ficus elastica* in Liberia, Ule *(Castilloa elastica)* in Mexico, and plants in the recently conquered Philippines. Those companies that failed to meet the requirements of the New York or London Stock Exchanges pursued the new and unsophisticated investor: the teacher, waiter, and widow. Journalists were hired to write copy that sold confidence instead of value. This became the age of the paid endorsement, as prominent businessmen, politicians, and even a former treasury secretary sold their names to promote bad deals. Enormous profits were promised. A monthly investment of $5–$150 assured an annual income of $500–$5,000, promoters said.

As in so many speculative bubbles, thousands were ruined. A typical hoax involved the Peru Pará Rubber Company, with a reported capital of $3 million and an "unlimited" number of virgin trees growing in undisclosed parts of the Amazon. In 1905, ads in Chicago newspapers promised dividends of 75 percent for life. Unsurprisingly, no one collected. Another company sold $250,000 in worthless stock to Philadelphia teachers, who lost everything. Pensions, trusts, and holding companies were weakened by their rubber-stock investments. Even managers of widows' funds, believed the most conservative investors, were fooled. A famous story concerned Lucille Wetherall, who, like thousands others, lost her life savings when in 1900 she invested seven thousand dollars in the Vista Hermosa plantation in Mexico, chartered in her state of Maine. Later that year the company went into receivership, ruining 1,800 stockholders while preserving the interests of company insiders holding A-grade bonds. But Lucy Wetherall was made of stern stuff and gained her place in investment history by showing up at the plantation and demanding residence on the property that had backed her securities. She managed the failing plantation until forced to flee the Revolution in 1914.

Nowhere was the money and madness more apparent than in Manaus, the rubber capital of the world. Unlike the coffee barons and other commodity tycoons, the rubber barons did not live on their estates. Instead, due to its central location, most lived in Manaus, and extraordinary concentrations of wealth came to what had been the small jungle town. In 1892, the republic's youngest-ever state governor, the diminutive Eduardo Gonçalves Ribeiro, swept to power and transformed the city with the profits of the Boom. A 20-percent export tax on every kilo of rubber enriched the state's treasury by as much as £1.6 million annually. From the malarial forests emerged a city of hospitals, banks, office blocks, a £500,000 Palace of Justice, and forty-five schools. Two million gallons of pure water flowed daily through the city's water system. Three hundred citizens were linked by the first telephone network in the Amazon.

The first order of business was to transform the harbor. The Manaus Harbor Co., Ltd., composed of Brazilian investors, English and Brazilian steamship companies, an English rubber firm, and others, was signed to build a customs house and quay along the Rio Negro—a problem, since that river rose and fell as much as sixty feet each year. The solution lay in building a floating dock equipped with huge iron air tanks. A four-hundred-foot platform connected the pontoons to the iron warehouse on shore, and merchandise moved back and forth along cables. Manaus's floating dock was the largest in the world, capable of unloading three tons of cargo per minute. The year 1910 set a record for ship movement at Manaus, unequaled for another fifteen years: Approximately 1,675 oceangoing steamships, river launches, and sailboats called at port that year.

But the floating dock was not the only sign of wealth and power. Bottle-green electric streetcars operated in Manaus before any other city in South America, looping sixteen miles from dawn to dusk through the city to the jungle, then back to the *praça*. The line was subsidized by Charles R. Flint, whose United States Rubber Company purchased a quarter of the city's rubber. Roads were built, but since no paving stones were quarried in the valley, special Plimsoll cobblestones were shipped from France. The customs house, or *Alfândaga,* was modeled after Delhi's, prefabricated in England, and assembled on the spot. Public gardens were sprinkled with fountains in the form of gold cherubs. Telephones, telegraphs, and electricity were installed.

The world was reflected here. English, French, German, and Portuguese managers directed rubber operations; Spaniards, Italians, Lebanese, and Syrians owned small businesses. One could buy Smith and Wesson revolvers, Omega watches, Scandinavian butter, Black and White whiskey, Underwood typewriters, and Parfum Lubin. A jeweler estimated that, in 1907, the city's per capita diamond consumption was the largest in the world. The British pound sterling was used as freely as the Brazilian *mil-reis,* but French style shaped the tastes of the rubber barons and their wives. The leading stores catering to women bore French names: La Ville de Paris, Au Bon Marché, Parc Royal. There were five "houses of diversion" for vaudeville and movies, projected by the latest Edison cameras. Every Sunday, the Derby Club held horse races at the Prado Amazonense, with five heats and heavy purses for winners. The rubber barons built private palaces out of Italian marble, then furnished them from England and France and hung the ceilings with crystal chandeliers. Their linen came from Ireland. Grand pianos stood in their salons. One baron bought a yacht, another a lion, a third watered his horse on champagne. Jewelry was imported in bulk, diamonds lavished on prostitutes imported from the best European bordellos. Police believed that two out of every three houses in Manaus was a brothel.

Stories of the barons' profligacy were legion. One "colonel," as they liked to be called, bought an entire consignment of thirty-six hats destined for a local merchant, chose five for himself and threw the rest in the river. Another paid four hundred pounds for a ride in the city's only Mercedes Benz limousine: he picked up his mistress, then rode three hundred yards from the theater to a bar. At the height of the Boom, 133 rubber firms and buyers were represented in the city's Rubber Exchange, and most were foreign firms, like Dusendchon, Zargas, & Co.; Kingdom & Co.; and Anderson Warehouses. The leader of this mercantile polyglot was Waldeman Scholz, president of the powerful Commercial Association. Scholz gazed over his empire with a round pale face, receding hairline and pince-nez. He was the city's second largest exporter of rubber and imported a wide variety of goods from European firms. A local paper described him as a man of "clear vision, incomparable energy, and extraordinary activity." He built for himself the finest house in the city, which is occupied today by the State Governor.

Manaus's crowning glory was the Teatro Amazonas, the famous opera house, inspired by the Opéra-Garnier in Paris and built completely of imported materials. Funded entirely from rubber profits, its construction lasted from 1891 to 1896 and cost $2 million, a cosmological sum for the time. Even Violet had heard of it. She would have loved to see it, but she was in the South Pacific, waking each morning to the smell of rotting sponge. Even for Manaus, the Opera House was out of scale, rising above the river like a huge Gothic cathedral that dwarfed some medieval town.

Its green, blue, and yellow dome was laid with thirty-six thousand Alsatian ceramic tiles acquired from Maison Koch Frères in Paris. The floor was paved in marble, and sixteen Corinthian columns lined the foyer. The theater itself was built in the shape of a lyre, with three rising tiers of box seats and a ceiling painted to resemble the base of the Eiffel Tower. There were 701 seats: since the city's turn-of-the century population was about forty thousand, this meant it could hold nearly 2 percent of Manaus's residents under its dome. For opening night, some of Europe's most famous performers were booked at vast fees to brave malaria and abandon all other engagements for months to perform in the middle of the Amazon. Tradition holds that Enrico Caruso sang in the fantastic auditorium, that Sarah Bernhardt performed, and Anna Pavlova danced, but all are unsupported by opera house records. They were apparently sought out, but could not be coaxed to the jungle for fear of death and disease.

By the turn of the century, two great streams of rubber flowed into American and European factories from the two greatest river basins in the world. Each year, approximately twenty-five thousand tons of rubber came from the Amazon and another five thousand tons from the Congo. The horrors in both regions were so similar that they seemed part of a continuum.

In 1885, King Leopold II of Belgium emerged, under the guise of philanthropy, as the sole owner and ruler for the next twenty years of the Congo Free State, an area as large as Europe. The purported reason for this masterstroke was to improve the moral and material conditions of the native, "a crusade worthy of this century of progress," Leopold said. The Congo's riches—its copper, ivory, diamonds, and, above all else, rubber—were the real targets. The rise of the bicycle and auto industries made rubber the most lucrative resource in Leopold's realm. Although the Congo produced a second-grade rubber, extracted by cutting down and smashing the plentiful *Landolphia* vine, so great was the need that manufacturers snapped it up immediately.

The method of collection was designed more for instant output than the continued life of the source, or the collector. "Each town and district is forced to bring in a certain quantity (of rubber) to the headquarters of the Commissaire every Sunday," wrote the American missionary John Murphy, in what would be a typical report.

> It is collected by force; the soldiers drive the people into the bush. If they will not go, they are shot down, and their left hands cut off and taken to the Commissaire. . . . [T]hese hands, the hands of men, women and children, are placed in rows before the Commissaire who counts them to see that the soldiers have not wasted cartridges. The Commissaire is paid a commission of about 1 d. a lb. on all the rubber he gets. It is therefore in his interest to get all the rubber he can.

In 1896, the British consul recounted how he'd been sitting with the commissaire in the Upper Congo one evening when a group of "sentries" passed, fresh from their pursuit of escapees. The sergeant held aloft a necklace of human ears to illustrate the results of the hunt; the commissaire congratulated his subordinate on a job well done.

Free trade became publicly equated with evil. It had been the battle cry for civilizing the Congo, but now, in the name of profit, mass murder was condoned. Immediately after Leopold took over, the missionary Holman Bentley exulted, "The most rigid injunctions enforcing free trades, absolute religious liberty and freedom of worship are guaranteed. We cannot fail to see the hand of God in this result." Within fifteen years, Sir Arthur Conan Doyle declared profit the True God:

> [I]t is the call to brutality which comes from above; the urgent call for rubber, more rubber, higher dividends, at any price of native labour and native coercion, driving the local agents on to torture and murder. . . . Is there anywhere any shadow of justification for the hard yoke which these helpless folk endure? Again we turn to the Treaty which regulates the situation. 'All the Powers . . . pledge themselves to watch over the preservation of the native populations and the improvement of their moral and material conditions of existence.' And this pledge is headed, 'In the name of Almighty God.'

The key to exposing the situation in the Congo was British diplomat Roger Casement's forty-page white paper, written for the government in 1903. Casement's grim tales of murder, mutilation, abduction, and beatings by soldiers of Leopold's Congo Administration were so incendiary that the British government tried to keep it secret; it was only released in 1904 after a struggle. Casement's investigations confirmed the horrific accounts of missionaries and journalists. The difference was that he took names. In one instance, he wrote, "Two cases (of mutilation) came to my actual notice while I was in the lake district:

> One, a young man, both of whose hands had been beaten off with the butt ends of rifles against a tree; the other a young lad of 11 or 12 years of age, whose right hand was cut off at the wrist. . . . In both these cases the Government soldiers had been accompanied by white officers whose names were given to me. Of six natives (one a girl, three little boys, one youth, and one old woman) who had been mutilated in this way during the rubber regime, all except one were dead at the date of my visit.

The climax came in 1908 when international pressure forced Belgium to wrest the Free State from the private ownership of King Leopold, but by then the damage was done. During the fifteen years of Leopold's stewardship, the population in the Congo Free State dropped from 25 million to 10 million—15 million dead for approximately 75,000 tons of rubber. That equalled one life per every 5 kilograms, a little more than the amount used in one automobile tire.

In 1907, similar evils came to light on the Upper Amazon. The Putumayo is a vast area around a river of the same name, which runs through territory that was disputed between Peru and Colombia; the river joins the Amazon near the western border of Brazil. In many ways, the revelations about this "Devil's Paradise" were merely part of a larger continuum. During the Boom of 1890–1912, small wars of conquest erupted throughout the Amazon, but now the conquerors were rubber men. The Boom struck the Indian tribes their most destructive blow since the arrival of the Jesuits. Slavers rounded up entire tribes and forced them to work on rubber plantations. When some rose up and killed their masters, massacres and atrocities exploded on both sides.

What made the Putumayo different was its extreme isolation. There was no competition or ameliorating influence when the family of rubber baron Julio Cesar Araña established its empire. Although many latex-bearing trees grew in the Putumayo, the main source was *Castilloa elastica,* a tree that is destroyed during tapping. Unlike the long-term process involved with hevea, the production of *caucho* from *Castilloa* was quick, efficient, and designed for maximum profit. Since a tapper destroyed one tree and proceeded to the next, production could go on all year—an assembly line of rubber more in keeping with modern demands. As a British Select Committee charged with judging the atrocity would later note, "The insatiable desire to obtain the greatest production in the least time and with the least possible expense was undoubtedly one of the causes of the crime."

The scandal began in the northwest Amazon boomtown of Iquitos with a series of articles in the muckraking newspaper *La Sanción.* Editor Benjamín Saldaña Rocca accused the Casa Arana of systematically employing terror and torture against its native work force for higher profits. The Indians worked day and night; they were starved and sold at market. They were beaten, mutilated, tortured, and killed as punishment for "laziness" or the amusement of bored overseers. Women and girls were raped, the elderly were killed when they could no longer work, and children's brains were bashed out against trees. Moreover, Arana registered his Peruvian Amazon Rubber Company in London, thus linking Britain, the world's leading antislavery nation, with a firm that was enslaving Indians.

La Sanción did not circulate outside Iquitos, however, and it would take American adventurer Walter E. Hardenburg to bring the story to the world. In 1907, while drifting through the Putumayo, he ended up in a gun battle between Peruvian and Colombian rubber traders. He was nearly killed, and a friend from upriver died. Hardenburg was taken prisoner by the Casa Arana and for the next few months taught English in Iquitos in an attempt to earn his passage home. During his stay, he read the paper's allegations and collected further horrors on his own.

When Hardenburg arrived in London, he contacted Rev. John H. Harris, secretary of the Quaker-sponsored Anti-Slavery and Aborigines Protection Society. Harris sent him to the crusading London magazine *Truth*, which carried Hardenburg's article under the title "The Devil's Paradise: A British Owned Congo." The reference to the Congo stung badly, but it was apt. Arana owned 8 million acres on the Putumayo. His 1,500 armed section chiefs and their 600 hired gunmen were paid only by commission on the amount of rubber they sent downstream, thus ensuring systematic brutality. Worse, Arana had manipulated the British cult of free trade like a maestro, equipping his company with a tame set of British directors who allowed easy access to London funding. This was a British company involved in the evil. Even its Barbadian overseers were British subjects.

Now the story circulated worldwide, not just in a remote part of the Amazon. Hardenburg's account was even more horrific than that in *La Sanción*. The Huitoto, Boras, Andokes, and Ocainas were flogged till their bones showed. They were denied medical treatment, left to die, then eaten by the company's dogs. They were castrated. They were tortured by fire, by water, by being tied head-down, and by crucifixion. Their ears, fingers, arms, and legs were lopped off with machetes. Managers used them for target practice and set them afire with kerosene on the Saturday before Easter as human fireworks for the Saturday of Glory. Whole tribal groups were exterminated if they failed to produce sufficient rubber. At one point, a manager called for hundreds of Indians to gather at his station.

> He grasped his carbine and machete and began the slaughter . . . leaving the ground covered with over 150 corpses, among them men, women, and children. Bathed in blood and appealing for mercy, the survivors were heaped with the dead and burned to death, while the manager shouted, "I want to exterminate all the Indians who do not obey my orders about the rubber that I require them to bring in."

The climax of Hardenburg's story focused on that crucial moment when Indians brought rubber in from the forest to be weighed. The

Indians were so terrified by their treatment that if the scale did not register the required ten kilos, they threw themselves on the ground to await
punishment. Then the manager or his lieutenant

> advances, bends down, takes the Indian by the hair, strikes him, raises
> his head, drops it face downward on the ground, and after the face is
> beaten and kicked and covered with blood, the Indian is scourged. This
> is when they are treated best, for often they cut them to pieces with
> machetes.

Those who fought back had their limbs hacked off, then were thrown
alive into the fire. In the rubber station of Matazas, Hardenburg said, "I
have seen Indians tied to a tree, their feet about half a yard above the
ground. Fuel is then placed below, and they are burnt alive. This is done
to pass the time."

Once more, Roger Casement was sent in by the British Government.
Once more, he confirmed the accounts, with more details. Of the 1,600
Indians he saw on his 1910 journey, 90 percent showed scars from being
flogged. Beating wasn't enough: The lash was combined with near-
drowning, "designed," Casement wrote, "to just stop short of taking life
while inspiring the acute mental fear and inflicting much of the physical
agony of death." Casement heard of little boys who watched as their mothers were beaten for "just a few strokes" to make them better workers.
When Casement asked a Barbadian whether he knew it was wrong to
torture Indians, he replied "that a man might be a man in Iquitos, but 'you
couldn't be a man up there.'"

Casement's report led to commissions. The Peruvian Amazon Rubber
Company was dissolved and an order issued for Arana's arrest. By now
the rubber baron had escaped to Iquitos, where he was untouchable, even
hailed for bringing prosperity to that remote land. When the company
resumed business, it cut back on the overt bloodshed but retained flog-
ging as a motivator, a practice that remains on the river today. From 1900
to 1912, the Putumayo's total output of four thousand tons of rubber cost
the lives of thousands of Indians. The population dropped by at least
thirty thousand.

Years later, an Indian tribesman would tell a reporter for the Viennese newspaper *Neus Wiener Tagblatt* that, "Rubber has taken the blood, the health, and the peace of our people." Tribes entered into debt and could never get out. Those who ran were beaten; those who didn't died. "Liquor consoled us, helped us to forget our troubles. The debt remained."

This was the world that Henry would not confront in his island hideaway. In early 1899, Violet could no longer bear the hermit's life and told him he must choose between her and his self-destructive dreams. He chose, and Violet left. After all the wandering, hardship, and danger, he could not see, or admit, how good a friend she'd been. In fact, she'd been his only constant friend. She left and eventually settled in Bermuda, never to see him again.

For a short while longer, he stayed in the Conflict Islands. He seemed to realize that, with Violet gone, he could no longer run his "plantation" alone. Soon after her departure, he left for London, seeking a new partner in his schemes. But no one was as patient as Violet. After four months of negotiation he agreed, in June 1899, to transfer all interest to the financial firm of L. F. Sachs of London. Sachs paid him £15,000 for the title and property, and Henry retained his post as resident manager, with payment of 10 percent of profits in addition to his salary. He returned to the Conflicts with a "Dr. Jameson," an expert on fisheries from the Science College of South Kensington. Henry was in charge of copra, Dr. Jameson in charge of pearl-shell.

But all did not go well. He quarreled with Jameson, and a final argument was so heated that the scientist left, declaring to Sachs that they could not be reconciled. In fact, the whole deal with Sachs had fallen through. In 1901, Henry returned to London and throughout the year tried to find a buyer for the islands, or at least someone to provide extra capital. In 1902, he returned to the Conflicts with a "Captain Holton" and three others, including Arthur Watts Allen, a distant cousin by marriage.

Allen was twenty-three, a freshly minted Cambridge graduate inspired by Henry's exotic tales. In August 1902, Henry borrowed money from Allen's family to finance his trading schemes, apparently with the caveat that he take young Arthur along. In his memoirs, written near the end of his life, Allen describes the beginning, at least, as an idyllic South Seas cruise. With the investment from the Allens, Henry was able to buy the freighter he'd always dreamed about. He filled the *Arthur* with trade goods like rice, bolts of calico, and Jews' harps, which his Papuan laborers loved. He brought staple foods, oakum for caulking decks, one hundred machetes, and a deep sea diver's suit with a hand-operated air pump and a complete set of hoses, all to reenergize his plans to harvest sponge, mother-of-pearl, and bêche-de-mer. They repaired and improved the houses on Itamarina and Panasesa. In addition to coconuts, he tried to cultivate bananas and papayas. In the evening, Henry sat on the deck, smoking pipefuls of trade tobacco as he regaled young Allen with tales of his travels, including that of the rubber theft. Henry's personality was unique, exhibiting an "absence of aggressiveness," Allen would write. "He envied nobody, coveted nothing, lived simply and frugally."

It seemed a dreamlike state, a paradise painted so long ago in Nicaragua when Captain Hill played the same role to the captivated young Wickham. But once again, there were problems. We do not know the details, only the results. Things were not as idyllic as young Allen described, and at the end of the year, a deal was finally made. Henry would receive three hundred pounds annually—provided he leave his appropriately named island paradise and never show his face there again.

CHAPTER 13

THE VINDICATED MAN

Henry wandered after his expulsion from the Conflicts, trying to prove his relevance, failing repeatedly. He would latch onto a forest product, promote it as the newest natural wonder, attract a few investors, and run aground. He made several attempts to buy back the Conflicts through direct offer and government intervention, but he was rebuffed every time. He bought ten thousand acres in Mombiri, the northeastern tip of Papua New Guinea, and discovered a forest vine that seemed to produce a good quality rubber. He also found a large rubber tree related to *Ficus elastica* that the natives called *maki.* Neither proved equal to hevea. At Mombiri he planted 650 acres of hevea at "half-chain intervals," or about forty trees per acre, and tested ideas about its growth that he'd mulled over through the years. He invented a machine for smoke-curing latex but only built thirteen and exported one. He invented a three-bladed tapping knife, but none ever sold. The land proved unwelcome in Mombiri for hevea, and in 1912, he sold out, with yet another great loss of money.

During this time he promoted piquiá, the tree that grew in abundance in the highlands behind Santarém. In 1876, shortly after his arrival with the seventy thousand rubber seeds, he'd written Hooker about this "most valuable," "quick-growing" tree, whose nut produced a "pure fat or butter." His sample sat forgotten in Kew's museum for forty years. Now he tried to respark interest at Kew, estimating that an acre of piquiá would yield up to one-half ton of fat, a product, he claimed, that could prove "of at least equal magnitude to that of Pará rubber." Kew did not comment, but obtained the leaf and seeds and identified piquiá as *Caryocar villosum,* a tree that grew in much of tropical South America. In 1918, as the Great War ended, Henry founded the Irai Company Ltd., with a partner, Joseph Cadman, a man twenty-five years his junior whom he'd met through a planters' society. Cadman invested £20,000, and with that Henry's company set up two experimental piquiá plantations in Malaya. Although Irai struggled along for a decade, the yields from piquiá were far lower than Henry had predicted and the production costs higher. In 1929, one year after Henry's death, the company dissolved.

There was arghan, a fiber plant known in British Honduras and Brazil as silk grass, which he promoted at the same time as piquiá. Later identified as Colombian pita fiber, or *Bromelia magdalemae*, each plant had twenty to forty leaves, each leaf ten feet high and four inches wide. Arghan's true identity was kept secret after the managing director of the Belfast Ropework Company said, "Its salt-water-resisting qualities are remarkable and its tensile strength is abnormal, giving a breaking strain of more than 50 per cent over that obtained in the finest Italian hemp or the best flax." A manager for the textile company of Hoyle and Smith told the *Times* that "there would be sufficient demand in Lancashire alone to take up the production for a long time to come." Arghan, the *Times* concluded, "had now become one of the greatest commercial factors in the history of Great Britain." By November 1919, the Arghan Co. Ltd. was formed with £40,000 seed money, this raised to £100,000 by 1922. Abraham Montefiore, a Jewish financier from Germany, was chairman; Henry was named technical adviser. In a general meeting of investors in 1922, Montefiore made this tribute to Wickham:

All of us know what a good thing he has done for this country in giving us the rubber plantation industry, but many of you may not know that he is possessed of a knowledge of tropical agriculture and forestry that is simply amazing, and you find that allied with almost undue modesty generally. He is so filled up with knowledge that he has not time for anything else. He has also given what we very strongly believe is going to be another great, and perhaps even a still greater industry than rubber to the country. I am certain there is no other man living with whom we would have embarked our reputation and capital. I think that is the greatest compliment we can pay him.

For all the public optimism, the company was in dire straits. By the following year, the firm, whose investors envisioned thirty thousand acres of miracle grass growing in Malaya, owned fewer than seven thousand young plants in a nursery covering half an acre. A company director and botanist set out for British Honduras and came back with twenty-five thousand seedlings and several ounces of seed, but there was an even more vital problem to contend with—how to take the tough sheathing off the leaves. It proved too time-consuming and costly to decorticate arghan, and no mechanical salvation was ever devised. In September 1924, two years after the glowing tributes to Henry, the Arghan Co. Ltd. went bankrupt, and its investors lost everything.

Henry's toxic reach extended to all things financial. What had started with Violet and his immediate family mushroomed to hundreds of investors buying thousands of shares. It wasn't stock fraud so much as misplaced faith and bad planning. Something had happened when Henry and Violet sequestered themselves on the Conflicts: Henry's rubber seeds were beginning to be noticed. He emerged reborn from the South Pacific without even knowing it. His name was a force; it seemed as good as gold.

Henry's star had risen because of time, panic, and disease. A coffee blight spread through the Eastern plantations in an unstoppable wave. Since the 1840s and the elimination of the ancient kings, coffee had ruled Ceylon as mahogany ruled British Honduras. The only thing that mattered was the price of coffee on the stock exchange and Mincing Lane, a short and dingy street where an enormous business in all kinds of tropical produce was enacted weekly. Rubber, tea, coffee, cocoa, gutta-percha, spices—all were spread on long tables in dark salesrooms along the streets or off narrow passages in dim warehouses. Every week the brokers mounted the rostrum at the Commercial Sales Room with their clerks, always known as "Charlie," and bargained from eleven to five. Attempts were made to introduce other crops in the East—sugar, tea, cotton, and finally rubber—but all failed when set beside coffee.

In 1869, the first signs of blight appeared, described as an orange-red splotch on the leaves. In a land where there was no autumn, the rolling hills of coffee burned with spots of autumnal color, but no one really paid heed. The selling price of coffee and its estates rose to record highs on Mincing Lane; astronomical sums were asked and paid. By 1876, the year Henry arrived at Kew with his stolen hevea, it was obvious that nothing could stop the blight, and in 1879 the ruin became complete. During those three years, as planters cast around in panic for a replacement, Henry's seedlings arrived at Heneratgoda Gardens in Ceylon. But speed of harvest was of primary essence to the planters, and several rubber species in addition to hevea were sampled. Ultimately the Ceylonese planters chose Ceará rubber, since it matured much faster than hevea.

The problems with Ceará took a decade to understand. It did not thrive in the climate and grew bushy, taking up too much space in ratio to its financial return. It was killed by continual rain. Planters who'd invested in Ceará lost badly; a few had sown hevea, but most who remained planted cinchona and tea. Hundreds of Britons were ruined and went back to England. The more enterprising went to Malaya to start again with coffee. They shrugged when they saw hevea trees flourishing in the Singapore Gardens. They'd washed their hands of rubber, they said.

Yet the coffee blight had spread throughout East Asia, and in Malaya, hevea's role as a replacement crop was handled differently. In 1888, Henry Ridley, a young protégée of Joseph Hooker, was headed to his new post as director of the Singapore Botanical Gardens when he passed through Ceylon. He observed the techniques used there for tapping rubber and was given 11,500 descendants of Henry's seeds. He stuffed them in gunnysacks, and most survived the 1,500-mile journey. In Malaya, he devoted most of his energy for the next twenty-four years to experimenting on and proselytizing hevea—only hevea—so much so that his work assumed a missionary flavor. Malayan planters called him Rubber Ridley, or Mad Ridley when his back was turned.

When he arrived in Singapore, twenty-two hevea trees grew at the garden. He planted the 11,500 seeds from Ceylon, experimented with new methods of tapping and coagulation, investigated plant diseases, and became famous for stuffing rubber seeds in the pockets of anyone who he thought might plant them. In later years, Ridley claimed most major discoveries for himself, but in truth his enthusiasm was the key factor, and by the end of the 1890s, an informal troop of botanists, gardeners, planters, and tappers had worked in tandem to resolve most of the mystery surrounding Wickham's tree. Planting in swampy ground proved to be a mistake, as Wickham had preached, but spacing 135 trees per acre permitted the most rapid growth, not the 40-tree-per-acre maximum that Wickham claimed. When Ridley proved that a tree could be tapped seven years after first planting, hevea began to look like a money crop—and for the planters, that was the most important consideration.

One problem with which Ridley struggled was a better way of tapping. By the end of the 1890s, he'd disseminated methods that proved superior than those of the Amazon. A new knife modeled on the farrier's gouge replaced the *seringuiero's faca*. It excised thin slices of bark from the trunk, and repeated excisions of the same portion of bark increased the latex flow. It was found that trees could be tapped on alternate days throughout the year, instead of during a six-month season. Annual yields of more than two pounds per tree were possible, with yields increasing as trees matured. The coagulation of latex with acetic acid proved to be more efficient for curing rubber than the toilsome method of smoking. When samples of acid-coagulated rubber were sent to Mincing Lane, they received grades that equaled the best smoked rubber from Brazil.

Hevea seeds were sent all over the world—to Selangor in Malaya, Malacca, British Borneo, India, Burma, German East Africa, Portuguese Mozambique, and Java. In 1883, five seedlings wrapped in brown paper and bound for the Buitzenzorg Botanical Gardens in the Dutch East Indies aboard the German paraffin carrier ship *Berbice* were suddenly caught in the fire and ash of Krakatoa's volcanic eruption. The *Berbice* hid behind an island for two days, and despite lightning strikes to the ship and volcanic dust covering the deck "at least eight English thumbs deep," both ship and seeds survived.

In the British colonies of Ceylon and Malaya, vast fields of hevea eventually covered the land, but the juggernaut took time to build. In 1898, the first commercial seed was sold in Ceylon, while Malaya registered its first commercial sale of "plantation rubber"—145 kilograms for £60. That same year, the Amazon sold 25,355 tons. A drop in the price of coffee paired with a sharp rise in that of rubber convinced the remaining hold-outs that hevea was their future. By 1905, these early groves still yielded no more than 230 tons of rubber, but seedlings were being planted on a tremendous scale. In 1905, nearly fifty thousand acres of rubber were planted in Malaya; in 1906, that figure doubled.

Much of this could be attributed to Ridley, but like his mentor Joseph Hooker, Rubber Ridley was a prickly, combative man. His jealous insistence that all advancements were through his effort alone tends to cloud the record today. Like Hooker, Ridley was reedy and slim, wound tight as a spring. According to one acquaintance, he had "little regard for those who did not share his views on botanical matters." He was, to say the least, single-minded. He carried rubber seeds in his pockets and would thrust a handful on a guest with the cryptic assurance, "These are worth a fortune to you." Once, as a friend discussed purchasing a grave plot, Ridley gave him a handful of seeds and admonished, "Never mind your body, man, plant these instead!"

As early as the 1890s, Ridley was echoing Hooker's prejudices—that Wickham was an amateur and that the foundations of the Eastern rubber industry lay in the seeds Robert Cross brought from Brazil, not Wickham's. In his old age, when Wickham had been knighted for his services and Ridley had not, when Wickham was called "the father of the rubber industry" and Ridley merely its mad shepherd, his tone turned condescending. "I looked on him as a 'failed' planter who had been lucky in that for merely traveling home with a lot of seeds had received a knighthood and enough money to live comfortably in his old age," Ridley said from his pensioned retirement home in Kew.

The statement, famous in its time, was a mix of truth and untruth, but most telling was its heavy-handed rancor. When Wickham entered the battle with Ridley, he was never so personal. He disparaged abandoning smoke-cured rubber in favor of that doused with coagulants "like junket" and baked like "toffee in vacuum driers." He decried the "fallacy" of planting more than forty trees per acre and ridiculed the new methods of tapping trees. Letter followed letter to newspaper editors: he identified himself as "sometime commissioner for the introduction of the Pará

(*Hevea*) Indian Rubber Tree for the Government of India." A favorite pose was to stand in the back of a meeting of rubber planters, "almost with the air of looking down paternally on his 'children' who were talking about 'his rubber,'" one acquaintance said. "He conveyed this idea in a general and benevolent way, never in an officious or assertive one." He'd played the role of Amazon planter and British spy; now he played the wise grandee. Even after Violet left the Conflicts and his finances barely sputtered along, he played that part and played it well. It was all he had.

The first time that Wickham seemed aware of the new attention paid to hevea was in 1901. Violet had been gone for two years; there is no evidence that they corresponded. Henry always focused on his next big discovery. This was the year he haunted the London investment firms and eventually descended on the Allens, his distant family. On September 3, 1901, he visited Kew, perhaps remembering the day of his sole triumph, almost exactly twenty-five years ago.

He was received graciously by William Thistelton-Dyer, who'd succeeded his temperamental father-in-law. The director seemed fascinated by the old wanderer's adventures as they walked along Kew's planted paths. Henry still limped from the wound to his foot, so they moved slowly. No doubt several gardeners and botanists came up to shake his hand. Thistelton-Dyer described Ridley's work in Malaya and suggested he take a look for himself when he returned to the South Seas.

The next day, on September 4, 1901, Henry thanked the director for his kindness:

Dear Sir Thistleton-Dyer:
Turning over what you said to me in the path-ways yesterday about the *Hevea* descendants, now in 3rd or 4th generation from those I introduced for the good of India, was glad to hear that after the question of their cultivation had remained in a state of suspended animation since the 'seventies they shall have at last been taken in hand.

Had those thousands of healthy young plants which were so well brought forward at Kew from my original introduction from the Amazon Valley been properly set out under suitable conditions, the ports of India would before this time have obtained very large returns therefrom. Now however as it seems nothing has been taken in hand, I hope you will use your best influence to avoid crowning error from planting out on unsuitable lands.

A very general error seems to obtain that swampy or wet land is the fitting locality.

This seems to have arisen from the "explorers" of a few weeks naturally in going up the rivers in boats would observe a group of these trees scattered along the margins. Whereas the true forests of the "Para" Indian rubber lies back in the high lands, and those commonly seen by the canoe traveler are but ill grown trees which have sprung up from seed brought down by freshets from the highlands—as matter of fact—the whole of those brought to Kew for the Indian Government were from large trees in the forest and were got away back in the forest on the broad plateau betwixt the Tapajos and the Madeira Rivers.

Perhaps it would be as well if I tried a talk at the India Office—if you think it would do any good? Who did you say was the right man to see?

Very faithfully yours,
H. A. Wickham

PS: If you should find occasion, will you kindly make my kind regards acceptable to Dr. Joseph Hooker?

When he visited the India Office, he was treated with respect. Bureaucrats valued his opinions. His 1876 report on rubber was dredged up: The handwriting was his, not Robert Cross's, they discovered.

For the next few years he crisscrossed the oceans—London to the Conflicts, Conflicts to New Guinea, once again to London, then back again. During this time he started the rubber plantation in New Guinea, where all his "practical" techniques for planting, tapping, and curing rubber would fail. At least once, in 1905, he visited the rubber fields in Singapore, then crossed to Ceylon.

He must have been amazed: The industry spawned by his stolen seeds had grown geometrically. In 1905, Ceylon was still the world leader for total acres planted in rubber, but within two years the Malay Peninsula would jump ahead. In all, some 5.32 million trees were growing in the colonies in 1905, about 56 million by 1910.

These were fantastic numbers, perhaps not as fantastic as the number of virgin trees said to lie deep in the Amazon, but these were real trees, scientifically tamed and chained to man's schedule. The Malay Peninsula stretched hundreds of miles south from what was then the Siamese border to the equator, a vast humid region of jungle, wild elephants, tribesmen, snakes, rice fields, tin mines, and now endless rows of hevea. Rubber was grown on British and Chinese estates and Asian small-holdings all along the western coast, field after field of bleeding trees fronting the Straits of Malacca. One reason for Malaya's jump-start was that a huge labor pool already existed for tin. By 1905, the Straits Settlements were known as the "melting pot" of Asia, with an industrial army of Europeans, Japanese, Tamils, Hindus, South Sea Islanders, and millions of Chinese.

Those who planted rubber in the 1890s were rich men by 1905; after that, they only grew richer. They strolled around Singapore in their golf brogues, white socks with clocks, khaki trousers, khaki *tutup* coats and *terai* hats sitting well back on the head. The future looked good for them.

The future for the indentured coolie laborer was another matter. Recruited primarily from the Tamil of South India, they served under a debt peonage system akin to that practiced on the Amazon and in Queensland. Tamil labor was in place when the plantations blossomed with coffee or tea. Between 1844 and 1910, some 250,000 indentured Indian laborers entered Malaya. During that time, their living conditions barely changed—low wages and isolation, accommodations on the rubber estates in "line rooms," and staples purchased on credit from the planter. From the planter's viewpoint, the system kept costs down. The *Selangar Journal*, a trade magazine for Malayan planters, noted that

Coolies lines, each room 12 ft. by 12 ft., can be built for dollars 25 to dollars 30 a room. Double lines—i.e., lines two rooms broad each facing on to a fixed feet verandah—will be found more economical than the long single lines, besides being dear to the heart of gregarious coolies. Not more than six coolies should be put into each room, but the planter need have no apprehensions on the subject of mixing the sexes as the Tamil coolie is most philosophical in this respect, a young unmarried woman not objecting in the least to reside with a family or even to sharing her quarters, if necessary, with quite a number of the opposite sex.

"Tamils are . . . cheap and easily managed," enthused the author of "Life on a Malayan Rubber Plantation." Although they worked hard for a capable "ganger," or shift boss, "the main task is suppressing their conversation. Off work they like drugs, and fight among themselves." A life of debt peonage on a rubber plantation, the writer concluded, "is a pleasant one."

Maybe so, for planters. In 1906, the future seemed rosy enough to convene East Asia's first Rubber Congress at the Perideniya Gardens in Ceylon. Results of pioneer plantings were reported, as were the best methods of tapping trees. Even at that early date, plantation rubber was beginning to catch the eye of industrialists in the West because of its low rate of impurity. While the best Brazilian rubber contained 16–20 percent foreign matter, plantation rubber contained less than 2 percent.

But the real attraction at the Rubber Congress was Henry, who'd received an invitation. In 1906, he was sixty. He'd been told never to return to the Conflict Islands, and he was sweating out his rubber experiments in New Guinea, yet another failed wilderness venture. He emerged from the jungle like some ancient warrior, and the planters treated him like a star. "The father of the rubber industry," they said. He'd refined his image: Instead of buttons on his coats, he used silver links. His tie was inserted through a nautilus shell. He carried a walking stick attached to his wrist by a rubber thong. His most prominent vanity was his walrus moustache: It was waxed and hung beneath his jaw, or was curled at the ends. "Mr. Wickham is no longer a young man," wrote the *Ceylon Observer* of May 1906, "[H]is grey hair, grizzled mustache and eyebrows, and suntanned face and hands bear witness to long exposure to sun and air. But he is upright and alert, and keen as many a young man who has not gone through his experiences." Said another: "Wickham never sermonized; he just talked." It was his debut as a legend, and he talked a lot, it seems.

Three months later, on August 10, 1906, Wickham wrote a letter in a spiky hand from his club at the Royal Colonial Institute on Northumberland Avenue in London, a place that was becoming his home. The letter was addressed to Joseph Hooker, now eighty-nine and retired. Henry was still convinced that Hooker had been his champion, and no one had ever bothered to inform him otherwise.

Dear Sir:

Having learned your address from the present director of Kew, will you permit me to congratulate you on the now, at last, after so long delays, development in systematic cultivation of the *hevea* (Para) Indian Rubber: remembering, as I do, your foresight and initiative in securing the free hand enabling me to bring away the original stock on which it is founded, from the forests of the Alto-Amazon. Permit me to remain,

> Ever faithfully yours,
> H. A. Wickham

This is apparently the last letter Henry wrote to Hooker, who died five years later. As with his very first, Joseph Dalton Hooker did not deign to reply.

In 1908, Wickham published his second book, *On the Plantation, Cultivation, and Curing of Pará Indian Rubber,* dedicated to "My Fellow Planters and Foresters." Claiming that he alone of all British planters and foresters had working experience with rubber in its native habitat, he railed against the new-fangled methods of tapping and curing. He spun his tale of the *Amazonas* and gave, for the first time, an account of the deception that allowed him to steal seventy thousand hevea seeds. "I was at that time, as one before my time—as one crying in the wilderness," he wrote. The idea of "cultivating" a "jungle forest tree" was seen as "visionary" and ridiculous. Time had proved him right, but the cycle was repeating with piquiá and arghan, and he was growing old.

Wickham was right about one thing: His seeds did change the world—with the help of Ford's Model T. The first one came off the assembly line in 1908, the same year as Wickham's book. It rolled along on four pneumatic tires made of rubber shipped straight from the Amazon. The "pool" of rubber barons in Pará and Manaus took notice and raised prices accordingly. In 1909, Amazon rubber sold for $2.22 a pound. In April 1910, it reached its peak at $3.06, and the world's rubber consumers let out a howl.

The madness gripping the world's rubber markets in 1909–1910 was like nothing previously seen. A proposal that the London Stock Exchange be closed for a week so that brokers and clerks could catch up with orders was drowned out in cries of derision. An office boy struck pay dirt with £19,000 in rubber shares. "The Rubber Market continued to astonish even its most enthusiastic supporters," wrote *The Times*. Across the Atlantic, the *New York Times* concurred:

> It is a maddening revel of speculation by the multitudes who are investing their small savings in rubber and oil shares. Rapid profits are made when two shilling shares rise to fifty or seventy shillings, and fresh investments are made in new issues.

According to the *New York Sun*, each morning's opening saw "riotous excitement" in that section of the floor devoted to rubber:

> New companies continue to be floated. The subscription list of three such companies will open tomorrow and they doubtless will be closed before the advertised time. . . . The brokers and clerks are becoming worn out. They rarely leave the City, snatching short spells of sleep at hotels.

The crash began in May 1910, though no one saw it initially. The price began to slip. Banks in Pará stopped accepting rubber land as collateral for loans, something that had never happened in the sixty years since Richard Spruce beheld the first effects of the world rubber trade. True, there were rumors from the East of possible competition, but the rubber barons in Manaus, Pará, and Iquitos were convinced that nothing could ever supplant wild rubber. This was the era of the fabulous wealth of Manaus, when Julio Arana's peak production of 1.42 million pounds of smoked Putumayo rubber cost thirty thousand lives. Yet the bankers had glimpsed some disturbing trends. All the easily accessed trees had already been worked; the more remote regions could be exploited only at greater cost and the import of more labor. In 1909 and 1910, the rubber barons withheld rubber and let American and European buyers feud among themselves for the limited supply: this drove the price sky-high. In the face of such need, it seemed to the barons that rubber profits could only soar.

At the same time, the Eastern plantations were coming into their own. It had taken a long time. In 1900, the Asian plantations sired from Henry's seeds produced 4 tons of domestic rubber against Brazil's 26,750 tons of wild; seven years later, they turned out 1,000 tons, still a drop in the bucket to Brazil's 38,000 tons. But a limit had been reached. The challenges of the jungle were too daunting, no matter how much one bribed, beat, flogged, or shot the *seringuieros*. By 1910, America produced 180,000 autos and her automakers prayed for a new source of rubber.

Their prayers were answered in 1913. That year, the British plantations turned the corner and produced 47,618 tons of high-quality, acetate-cured rubber compared to Brazil's 39,370 tons. In 1916, Brazil produced as much as ever, but the game had now changed. In three more years, British plantations would produce enough hevea to fill 95 percent of the world's need for high-quality rubber. Such fantastic supply seemed unimaginable just a few years earlier, and the price plummeted from the $3.06 high in 1910 to 66 cents per pound in 1915. By 1921, when Great Britain controlled the world market, plantation rubber sold for 12–21 cents a pound.

The indifference with which the rubber barons watched domestic rubber catch up seems suicidal today. By 1912, the world had 1.085 million acres planted in rubber, most of that in the East; each year saw increasing yields. But no one on the Amazon or in Rio de Janeiro seemed to care. Brazil had made millions at almost no expense. The region, which held one twenty-fifth of Brazil's population, produced one sixth of its revenues. From 1890 to 1912, the federal government collected 656 million *millreis* more than it spent in the rubber states of Pará and Amazonas; Brazil collected 241 million *millreis* in rubber export taxes and spent it all on grandiose palaces and payoffs to politicians.

Credit was the system's Achilles' heel. The entire Amazon rubber trade was built on perilously overextended credit, a fact that the bankers recognized when rejecting rubber land as collateral in May 1910. In 1913, when the bottom fell out, the truth was revealed to all. The business was based on a forward billing on the expectation of continuously rising prices. When the price plummeted, the import houses demanded payment from owners and *aviadors,* but there was no actual money, just a paper promise of "inexhaustible natural supplies" and the "unrivaled quality" of "Pará fine." The houses tried to pay in rubber, but as its value shrank, week by week, then day by day, no bank would take it. Although the *aviador* houses were supposedly worth hundreds of thousands of dollars, they folded overnight when the loans were called. The entire rubber system, from Pará to the remotest corners of the Amazon, was based on irreversible debt, and in 1913, it collapsed like an enormous house of cards.

Imagine if, in one month, the oil-consuming world switched suddenly from its addiction to petroleum to the use of solar power or some clean and cheap hydrogen-based fuel. Suddenly the OPEC nations would find that their natural riches meant nothing. The Amazon powers were loathe to blame themselves for their ruination and accused the federal government of lavishing its attention on coffee, or blamed "Yankee speculators" for causing the disaster. Amazon banks blamed American manufacturers for engineering a price drop. Even in their distress, they disparaged plantation rubber. Not until later would they turn their wrath on Wickham.

In the resulting panic, ruin rolled up the Amazon Valley like a giant tidal bore. Except for the floating docks in Manaus, British capital was mobile. The ubiquitous steamships were the first to vanish, gone in a puff of smoke over the horizon as they were rerouted east to where all the new money lay. With the end of shipping, the little trading towns along the rivers sank back into the jungle. The Sephardic houses in Boim closed one by one; the families moved on. Although David Riker had wisely sold his rubber plantation at Diamantino to an "English firm of gilt-edged bond folk" before the bubble burst, he lost a Santarém trading house inherited from his father. The town of Obidos, west of Santarém on the Amazon, dropped from thirty thousand residents in 1907 to three thousand in 1920. There were tales of *seringuieros* starving upriver because supplies never reached them, of others retreating to their hammocks to drink and commit suicide. Production continued at a trickle during the war years, 1914–17, but after that, the Amazon was finished as the world's rubber center. Singapore was king.

Manaus became a shadow of its former self. The commercial houses went bankrupt; the Opera House closed. The docks and warehouses deteriorated; the foreign merchants moved away. Before they left, the yachts and diamonds, the thoroughbred race horses, and Steinway grand pianos were sold for what they could bring. At the Zoological Gardens, the cages stood empty. The trolley tracks meandered through an almost unbroken jungle. The suburbs at its edges were empty and soon overgrown.

At the end of the line sat Iquitos, the last great rubber metropolis. The Malecon Palace—a three-story hotel erected for Julio Arana and his guests—was nearly vacant. Warehouses stood empty, and cows grazed in the boulevards near rusting fleets of autos. Most of their owners drifted away.

Sometimes those who remained tried to recapture the ostentatious dream. In the evenings, when the tropical sun dipped beneath the horizon, the few remaining café owners on the central plaza dragged their tables to the middle of the street, safe now that it was devoid of cars. They sat and sipped an aperitif, always donning a stiff white collar for the evening ritual. In the barren plaza, a band played romantic ballads. The café remnant would think of how yet another cycle of boom and bust had played itself out in the Amazon Valley, and how rubber was merely the last in a procession starting with dyewood and promenading through sugar, gold, tobacco, cotton, and cacao. They remained loyal to the memory, like survivors bearing witness to the end of the world. On Sundays, they mounted the town's one remaining trolley. It took them around and around the plaza, along the riverbank's precipice, and to the jungle at the far edge of town.

If Henry's theft ruined the economy of one of the lushest regions on earth, it seemed to prove an oft-debated point during the runup to the Great War.

In the years of Victoria's reign, Britain's foreign policy was based upon the assumption that international amity could exist between the nations that mattered most to her—the other European powers. But with Victoria's 1901 death, Britain began to realize she had hardly a friend in the old coterie. Some blamed it on the Boer War. Outside Britain and her colonies, the war in South Africa was almost universally despised. Her aggression was a threat to others, warned critics, and it was what had turned others against her. But this was probably unlikely. As historian Bernard Porter pointed out, nations do not generally raise more than perfunctory complaints about the aggressions of others unless the offending state is already regarded with hostility—as became clear when the United States invaded Iraq in March 2003.

The "new imperialists" who rose to power in Britain in the early twentieth century sound very much like the American neocons who rose in the early twenty-first. Both saw threats from without, believed that control of the world's resources was a battleground, and concluded that the means of survival was a preemptive will to empire. Great Britain's naval power was no longer exclusive: she was vulnerable, even at sea, especially if two or more great powers became allies. The "scrambles" in Africa and Asia were bellwethers of the coming conflict, which at its heart would be a war over raw materials. "The fight for raw materials plays the most important part in world politics," the president of the German Reichsbank would declare in 1926. "Germany's only salvation is her acquisition of colonies." In effect, he stated his nation's rationale for two world wars. The world was an unfriendly, predatory place, warned the new imperialists: Maintaining the resources and allies intrinsic to empire was the only way to survive.

When war came in 1914, the increased use of material transport by combatants made rubber an essential war material. The European armies depended on cars, motorcycles, and insulated cable for lines of communication, and on railroads and trucks to move troops and keep open the lines of supply. When Lawrence of Arabia was not riding around on a camel, he sped across the desert in a fleet of nine armor-plated Rolls-Royces, blowing up trains. This was total war, each side trying to strangle the other to death by cutting off sea routes that brought war material for the forces and food for the populace—and the German submarines and British dreadnoughts engaged in the fight needed rubber for the seals and gaskets of their engines. The first Battle of the Somme saw the introduction of gas, as a million French, German, and British soldiers died in the mud. By the end of the war, American factories would churn out 3.9 million gas masks; French factories, 35 million; English industries, 50 million. One of the prime materials was rubber. The face piece, the gaskets around the goggles, the breathing tube and exhalation valve, the hose feeding into the canvas bag that contained the canister of chemicals filtering gas from the air—all were made of rubber. Observers in balloons watched overhead as men were being gassed. They were held aloft by hydrogen trapped in giant bags of rubberized fabric. Their telephone wires, insulated in vulcanized cables, passed through the center of steel umbilicals tying the balloons to the ground.

One new weapon was the embargo. In December 1914, the British government banned the shipment of rubber to any country not allied to Britain, the United States being the sole exception. The target was Germany, for whom the embargo turned into an emergency of the greatest scale. In 1901, the Russian chemist Ivan Kondakov had made a synthetic rubber by heating potash with dimethyl butadiene, which could be produced on an industrial scale from acetate, made from potatoes. But civilians needed potatoes, or they would starve. Instead, German chemists found they could synthesize dimethyl butadiene from calcium carbide. If they left the prepared dimethyl butadiene in huge drums in cold storage for several months, the resulting ersatz rubber was about a third as stretchable as natural rubber; if they kept the drums hot, the synthetic product was as soft as its natural cousin, but only a tenth as strong. The synthetic was spongy and cheesy but could still be used for insulation, observation balloons, and dirigibles, and for an inferior but usable tire. By the end of the war, Germany had set up plants making eight thousand tons of ersatz rubber per year. If not for the discovery, Germany would have been forced to surrender earlier.

The world came out of the war crippled and diminished, but not the rubber industry. In 1918, the United States imported 333.8 million pounds of rubber from the Far East at a price tag of $181.6 million; that year, the ten leading plantation companies paid enormous cash dividends, ranging from 30 to 150 percent. In the years after the Armistice, the United States realized it had entered its own Rubber Age. It took a war for Americans to grasp how big a part rubber played in their lives. The center of the domestic rubber industry was Akron, Ohio, a brash, noisy city perched above the Little Cuyahoga River. By 1920, half the rubber manufacturing in America was located here: B.F. Goodrich Tire and Rubber, followed by Goodyear, Firestone, and others. Since Akron's economy was hooked to one commodity, its booms and busts were larger than life, just like the fortunes of the one-crop Amazon.

As early as 1914, the public relations machines of Detroit and Akron began spinning out films and copy they grouped under the rubric "The Romance of Rubber." This was a purely male mythology set in an infernal world of sulfur and sweat, where sixty thousand muscular modern-day Vulcans wrestled great rings of rubber into submission amidst suffocating heat and chemical fumes. In 1914, Detroit required 1.8 million tires, and most came from Akron. In April 1914, the Ford Motor Company alone received 87 railroad cars filled with tires, each car carrying 400 sets; the following month, this increased to 110 cars. This meant that in two months alone, Ford required 78,800 sets of tires. Akron had the same boom-town flavor as Singapore or Manaus—bars, brothels, silk suits, and music halls.

The United States and Great Britain were joined at the hip in the rubber business, but they pulled in opposite directions by the forces of supply and demand. In 1921–27, Great Britain produced 67–71 percent of the world's total rubber supply and almost all of the high-quality rubber from hevea. At the same time, according to different estimates, the United States consumed two thirds to three quarters of exports from Great

Britain and the world, and most of that went directly to the tire and auto industries.

British producers believed they were in a position to control the world market—an echo of the Brazilian "pool," though the price per pound would be much lower. In November 1922, they took a step in that direction by introducing the Stevenson Rubber Restriction Plan, a unilateral British response to the falling fortunes of the plantation rubber industry after World War I. The withdrawal of ships from the East Asian ports in 1918 to add to the war effort curtailed the shipping of rubber. The resultant glut in the Singapore market drove the price down to fifteen cents per pound. The planters screamed that they could not survive under such conditions, and they were probably right. Therefore the act restricted output to 60 percent of 1920s production, permitting an increase of 5 percent if the price averaged thirty cents per pound and a decrease of 5 percent if the price was lower. Consuming nations might not like the Stevenson Plan, but there was no one else to turn to but the British for vast quantities of hevea. By now, the once proud Amazon was so decimated that it barely supplied 4.6 percent of the world's rubber.

Most Americans were not worried, but angry voices rose out of the tire and auto industry. Harvey Firestone was the loudest: No government had the "moral right" to withhold exports of a vital world commodity "for the benefit of a few stockholders," he roared to his senior executives in Akron:

> I am going to fight this law with all the strength that is in me. . . . It is a vicious plan, which will result in making Americans pay exorbitant prices for their automobile tires. If we submit meekly, it will cost the car owners of this country millions of dollars and do the rubber industry irreparable harm. The time to break up this monopoly is now.

The quest to find a new rubber source was wrapped in the flag. Early in 1923, Firestone called a conference of rubber- and auto-industry captains to meet in Washington and organize the opposition. He turned to President Warren Harding and Secretary of Commerce Herbert Hoover, both personal friends. Firestone wanted to whip up public opinion.

Although Harding was lukewarm, Hoover was sold on Firestone's scheme. In his letter of invitation to Henry Ford, Firestone ranted that the price of rubber had risen from fifteen to thirty-seven cents a pound in three months: That meant a $150 million increase "to the crude rubber bill to the United States for 1923," he warned. The only recourse was to meet "an invading nationalism with a defending nationalism—and by taking steps to make sure that in the future Americans can produce their own rubber."

Rubber grown under the flag spelled "resource independence," a theme that echoes from Clements Markham's theft of cinchona to national oil politics today. Firestone vowed he would give "the play an Olympian cast of characters, mustering the household gods of the nation, Hoover, Ford, and Edison, to the defense of America in distress."

And so the quest began. Thomas Edison tested 2,300 different plants in his search for a high-quality domestic rubber. He focused on Mexican guayule, a twiggy, knee-high shrub that is a distant cousin of the sunflower, then turned his attention to a new strain of goldenrod that yielded 7 percent rubber. But both were inferior to hevea, and the research ended with Edison's death in 1931. Harvey Firestone sent his son to Liberia, where plantations were started that met with modest success but nothing that could meet the demands of the auto industry. Plantations were begun in the Philippines but never on a grand scale. Filipino leaders feared that a giant American industry on the islands would act as a deterrent to independence. Hoover sent scouting expeditions to South America. And Henry Ford began to dream of vast rubber fields in the Amazon.

Henry Wickham stood on the sidelines as the feeding frenzy grew, an iconic observer of what he'd started, amazed by the stupendous fortunes—none of which he shared. From 1913, when the British plantations took over, until 1922, the United States alone had imported 2.7 billion pounds of rubber for $1.16 billion. It was an inconceivable figure, beyond Henry's greatest imagining, and in all that time, he told acquaintances, he hadn't made a dime from his stolen seeds.

That was not quite true. He *had* been paid his £700 bounty—not a fortune, but enough to take him to Queensland. In 1911, the British Rubber Growers' Association and the Planters' Association of Ceylon and Malaya presented him with a check for one thousand pounds, a small annuity, and a silver salver at London's Second International Rubber Exhibition. That year he made his last trip out east. He was photographed leaning against the largest rubber tree in Ceylon, a giant planted in 1876 from his seeds at Heneratgoda Gardens, which yielded 371 pounds of rubber from 1909 to 1913. He wore a khaki jacket and white Captain's hat. The V's of the herringbone scars crept up the tree like chevrons. Henry rested his hand against the trunk like an old friend.

After that, he was poor, frightfully poor, spending most of his time at the Royal Colonial Club, surrounded by fellow imperialists, each spinning their separate tales. Somehow the annual payment of three hundred pounds from the Conflicts dried up. He ventured out to promote arghan and piquiá, but these efforts were hideous failures. He was sixty-eight when the Great War started; he joined the City of London National Guard, where he was one of his battalion's crack shots, a skill he'd learned potting birds in Nicaragua. On June 3, 1920, he was made Knight's Bachelor on the King's fifty-fifth birthday "for services in connexion with the rubber plantation industry in the Far East," according to a brief notice tucked far back in *The Times*.

He was now Sir Henry Wickham, and around this time he began embellishing his legend for anyone who cared to listen. At an unspecified date after his knighthood he told a reporter for *The Planter* that, "The first natural seeds of the rubber tree to be brought to Britain from South America were loaded by stealth in a small steamer under the nose of a gunboat which would have blown us out of the water had her commander ever suspected what we were doing!" Not *could* blow them out of the water, but *would*, as if what Henry did was espionage at its most ruthless, as if both sides engaged in a deadly game of resource management and the stakes were paid in lives. By then he was eighty-three. It was a detail he'd never revealed before, the final flourish to the imperial legend that probably turned more scholars against him than any others. No one likes a blowhard, especially an *old* blowhard. But could it have been true?

The idea of a Brazilian gunboat lingering off Boim is a little farfetched, since in 1876 the Brazilian Navy was stretched thin, assigned to major ports along the Atlantic coast and on the Amazon. That means, however, that one could have been at anchor in Pará, and that Henry eyed her as the *Amazonas* passed by. In 1876, the Brazilian Navy had seventy vessels of war, including nine steam launches; of those, fifteen were ironclad and fifty-five were wooden vessels. They carried 72 rifled guns, 65 smooth-bore guns, an aggregate power of 11,188 horsepower, and a compliment of 338 officers commanding 1,122 enlisted men. One of the navy's principal duties was to uphold customs regulations in the shipping lanes. Such a gunboat *could* have blown the *Amazonas* out of the water, but consider the consequences: such an action taken against Brazil's principal foreign investor would not bode well for business, foreign relations, or the career of the gunboat's commanding officer.

So the threat probably never existed, and if it did, the threat was primarily in Henry's mind. It shows that he knew what he was doing and that he may have felt guilty, knowing what his transplanted seeds could do to the livelihoods of his old friends. He'd brooded over the fact that he'd left the tattered remnants of his family back in Santarém without so much as a good-bye. It speaks to legend as well. The idea of duplicity was essential to his aura: without danger, there would be no triumph; without sleight of hand, no battle of wits was involved. Henry told his tale in his club among the retired servants of Empire. His victory over the natives, and thus the harsh world outside Britain, was more by wit than by force. It was the stuff of imperial legend, a confirmation of national and moral superiority while civilizing the world.

By then, no one remained to dispute Henry's tweaks to his legend. The adopted Indian boy disappeared after Queensland. Violet and Henry had been separated for over twenty years. After leaving the Conflicts, she dropped from the records almost as totally as the adopted boy. According to family, she moved to Bermuda to be near friends and kin, but that is all we know. She inherited with Henry's knighthood the title of Lady, and she would have liked the measure of respect it brought in her small colony of friends. But it would have turned her thoughts to Henry, and those memories would be sad. Rubber had made him a disappointed man who'd driven her away. We know they loved each other. There are secrets we'll never know, but she'd stood by him through the worst, and he'd loved her so recklessly that he swam a shark-filled channel in the middle of the night just to be with her. He was often a fool, but he was *her* fool, and for all his faults, he remained brashly, outrageously charming. Now he was simply a character, a bitter, comical figure with a mane of white hair and a walrus mustache who railed against the newfangled developments of the Malayan rubber growers whose pockets he had lined.

He was penniless, while his seeds made billions. If Victoria had ruled for him on the Temash River, perhaps there would have been a modicum of justice, but she'd ruled for the rulers and sent Henry on his doomed quest in the South Seas. Henry kept his financial straits to himself, reluctant to relinquish his fierce pride. But in 1923, his situation must have reached that point where he simply could not go on. His former brother-in-law, Frank Pilditch, lived in London, and his brother John lived in Texas as a cattle rancher, but he'd left them to their own fates when he'd shipped out on the *Amazonas,* and he could not turn to them now. That year he must have revealed his desperation to Quincy Tucker, an American who'd prospected for rubber in the Bolivian wilderness and who'd met Wickham in September 1908 at the International Rubber Exposition in London. Tucker revealed Henry's plight to Fordyce Jones, Wickham's closest friend and owner of the Reliance Rubber Company Ltd., the world's largest manufacturer of "molded seamless hot water bottles." The two tracked down a strange American savior to ride over the hill like the cavalry.

Edgar Byrum Davis was odd even by American standards. An oversized block of a man with a wistful smile and a mystical faith in his own greatness, Davis made and lost several fortunes before striking oil in Texas, the fortune that made his name. He'd made his first million as an officer and stockholder in the Charles A. Eaton Shoe Company of Massachusetts, then had a nervous breakdown and commenced an around-the-world voyage, which depleted his funds. In 1907, during his travels, he became interested in rubber and tried to convince the United States Rubber Company to start their own Far Eastern plantations, but the officers would not listen until 1909. Davis was made vice president in charge of plantations and given $1.5 million for development. Though paid only twelve thousand dollars annually, he wrote into his contract a codicil that he'd be paid 8⅔ percent of the company's value in excess of its investment when the plantation reached full production. In 1913, he was rich again, and switched his attention to oil.

In March 1921, Davis started the United North and South Oil Company and came to Luling, Texas, convinced that God directed him to deliver Texans from cotton, the one-crop economy that ruled their lives. He spent $600,000 drilling the first six wells, and not a drop of oil was found. Creditors took his furniture; banks refused to renew his loans. In despair, he sought out Edgar Cayce, the "sleeping prophet," who would go into a trance and see the most remarkable things. Although most of Cayce's pronouncements were failures, enough were successful that by 1922 he had a legion of believers. In a trance, Cayce described an underground geological structure that Davis believed lay beneath the 126-acre farm of Rafael Rios outside Luling. He called it Rafael Rios No. 1. On August 9, 1922, the same day his bank returned a $7.40 check marked "insufficient funds," the rotary drill at Rafael Rios ground away at 2,100 feet and struck oil. He had, in fact, opened up an oil field that was 12 miles long and 2 miles wide, and which by December 1924 was pumping 43,000 barrels a day. His wistful smile grew wider as oil rained on his suit. The Lord, through Edgar Cayce, had made Davis rich for a third time.

This was when Fordyce Jones and Quincy Tucker caught up with him. Davis would have known about Wickham during his days with U.S. Rubber in Sumatra. He might have seen the old man as a spiritual twin. There was something very American about Wickham's saga of repeated failure punctuated by the lucky strike—success wasn't a matter of intelligence, or even common sense, but *faith*, as Davis liked to proclaim. Davis promised Fordyce Jones and Quincy Tucker that if he "made a killing in oil," Henry would be comfortable the rest of his days.

In June 1926, Davis made that killing, selling his leases at Luling for $12 million. He went on an orgy of charitable giving, holding a picnic in Luling for 35,000 locals, giving employees 25 to 100 percent of their salaries as bonuses, giving away millions more. In 1926, he sent Henry a check for five thousand pounds, quickly followed by another thousand pound check from other American oil kings.

Henry had just turned eighty when Davis wrote his check. He seemed set for life, and more money rolled his way. It was apparently the result of international jealousy, if not simple shame. Some of the planters in Malaya remembered Davis, and the memories weren't pleasant ones. While Davis worked for U.S. Rubber, some British and Dutch planters made wealthy by rubber's high price gloated that "America is paying the bills." Their "cocksure attitude that America is left at the post only acts as spurs to my determination to secure for our company and America their just share of this good thing," Davis later said. His frustration with British rubber interests apparently turned to disgust when, in 1922, he met with a group of openly hostile London investors during a meeting in which he proposed a $250 million company formed to absorb a number of small British and Dutch rubber companies. When he suggested that American capital might be raised to support the undertaking, the financiers saw it as an American attempt to break the British monopoly during the start of the Stevenson price-fixing plan. When they voted him down, Davis exploded: "If you men think you are doing the best for the industry by merely consulting what seems to be your own immediate interest in a blind or near-sighted patriotism, you are wrong."

When it came to rubber, there was no love lost on either side of the Atlantic, and now the Americans were saving Wickham, "the father of the British rubber industry," whom the Britons had ignored for so long. Soon afterwards, the governments of the Straits Settlements and the Federated Malay States gave Henry a check for eight thousand pounds in recognition for his services to rubber.

He was well off for the first time in his life, as well off as he'd been as a child on Haverstock Hill. He should have been at peace, but one wonders. The world praised him, but the praise was never enough. The craving never ends.

He did not have long to enjoy the money or accolades. Four months after his eighty second birthday, on Monday, September 24, 1928, he suddenly fell ill. His condition deteriorated, and he passed away quietly three days later, on Thursday, September 27, "of senile decay." He'd always believed he was descended from William of Wykeham, the medieval bishop of Winchester, and asked to be buried at the Hampshire village of Wickham. Although the belief was probably mistaken, his wish was fulfilled.

Violet fell ill within a month of him. She was in London and heard of his sickness. There is some suggestion that Henry knew she was in town. But both were too weak to make the trip and reunite. They'd parted over Henry's insatiable attempts for a second success and Violet's unending loneliness in the world's worst places. They'd loved each other deeply, but rubber and the madness and greed it spawned finally came between them. Both died alone.

On the day after his death, Henry's obituary ran inside the *Times:*

Sir Henry Alexander Wickham . . . was the man who, in the face of extraordinary difficulties, succeeded in smuggling seeds of the Hevea tree from the Upper Amazon, and so laid the foundation of the vast plantation rubber industry. Looking every inch a pioneer, broad-shouldered and heavily built, with an extraordinarily long and wavy moustache, his physical strength was as great as his resolution. To these qualities he owed his escape from the many adventures which he encountered in his wanderings. In conversation he was most entertaining, and his stories were not only interesting but also instructive.

He told stories, all right: stories to everyone. To his mother, his siblings, his wife. To those who wanted to carve a future out of Nature; to those who invested thousands in his name. There were many like him in the world wanting to grow an empire. Rubber was the seed, the starting point, but like the waters of the Amazon, the stories and promises never ended.

During his last days, he had a housekeeper who was also a trained nurse. Her name seems forgotten too. "She showed her heroism," Quincy Tucker later wrote, "by remaining on the job without pay when his cash ran low." He was supposed to be rich, but he'd run through his funds again. Whether the money went to creditors or more bad investments is uncertain today. No doubt the nurse listened to Henry's tales of the jungle, about the obstacles that were placed in his path, and of the people who failed to trust him, but he'd proved himself right in the end. He was, after all, a vindicated man.

Henry willed to her his house and furniture, and some shares in a Burmese rubber plantation he'd bought long ago. After all, no one could go wrong by investing in rubber. It was a necessity in this modern age. But the British government demanded inheritance taxes in cash, and she sold the house and furniture to pay the debt.

And when she sold the rubber shares, they were worthless too.

EPILOGUE

THE MONUMENT OF NEED

What was it about the jungle that robbed a man of his judgment and fueled the most grandiose dreams? In 1928, the year of Wickham's death, it was Henry Ford's turn. Like his predecessors, he saw the Amazon as a limitless, untapped treasure. Of all the schemes in progress to wean America from her dependence on foreign rubber, his seemed most likely to succeed. The deed to a 2.5-million-acre swath of land—four fifths the size of Connecticut—was signed, sealed, and delivered. All deemed necessary to conquer the jungle was loaded into the hold of the white *Lake Ormoc* and her barge *Lake LaFarge*. Someone from Ford's legal department traveled to Pará and the Tapajós to review preliminaries. Ford was returning to where the rubber industry started, but he would do it better. Nothing could go wrong.

Ford's name inspired confidence. He had a compelling vision of a bright new age, and he aimed to spread it to every dark corner of the world. Mass production would create a world economy of wonders. Rising wages and falling prices would increase society's buying power, and abundance would change the world. Scale models of cars, factories, and towns were designed in his headquarters and turned into reality whenever he saw fit. If Fordism had conquered the jungles of American capitalism, it could subdue the Amazon.

David Riker was sixty-seven when, in December 1928, the *Lake Ormoc* anchored outside Santarém, and Ford's managers hired him as guide. By then, he'd built a large family estate overlooking the Tapajós. It was fashioned in the antebellum style of the Lower Mississippi Valley and built of wood in defiance of the white ants. A parrot squawked from its perch on the veranda. But this was nothing compared to the *Lake Ormoc*. The vessel had been refitted to serve as Fordlandia's initial headquarters: a hospital, laboratory, and refrigeration unit were squeezed into its interior. The ship lay offshore like some industrial odalisque, awing the locals like the blue-lit *Amazonas* fifty-two years earlier.

The birth of Fordlandia progressed with assembly-line efficiency. The little village of Boa Vista was a beautiful site: The shore mounted fifty feet above the clear river. The land rose as one progressed inland, the hills forested with towering, lovely trees—*castanheira,* or Brazil nut, Spanish cedar, uxy, and itauba. All were burned clear and bulldozed flat. In their place rose a modern suburb with rows of white, green-shuttered bungalows. The main street was paved and ran uphill. Residents collected well-water from spigots in front of their houses, while the American staff and few Brazilian managers enjoyed running water inside. Screens were installed in the windows to keep out mosquitoes. A modern hospital was staffed with tropical specialists and equipped to produce quinine. Schools for the children of workers and managers were staffed with teachers from Pará. There was a private club and pool in the "American village"; the *caboclo* workers had their own separate pool. The "Villa Brasiliena" boasted tailors, shops, bodegas, and a butcher; the smell of fried bread wafted from the bakery. Until the first tapping, scheduled for 1936, the sawmill would process and export hardwoods felled on the property. U.S. newspapers called it "The Miracle City of the Amazon."

At first it seemed Fordlandia would bring prosperity to the valley as advertised. Over three thousand Brazilians were hired to clear land, plant rubber seedlings, expand the physical plant, and run timber through the mill. Several old *confederado* families leased their land to Ford at profitable rates, while their sons were recruited into a variety of white-collar jobs. Three of Riker's sons were hired soon after their father: Robert and Rubim, his oldest, would go to Dearborn, while Ditmar, the third boy, would stay as a field manager, first for Ford and later for Goodyear. Ford's advance guard had arranged with planters across the river to set out one thousand rubber seedlings from the highlands behind Boim to be used when Fordlandia was cleared. By August 28, 1929, less than a year after the *Ormoc*'s arrival, 1,440 acres had been flattened, much of it planted with young rubber trees sown six inches apart on the naked hills. Riker would later say he planted the first rubber tree for Henry Ford.

But these trees were harbingers of things to come. By September 1929, most of the seedlings were dead or dying. Hevea was a jungle tree, used to shelter, and Ford had exposed the seedlings to the burning sun and pelting rain. His Dearborn managers made two great mistakes, ones that ran counter to all their training in the latest time-and-motion technologies. Although Ford insisted that they industrialize the forest, nature could not be forced into a schedule. And by clearing the forest, he'd depleted the thin layer of soil and nutrients, and destroyed the natural weather machine. Ford had created a desert. His seedlings withered and died.

Fortunately, there were still those thousand seedlings planted across the river—until the managers learned they could not be used. The Tapajós divides the states of Amazonas and Pará, and old rivalries divided them more than the river. In 1927, before Ford concluded the deal with Pará, Amazonas officials had tried to woo him away. When he moved to Boa Vista, they held a grudge. When Ford needed those one thousand seedlings, Amazonas forbade their export; in retaliation, Pará refused to permit their entry. The case went to court, but in vain. Good seedlings were suddenly in rare supply just when Ford needed them most, so his men traveled to Malaya and bought seeds that descended from Wickham's stolen hevea.

So Henry's seeds completed their global journey, a great circle from the Amazon, to Asia, and now back home. Much was made of it, but as Thomas Wolfe would write, you can't go home again. Henry's seeds had been gone too long: they'd lost their resistance to local conditions. Almost as soon as they were planted, Henry's hevea wilted from the climate and local disease.

In late 1929–early 1930, an expedition was mounted to the Acré territory, with Riker along as translator. At the time, American and Brazilian scientists were convinced that the Acré would yield a mother lode of mythic trees, whose latex flow would surpass all others and whose natural resistance would defeat disease and predators. The Acré was considered an unofficial state devoted to rubber. In the 1890s, thousands of Cearense tappers rose up to wrest it from Bolivia and claim the land for Brazil. Afterward, the *seringueiros* returned to their tapping, only to rise again in 1907 when Brazil tried to "regularize" land titles. The Acré was the only place in the Amazon Valley where tappers had not been reduced to slaves, the only place where the romance of rubber had evolved into effective grassroots power. Riker and his group came to the Acré expecting miracles but instead encountered problems. They'd arrived six months before fruiting and had to wait for the season to start. Even then, the seed they collected didn't do well.

When Riker returned, he saw Fordlandia through different eyes. It had reached impressive proportions, having become the Amazon's third-largest town. It had comfortable housing, a school, a sanitary water supply, thirty miles of road, a cinema, machine shop, and a sawmill with a capacity of twenty-five thousand board-feet, the largest in Brazil. Plans were made to produce wooden auto parts for use in the United States. A new manager revealed plans for a tire factory and envisioned Fordlandia as a city of ten thousand within a decade. Ford's workers received free housing, free medical and dental care, recreational facilities, and a wage of thirty-three to sixty-six cents a day, twice that paid anywhere else in the region. The workers bought food and supplies at controlled rates; the debt system of the *patrão* did not exist here. Babies born in Fordlandia got free pasteurized milk. Workers who died on the job received free burial in the Fordlandia cemetery. From birth to death, the company enfolded the worker in Ford's vision of a protective, paternal industry.

Very little of this had anything to do with rubber, however, and it seemed to Riker that Fordlandia was drifting in the same way that Henry Wickham had drifted fifty-four years earlier. Purpose dissipated like that in this country. A man woke up one morning and wondered what had become of his life. He came up the river with hope, but soon the bright plans vanished, replaced by the daily fight for survival.

By 1930, it was obvious that Fordlandia's workers were not about to embrace Fordism in the way Henry Ford desired. Ford's "humanity in industry" was a distant benevolence, unlike the *patrão* system, which the *caboclos* knew. True, the *patrão* held the worker in debt, but the *patrão* also served as protector, advisor, and godfather. Direct obligations existed that were familial and human. In contrast, Ford's approach was characterized by his statement: "A great business is really too big to be human. It supplants the personality of man."

It was growing obvious that Henry Ford would never visit his Miracle City in the Amazon. He planned to run his empire by remote control. Everything was done by American standards, and Ford made the decisions from Dearborn. In exchange for health care, housing, and wages, Ford expected his plantation workers to adjust to company-imposed routines. Ford's temperance rule was a failure: On paydays, boats filled with potent *cachaca,* brewed from sugar cane, pulled up to the docks. Between paydays, workers motored down to Aveiro, and the little town of fire ants experienced a firewater boom. A people accustomed to living in open-sided houses felt suffocated by Ford's four walls; they found the two-family houses hot and ugly, the idea of indoor bathrooms repulsive. The six A.M. to three P.M. workday was unpopular with tappers accustomed to slashing trees several hours before dawn. Planting trees in a man-made desert, just to watch them die, was especially perplexing and discouraging. Said one old *confederado* who watched from the sidelines: "They tried to do to these Brazilians what Northerners had always wanted to do to the South— Yankeefy it!—and it didn't work there either." Within two years, two thousand of Ford's original three thousand workers were fired or had left on their own.

In 1930, there was a riot. Actually, there were two. The first ignited over the rigid work regime and appeared in the least likely place—the new cafeteria during the noon meal. Ford had introduced the cafeteria as a way to reduce time spent on lunch breaks. As the workers filed down the serving line for the first time, one stopped and shouted, "I'm a worker, not a waiter!" The foremen were already angry after learning they'd be eating in the same manner as their subordinates. There'd been complaints. The Midwestern cuisine gave Brazilians indigestion. Others grumbled that they were not dogs and deserved to be served.

With that cry, the *caboclos* Ford officials saw as their "little brown brothers" turned wild. They smashed the cafeteria, tore down everything loose, then armed themselves with rifles, shotguns, and machetes and rampaged through the plantation, breaking tractors, smashing windows, and gutting cars. The North American staff and their families escaped on company boats to the middle of the Tapajós. In the panic, the young daughter of a boss fell into the river and was swept by the currents into a high stanchion by the pier. One of the rioters dived in, pulled her to safety, and took her to her family. As the rioters tore through Fordlandia, the managers heard what sounded like slogans. At first they thought it was "Down with Ford!", but it turned out to be "Down with spinach! No more spinach!" The *caboclos* were sick of spinach and other "well-vitaminized" foods; they could not even look at the stuff, while corn flakes made them gag.

The managers remained in the middle of the river until Brazilian troops arrived and squelched the riot, but by then the violence had run its course. After that, recalled Doña Olinda Pereira Branco, then a twenty-two-year-old laundress for the managers who still lives at Fordlandia, people began to leave. The riot had opened a chasm that seemed hard to bridge. All sensed a void beneath the bright surface. "Everyone was sad," she recalled.

"In one night," observed Brazilian writer Vianna Moog, "the officials of the Ford Motor Company learned more sociology than in years at the university.

They learned that the *caboclos* detested the tiled houses in which they lived and the Puritan way of life the officials wished to impose upon them . . . the houses were veritable ovens, as is easily imaginable when one considers how hot the majority of American houses are in the summer. . . . Mr. Ford understood assembly lines and the designs of Divine Providence. He did not, absolutely could not, understand the psychology of the *caboclo*.

The second riot started later in the year. The *caboclos* complained that black Barbadians who lived across the Tapajós would take their jobs and be paid a higher wage. The violence erupted on payday as everyone stood in line. Three West Indian workers were injured, and afterward, Ford managers ended the employment of all Barbadians at Fordlandia. What happened to the immigrants after their dismissal remains a mystery. Brazilians in Santarém later told writer Mary Dempsey that they never made it home. They drifted to the cities of the Amazon, looking for work or entering the latest rush for riches. Today, their descendants are sprinkled throughout the valley.

Labor trouble ceased in 1930, replaced by assaults from the jungle. By 1932, nearly seven thousand acres of forest had been cleared; nine thousand by 1934. The timber proved useless because some of it absorbed too much moisture, while other trees were hard as stone. Waves of insects and fungi advanced from the tree line to attack the young hevea: five species of fungi were counted and two dozen types of insects. The pests included black crust, yellow scale, white root disease, canker, red mites, hawk moths, sauva ants, lace bugs, and army caterpillars. Olinda Branco remembered how the Ford managers paid residents a bounty per bug to pick army worms off the leaves and stems: "I was really scared of them. They were gray and flat, with red stripes, so they were easy to find. But they had pincers that went right through your skin." The caterpillar problem was temporarily solved when the sauva ant attacked, but after the ants had exterminated the caterpillars, they turned on the young rubber trees.

Most devastating was the South American leaf blight, caused by the fungal parasite *Dothidella ulei*. The spores floated in with the breeze and lodged on the leaves. Isolated trees in the forest stood a better chance, since they could not pass on the blight to close neighbors, and those that survived developed immunity. Since the fungus did not attack seeds, it had not been a hidden passenger in Wickham's smuggled shipment to Kew and thus did not spread to Asia. Riker had seen traces of rust on his small plantation before selling out, and there'd been rumors that a similar disease had wiped out the entire plantation rubber industry of Surinam and Guyana. It struck quickly, spreading through a stand of trees like cholera. In one case, a plantation of two hundred to three hundred acres lost one third of its young trees in a matter of days, while the rest were in such a hopeless state that the plantation was abandoned and the planter ruined.

Dothidella had been at Fordlandia from the beginning, but since the first groves were planted far apart, the conditions mimicked hevea's isolated behavior in the forest. In time, however, the trees were planted close, and after four to six years, as the rows of trees matured, the canopy closed overhead. This made it easy for the fungus to travel from one tree to another. By 1934, the blight had swept through the plantation, denuding almost every tree. "Practically all the branches of the trees throughout the estate," said Fordlandia's chief biologist, "terminate in naked stems." Each time buds appeared, they were attacked and eaten through. Soon, the hills were covered with skeletal trees or young saplings—and all knew the saplings would die.

In a very real sense, Ford had been given prior warning. Others before him had been ruined. Something had swept through Guyana and Surinam, killing a thousand plants and crippling others. The reason for hevea's isolated spread through the forest could now be understood. The trees did not grow in dense stands because the fungus feasted in lush conditions and reproduced at a spectacular rate, overwhelming the tree's defenses. Hevea's scattered distribution was its best protection. Plantation conditions, meant to encourage growth, instead destroyed them all. Ford's assault on the Amazon was foreordained to fail.

To many of that era, such an end seemed impossible, so pervasive was the myth of Ford's invincibility. In 1932, when Ford still dared dream of an Amazon Utopia, a German journalist visited Fordlandia. "Henry Ford has never yet seen one of his big plans fail," he wrote. "If this one succeeds, if the machine, the tractor, can open a breach in the great green wall of the Amazon jungle, if Ford plants millions of rubber plants where there used to be nothing but jungle solitude, then the romantic history of rubber will have a great new chapter. A new and titanic fight between nature and modern man is beginning." The Germans had endured the rubber embargo of the First World War and were captivated by the mechanized struggles of "Heinrich Ford," the American industrialist whom Adolf Hitler most admired.

The industrialized nations had wakened to the strategic danger of resource dependence. Such vulnerability "threatens not only the sane progress of the world," said Herbert Hoover, "but contains in it great dangers to international goodwill." Many believed that if an essential resource were held back, a nation-state had the right to go and take it. The Japanese turned the abstract argument into strategy when they went to war for resources: they took the Eastern rubber plantations from the British early during World War II.

In 1942, President Franklin Delano Roosevelt bluntly warned that "modern wars cannot be won without rubber." During World War II, rubber proved indispensable for the manufacture of motor transport, airplanes, submarines, antiaircraft balloons, gas masks, electric motors, ships, trains, electric lights, telephones, telegraph equipment, typewriters, wireless sets, radios, medical goods, fire hoses, shoes and boots, bulletproof gas tanks, deicers on airplane wings, rubberized flying suits, rubber track blocks for tanks and halftracks, inflatable rubber boats, self-sealing airplane fuel line hoses, airfoam pads for wounded limbs, and surgeon's gloves. Roosevelt asked Americans to undertake a huge rubber recycling program to make up for the lost Eastern plantations. Within three years, the formula for synthetic rubber was perfected, making raw hevea, at least for a time, as obsolete as "Pará fine."

Before this, however, Henry Ford had realized that he would never be king of an American rubber empire. He would not do for his nation what Henry Wickham had done for Great Britain. In 1934, after *Dothidella* laid his dream to waste, he traded a third of his property at Fordlandia for a new site thirty miles closer to the mouth of the Tapajós and called it Belterra. He abandoned all essential operations in Fordlandia, designating it a "research station." Belterra was laid out in squares, and planting went slowly. By 1937, twelve thousand acres were cleared and 2.2 million seedlings sown. By 1941, that number rose to 3.6 million. But in 1942, the harvest yielded a mere 750 tons of rubber, a fraction of the Amazon's 45,000-ton output at the height of the Boom. By 1945, after spending $10 million, Ford sold everything to Brazil for $500,000.

Nature had won.

The fear of biopiracy that haunts Brazil today is one legacy of Henry Wickham's dream. The Amazon rain forest is thought to contain forty percent of all plant and animal species on Earth, many still undiscovered, while loggers and farmers slash the forest at a rate of six football fields a minute. Yet the search for new species—and development of new drugs from them—is hampered by a deep suspicion among Brazilians that the rain forest crawls with biopirates scooping up animal blood samples, leaves, and seeds. Thomas Lovejoy, the U.S. scientist credited with bringing the Amazon's deforestation to the world's attention in the 1980s, was accused of being a CIA agent while doing rain-forest research for the Smithsonian Institute. Marc Van Roosmiler, a Dutch researcher who'd discovered twenty new monkey species, was accused of trying to smuggle primates when twenty-seven rare monkeys were found in his home in Manaus. Although Van Roosmiler insisted he was studying the animals, the case showed how former friends of the forest were now branded their enemy. General Luiz Gonzaga Schroeder Lessa, former chief of the Amazon Military Council, claimed that collectors disguised as missionaries or scientists were stealing biological samples. Amazonas State Governor Eduardo Braga warned that foreigners planned to "take from us our flora and fauna." Even Brazilian scientists feel the squeeze. "I've all but given up," said Paulo Buckup, a professor of ichthyology at the Federal University of Rio de Janeiro who collects river fish for research. "Brazil has lost the capacity to control its own resources because it doesn't know what it has."

Fordlandia still stands along the river, a barely inhabited ghost town. In October 2005, I journeyed up the Tapajós and landed on the spit that Old Man Franco sold for all the money in the world. Zebu cattle imported from India for their mosquito resistance wander the dusty streets. The American Villa still stands on the hill. The green and white cottages line the shady lane, but the only residents now are fruit bats and trap-door tarantulas. The state-of-the-art hospital shipped from Michigan is deserted. Broken bottles and patient records litter the floor. A towering machine shop houses a 1940s-era ambulance, now on blocks. A riverside warehouse built to hold huge sheets of processed rubber holds six empty coffins arranged in a circle around the ashes of a small campfire.

A few aging veterans from the Ford era still remain in the dusty town. The dirt streets climb the red hillside. The fastest way up is also the bumpiest, crammed into the town's one taxi, an aging, rust-colored

Toyota Corolla. Although its passenger door is held shut by wire and the windows are stuck, "that doesn't matter," said the owner, "because it still gets around." At the top of the hill lives Biamor de Sousa Pessoa, whose father was a company rubber tapper. Now seventy-five, he can still remember the sounds of the clarinet played between ten and ten-thirty A.M. to call the rubber workers from the fields. It was a haunting sound, he said, like something ancient that drifted from the forest and settled like mist over the young trees. Doña Olinda Pereira Branco lives close to him and is even older. In 2005 she was ninety-two and could still recall being paid for each ugly, red-striped caterpillar she plucked from the trees. The thought still makes her shudder. There were no Brazilians promoted into management, she said, and plenty of resentment grew up around that fact. A few Americans died of fever, as did many Brazilians, but no American bosses were buried at Fordlandia. "There were bosses for everything," she remembered. She paused, and gazed out the window. "Too many bosses," she said.

And what of the thousands of rubber trees that once crept up the hills? She smiled and shook her head. Until recently, a few of the old giants still stood. "But they were getting rotten and started to fall. So we cut them down."

Two decades would pass before new developers dared take on the jungle again. When they did, they made the same mistakes, but on a more massive scale. In 1967, American billionaire and shipping magnate Daniel K. Ludwig bought 1.6 million acres of Amazon land along the Jari and Paru rivers for $3 million, then expanded that to 3.7 million acres, an area equivalent to half of Belgium. Expecting the world to run out of wood fiber, he cleared 250,000 acres of rain forest and planted *Gmelina arborea*, a fast-growing Burmese species—his own "miracle tree." He brought in by barge a seventeen-story preassembled pulp mill made in Japan. He also envisioned mining and cattle empires, the world's largest rice plantation, and modern suburbs in the jungle. He built a settlement, Monte Dourado, with houses, schools, nurseries, bridges, and community centers. Given enough money, he thought, any place could be civilized.

Instead he created a wasteland. His heavy land-clearing equipment scraped and compacted the soil, repeating the same mistakes Ford had made. The transplanted *Gmelia* grew slower than in its native environment, and the pulp had to be supplemented by local wood—which he had judged useless before. He replaced his plantations of *Gmelia* with equally disappointing eucalyptus and pine. Although the mill processed pulp, it was never enough to be profitable, and Ludwig's dream began to fade. Workers contracted malaria; insects destroyed the wood and supplies. After fourteen years, Daniel Ludwig gave up, losing an estimated two thirds of a billion dollars to the same mistakes Ford had made.

Today it is soy that drives such dreams. In 2003, Cargill Inc., the Minneapolis commodities giant, opened a new $20-million soybean terminal on the river at Santarém. The draft at the Amazon port is forty to fifty-five feet, deep enough to load oceangoing vessels directly and avoid the transfer of freight once the ship reaches sea. Cargill bet that Brazil would pave the impassable 1,071-mile mud and rock path optimistically known as BR–163. This would open the Amazon port to the vast soybean fields in the south that cover the state of Mato Grosso. During the 2003–2004 harvest, Brazil exported 20.5 million tons of soybeans, ranking it second in the world to the United States, which exported 24.5 million tons. Brazil hoped to rank first: While U.S. producers have run out of land, Brazil has millions of acres of Amazon jungle that could be cleared.

In March 2007, the federal government closed the Cargill plant for failure to comply with environmental laws, but by then the damage was done. Tens of thousands of acres around Santarém had been cleared of forest for rice and soybeans, in a region that had already lost 20 percent of its trees to loggers. It didn't rain once in Santarém during my trip, and only once in the south around Fordlandia. A day without rain in the Amazon is extremely rare, even during the "dry" season. Older residents called it the greatest drought they remembered in fifty years. A few hundred miles west in Amazonas the water level sank to unprecedented levels, closing ports and crimping river traffic, resulting in the declaration of a state of emergency. The jungle's weather machine seemed broken, and even the most hardened Amazonian seemed scared. The newest dream of riches may have gone the way of Fordlandia, but this time something essential to the health of the region seemed to go with it, also.

There is an old saying on the Amazon, often uttered with a cynical tilt of the head. *Deus é grande, mas o mato é maior:* "God is great, but the forest is greater." Brazilians also say, "God sees the truth, but sometimes forgets," but the jungle never forgets. One way or another, nature always wins.

Henry Wickham could have warned them all of this—the soybean prophets, the Daniel Ludwigs and the Henry Fords. A man is never greater than the jungle, no matter what his worth back home. But Henry understood their mania, too. There was something in the jungle that drove a man beyond all wisdom: He chased his El Dorado until the fever killed him, until he was broken and humiliated, until his obsessions destroyed those who dared believe in his dreams. It was an enslavement unlike anything on earth, because he forged the shackles himself. Joseph Conrad felt it, in the site of another rubber madness half a world away. "The brown current ran swiftly out of the heart of darkness bearing us towards the sea," he wrote in his most famous work. The only hope was to leave the jungle for the ocean's far horizon.

But once seduced, one was never free.

APPENDIX I

World Rubber Production

1905–1922 (percent)

Year	Plantation	Wild Rubber
1905	0.3	99.7
1906	0.9	99.1
1907	1.7	98.3
1908	2.7	97.3
1909	4.6	95.4
1910	9.0	91.0
1911	17.4	82.6
1912	29.0	71.0
1913	45.0	55.0
1914	60.4	39.6
1915	67.6	32.4
1917	79.6	20.4
1918	83.1	16.9
1919	87.4	12.6
1920	89.3	10.7
1921	92.0	8.0
1922	93.1	6.9

Source: James Cooper Lawrence, *The World's Struggle with Rubber* (New York & London, Harper & Bros., 1931), p. 31.

APPENDIX II

The World's Rubber Requirements, 1922
(a conservative estimate, in tons)

U.S.A.	290,000
U.K.	15,000
Canada	10,000
France	25,000
Germany	25,000
Japan	15,000
Australia	4,000
Russia	3,000
Italy	5,000
Spain	3,000
Others	5,000
Total	**400,000**

Source: Harvey S. Firestone, *America Should Produce Its Own Rubber* (Akron, OH: Harvey S. Firestone, 1923), p. 13.

APPENDIX III

New York Price Quotations for Crude Rubber
(price per pound)

Year	Market price	
	low	high
1890	.73	.74
1900	.92	.98
1905	1.18	1.33
1909	1.26	2.22
1910	1.34	3.06
1911	1.08	1.74
1912	.98	1.38
1913	.48	1.09
1914	.47	.72
1915	.475	.99
1916	.995	1.03
1917	.535	.81
1918	.49	.62
1919	.38	.58
1920	.16	.55
1921	.12	.21
1922	.14	.295

Sources: Bradford L. Barham and Oliver T. Coomes, *Prosperity's Promise: The Amazon Rubber Boom and Distorted Economic Development,* Dellplain Latin American Studies, no. 34, David J. Robinson, ed. (Boulder, CO: Westview Press, 1996), p. 31; Harvey S. Firestone, *America Should Produce Its Own Rubber* (Akron, OH: Harvey S. Firestone, 1923), p. 17.

GLOSSARY

aguardiente. A local rum brewed in the Americas from sugar cane.

aldeira. Village.

aviador. The wholesale rubber dealer, or "forwarder," who sat at the top of the chain in the Amazon rubber business. He bought rubber from the *patrão* and sold it to the exporter.

balatá. A type of rubber made from the latex of the *Manilkara bidentata* tree native to northern South America, Central America, and the Caribbean. The rubber produced from this tree is almost identical to gutta-percha.

borracha. Rubber (latex of the hevea tree).

caboclo. The peasant and subsistence farmer of the Amazon Valley.

cacau. Cocoa.

campos. Prairie-like pasture leading up to or on the high plateaus and mountaintops.

candiru. The "toothpick fish," a tiny catfish that swims up the stream of uric acid released by bigger fish (or animals), then lodges in the gills or cloaca (or urethra) with the help of some sturdy, sharp spines.

caoutchouc. Early name for rubber, still current in France.

caucho. Rubber from the *Castilla elastica,* or Panama rubber tree, a genus of the Moraceae (or mulberry family). *Caucho* was the rubber reported in his travels by la Condamine.

caudillo. A type of landed strongman or populist dictator, a social force in revolutionary South America of the early nineteenth century.

cidade. City. Nearly every Amazon town was divided into *cidade* and *aldeira,* the city and the village.

cinchona. A genus (*Cinchona*) of about twenty-five species of the Rubiaceae family, native to tropical South America. They are large shrubs or small trees, and the bark is the source of a variety of alkaloids, the most famous and sought after of which was the antimalarial quinine.

confederado. Former Confederates of the defeated American South, who came to Latin America after the Civil War in hopes of starting a new slave society.

Cruzob. During the Caste War of the Yucatán peninsula, followers of the "Talking Cross."

curiara. Small fishing canoe.

curupira. The "pale man of the forest," a lethal spirit in Indian myth that kills forest wanderers, inhabits their bodies, then kills their families.

Dothidella ulei. Fungal parasite known to destroy vast stands of maturing *Hevea brasiliensis* in Central and South America. It does not naturally occur in the rubber plantations of Southeast Asia. *Dothidella* destroyed Henry Ford's dream of becoming a rubber baron in the Amazon Valley.

engenho. Sugarcane plantation.

estrada. The jungle path cut between trees in a rain-forest rubber "estate."

faca. The small, curved knife used by rubber tappers in Central and South America to make a thin cut through the bark of the rubber tree.

farinha. Manioc flour, toasted and often popped in the mouth like popcorn. Also *farinha de mandioca*.

feiticeira. "Witches" living at the edge of the forest in solitary hovels. Most of them specialized in jungle cures and love potions, though some had knowledge of native poisons.

Ficus elastica. A species of plant in the banyan group of figs, native from northeast India south to Indonesia. Grown today around the world as a houseplant, its milky latex sap could be used to produce an inferior grade of rubber.

formigas de fogo. Fire ants.

goma. On the Orinoco River, the local term for rubber.

gaurapo. The heated juice of sugarcane.

guayule. A shrub of the *Parthenium* genus, native to the southwestern United States and northern Mexico. It can be used as an alternate source of latex that is hypoallergenic, unlike the normal hevea rubber.

gutta-percha. The *Palaquium* genus of tropical trees native to southeast

Asia and northern Australasia, which produces a latex that is a good electrical insulator. By 1845, telegraph wires and transatlantic telegraph cables were being coated with gutta-percha, since this rubber was not attacked by marine plants or animals.

Hevea brasiliensis. Also known as the Pará rubber tree and referred to simply as hevea throughout this book, a tree of the Euphorbiaceae family that initially grew only in the Amazon rain forest and is still the world's primary source of natural rubber. Today most rubber-tree plantations are grown with hevea in Southeast Asia, though some are also found in tropical Africa. In 1876, Henry Wickham smuggled seventy thousand hevea seeds from the Amazon Valley to the Royal Botanic Gardens at Kew, a disputed act of biopiracy that began the British rubber plantation industry.

jararaca. The fer-de-lance, the most common, and deadliest, venomous snake in Amazonia.

lancha. Portuguese term for a larger trading vessel used in the rain-forest rivers, often covered with a shelter, or *toldo,* a half-open deck.

Landolphia. A genus of rubber vine, also known as white rubber vine and *mbungu* vine, particularly indigenous to tropical Africa.

Lingoa Geral. European missionary script form of Tupi Guarani language.

llanos. The vast tropical grasslands situated east of the Andes in the Orinoco Basin.

matapalo. The parasitic "tree-killer," or strangler fig, native to tropical America.

mestizo, mestiza. A term of Spanish origin used to describe people of mixed European and Indian ancestry.

mishla. A native beer in Nicaragua made from fermented cassava and other fruits and vegetables.

Paranáquuausú. "Great River," Indian name for the Amazon River.

Paraponera clavata. An inch-long, glistening black ant native to Central and South America that sports a massive hypodermic syringe at the end of its abdomen and large venom reservoirs.

patrão. The rubber boss, or patron, who often owned the rubber stands in the jungle, hired the *seringueiro,* and supplied him with food and essential supplies during the tapping season.

perros de agua. "Water dogs," the large river otters of South America.

pitpan. A long, flat-bottomed canoe used in the rivers and lagoons of Central America.

praça. Plaza or square.

rancho. A rude thatch hut or open-air shed.

rede. Hammock used for sleeping in the Amazon Valley.

saudade. A Portuguese term considered untranslatable: a sadness of character best thought of as the longing that infiltrates many literary descriptions of life in the Amazon Valley. *Weltschmerz.*

seringa. Early name for "syringe" rubber.

seringal, ciringal. A stand of rubber trees.

seringueiro. A rubber tapper, rubber collector.

sucuruju. Anaconda.

temblador. *Electrophorus electricus,* the South American electric eel.

terra firme. Forest on land above the flood plains, terra firma.

terra preta do Indio. "Indian black earth," a rich compost built up over the generations by ancient Indian farmers, still found around ancient sites in the Amazon Valley.

timbó. Vine from which insecticides are produced.

toldo. Shelter, awning.

tonina. The fresh-water dolphin of the Amazon basin.

ubá. A fire-hollowed dugout canoe.

vulcanization. A chemical curing process for rubber discovered accidentally in 1839 by American chemist Charles Goodyear. Individual polymer molecules are linked to other polymer molecules by atomic bridges, resulting in a springy rubber that is hard, durable, and more resistant to wear.

Xtabay. In Mayan mythology, the sinister and seductive temptress of the rain forest who lured men to their doom.

zamora. The ubiquitous "turkey buzzard" found throughout tropical and semitropical America.

NOTES

Prologue: Henry's Dream

1. **Deep in the forest grew a ruinous tree** Several Indian myths of the Tree of Life in the South American rain forest are summarized and compared in Claude Lévi-Strauss, *The Raw and the Cooked: Introduction to a Science of Mythology*, vol. 1, trans. John and Doreen Weightman (New York: Harper Colophon, 1975, first published in France as *Le Cru et le Cuit*, 1964, Librairie Plon), pp. 184–186.

2. **where another wrecked and the captain blew out his brains** Henry C. Pearson, *The Rubber Country of the Amazon; a detailed description of the great rubber industry of the Amazon Valley, etc.* (New York: India Rubber World, 1911), pp. 18–19.

2. **increased to 216,000 cubic yards** Alex Shoumatoff, *The Rivers Amazon* (San Francisco: Sierra Club Books, 1978), p. 89.

2. **"So great was the stench of their decomposing carcasses"** Richard Spruce, *Notes of a Botanist on the Amazon and the Andes*, Alfred Russel Wallace, ed., vol. 1, (Cleveland, OH: Arthur H. Clark, 1908), p. 114.

4. **In 1914 alone, Detroit consumed 1.8 million tires** "Automobiles and Rubber: How the Automobile, and Especially Ford Cars, Has Revolutionized the Rubber Industry," *Ford Times*, July 1914, vol. 7, no. 10, p. 473.

4. **"agro-industrial utopia . . . one foot on the land"** John Galey, "Industrialist in the Wilderness: Henry Ford's Amazon Venture," *Journal of Interamerican Studies and World Affairs* vol. 21, no. 2 (May 1979), p. 273.

6. **"A million Chinese in the rubber section of Brazil"** Carl LaRue is quoted in Susanna Hecht and Alexander Cockburn, *The Fate of the Forest: Developers, Destroyers and Defenders of the Amazon* (New York: HarperPerennial, 1990), p. 97.

6. **to buy Villares's concession for $125,000** Hecht and Cockburn, *The Fate of the Forest*, p. 98.

6. **Old Man Franco** Interview with Cristovao Sena, regional rubber historian, Santarém, Pará, Brazil, October 18, 2005.

6. **"While there may be a difference of opinion"** Roger D. Stone, *Dreams of Amazonia* (New York: Viking, 1985), p. 86.

7. **rumors of a kickback began to taint the venture** Hecht and Cockburn, *The Fate of the Forest*, p. 98. No charges were ever filed against LaRue, but the Michigan professor would have been the natural choice to head the plantation. Yet after the purchase, Ford reportedly never had anything more to do with LaRue, despite overtures by the professor.

8. **"take root almost without fail"** Hecht and Cockburn, *The Fate of the Forest*, p. 99.

10. **A 1976 article in the *Times*** Michael Frenchman, "Unique Link with Amazon," *Times* (London), no. 59694 (May 3, 1976), p. 2, col. 6.

10. **A British cruise-boat tourist recently claimed** Interview with Gil Sérique, guide, Santarém, Pará, Brazil. December 15, 2005.

Chapter 1: Fortunate Son

19. **"I will tell you what I believe"** Joseph Conrad, "The Planter of Malata," first published in *Empire Magazine* (January 1914), then as a novella in the collection *Within the Tides* (London, 1915).

20. **"a crowded island where towns and cities rub up against one another"** John Cassidy, "The Red Devil," *New Yorker*, Feb. 6, 2006, p. 48.

20. **"province covered with houses," "a state"** Lynda Nead, *Victorian Babylon: People, Streets and Images in Nineteenth-Century London* (New Haven: Yale University Press, 2000), p. 15.

20. **"No one will ever understand Victorian England"** John Gardiner, *The Victorians: An Age in Retrospect* (London: Hambledon and London, 2002), p. 4.

21. **"For them . . . the empire was hazily exotic"** Gardiner, *The Victorians*, p. 6.

21. **"being need for any inflation"** Anthony Smith, *Explorers of the Amazon* (New York: Viking, 1990), p. 268.

21–22. **"from *batey* to *Vok-a-tok*"** Peter Mason, *Cauchu, the Weeping Wood: A History of Rubber* (Sydney, Australia: The Australian Broadcasting Commission, 1979), p. 15.

22. **The English discoverer of oxygen, Joseph Priestley** Priestley gave rubber its name in the Preface of his *Familiar Introduction to the Theory and Practice of Perspective* (1770), which is quoted by Howard Wolf and Ralph Wolf in *Rubber: A Story of Glory and Greed* (New York: Covici, Friede, Publishers, 1936), pp. 288–89.

22. **"all kinds of leather, cotton, linen and woolen cloths, silk stuffs, paper, wood"** Peal's patent application is quoted in Wolf and Wolf, *Rubber: A Story of Glory and Greed*, p. 289.

22. **"perfectly waterproof"** Ibid., p. 269.

23. **In 1827, the first rubber fire hose was used** Ibid., p. 295.

23. **"vile taste"** Ibid.

24. **"instrument in the hands of his Maker"** Ibid., p. 300.

24. **"While yet a schoolboy"** Ibid., quoting from Charles Goodyear's *Gum-Elastic*.

24. **"The most remarkable quality of this gum, is its wonderful elasticity"** Ibid., quoting from Goodyear's *Gum-Elastic*, p. 299.

25. **"In time, the process would be dubbed"** The term *vulcanization* was actually dreamed up by one of industrialist Thomas Hancock's friends, a "Mr. Brockedon," when trying to come up with a better term than just "the change." Thomas Hancock, *Personal narrative of the origin and progress of the caoutchouc or india-rubber manufacture in England* (London: Longman, 1857), p. 107. Hancock repeats the story again on p. 144.

25. **"one of those cases where the leading of the Creator"** Ibid., p. 311.

26. **His mother, Harriette Johnson** This and subsequent information about Wickham's family comes primarily from two sources: Edward V. Lane, "The Life and Work of Sir Henry Wickham: Part 1—Ancestry and Early Years," *India Rubber Journal* 125

(Dec. 5, 1953), pp. 962–65; and Anthony Campbell, "Descendants of Benjamin Wickham, a Genealogy" (self-published, Jan. 30, 2005).

27. **fought and governed in the American Revolution** Campbell, "Descendants of Benjamin Wickham," quoting the Barlow Genealogy Papers, "unpublished ms. in possession of A.S.C. (Sallie) Campbell (born 1931.)"

27. **"richest and most populous metropolitan parish"** From "St. Marylebone: Description and History from *1868 Gazetteer*," in *Genuki: St. Marylebone History,* http://homepages.gold.ac.uk/genuki/MDX/StMarylebone/StMaryleboneHistory.html.

27. **Seven years later, in 1845, he married Harriette** They may have married in Muthill, near Crieff in Perthshire, where Henry Wickham's brother-in-law, Alexander Lendrum (1811–1890) was a minister of the Scottish Episcopal Church. Much family lore seems to have passed down through the Lendrums, especially since Henry and Violet Wickham would not have children. It was presumably through the Lendrums that Edward Valentine Lane would collect much of the undocumented stories of Wickham's adventures. Since Wickham's mother and father married before statutory registration was enacted in Scotland, there seems to be no record of their wedding. Campbell, Anthony. "Descendants of Benjamin Wickham, a Genealogy" (self-published, Jan. 30, 2005), p. 3.

27. **By then, the Wickhams were firmly ensconced in the country** Information on Hampstead Heath and Haverstock Hill came from the following sources: "Genuki: Hampstead History, Description and History from *1868 Gazetteer*," http://homepages .gold.ac.uk/genuki/MDX/Hampstead/HampsteadHistory-html; "Finchley Road and Haverstock Hill," www.gardenvisit.com/travel/london/finchleyroadhaverstockhill .html; and "Genuki: Middlesex, Hampstead," http://homepages.gold.ac.uk/genuki/MDX/ Hampstead/index.html. Development at Haverstock Hill and Hampstead Heath appears to have progressed in a surprisingly modern manner; a landowner would divide his estate into parcels, and then speculators would build houses, or at least a couple of models, and promote the advantages of living in the healthy suburbs. Details are found on the following Web sites: "Hampstead: Social and Cultural Activities/British History Online," www.british-history.ac.uk report.asp?compid-22645; "Hampstead: Chalcots/British History Online," *Ibid.;* "Hampstead—MDX ENG," http://privatewww.essex.ac.uk/~alan/ family/G-Hampstead.html; and "Genuki: Hampstead History," http://homepages.gold .ac.uk/genuki/MDX/Hampstead/HampsteadHistory.html.

27. **Several rich courtesans built retirement homes** Dan Cruikshank, "The Wages of Sin," in BBC Online-History, www.bbc.co.uk/history/programmes/zone/georgiansex.shtml.

27. **"open and airy slope of Hampstead"** *Lancet,* Nov. 5, 1881, quoted in "Sanitation, Not Vaccination the True Protection Against Small-Pox," www.whale.to/vaccine/tcbbl.html.

28. **"occupied by the very lowest class of society"** Pamela K. Gilbert, *Mapping the Victorian Social Body* (Albany, NY: State University of New York Press, 2004), p. 92.

28. **"wretched houses with broken windows [and] starvation in the alleys"** Dickens is quoted in Gilbert, *Mapping the Victorian Social Body,* p. 92.

28. **Hampstead was an urban center in itself** "Genuki: Hampstead History," http:// homepages.gold.ac.uk/genuki/MDX/Hampstead/HampsteadHistory.html.

29. **due to give birth in the summer** John Joseph Edward Wickham, the third of the

children of Henry and Harriette Wickham, was born in Croydon, Surrey, in 1850, according to the Census of 1871. Quoted in Campbell, "Descendants of Benjamin Wickham," pp. 6 and 14, ff42.

29. 25 Fitzroy Road in Marylebone The 1871 census is quoted in Campbell, "Descendants of Benjamin Wickham," pp. 5–6.

29. "set up a not very successful millinery business in Sackville Street" Edward V. Lane, "The Life and Work of Sir Henry Wickham: Part I,—Ancestry and Early Years" p. 15. Lane is the only historian to write at length about Wickham, and this in a nine-part series spanning his life, in 1953–54, in a trade journal called the *India Rubber Journal*. Unfortunately, according to the fashion of the time, he did not cite sources. It is obvious that much of his material came from Wickham's two works and also from Violet Wickham's memoir. Much that is quoted, however, was probably gotten from surviving extended family. Lane was writing thirty years after Wickham's death, so although he neglects to say it, there still would have been people around who knew him personally.

30. the case of a destitute, seventy-two-year-old milliner Jack London, *The People of the Abyss* (London: Pluto Press, 2001, first published in Great Britain in 1903), p. 134. The economics of millinery are discussed in Helena Wojtczak's *Women of Victorian Sussex—Their Status, Occupations, and Dealings with the Law, 1830–1870*, reviewed on the page "Female Occupations C19th Victorian Social History" at www.fashion-era.com/victorian_occupations_wojtczak.htm.

30. Henry Alexander, deprived of a father's guidance Lane, "The Life and Work of Sir Henry Wickham: I—Ancestry and Early Years," p. 16.

30. "The Rape of the Glances" Lynda Nead, *Victorian Babylon: People, Streets and Images in Nineteenth-Century London* (New Haven: Yale University Press, 2000), pp. 62–65.

31. The voices of critics rose with it Lytton Strachey, *Queen Victoria* (London: Penguin, 1971, first published 1921), p. 119.

31. "the working bees of the world's hive" Asa Briggs, *Victorian People* (London: B. T. Batsford, 1988), p. 41.

31. "industry and skill, countries would find a new brotherhood" Briggs, *Victorian People*, p. 41.

31. "a vegetable wonder" Susan Orlean, *The Orchid Thief* (New York: Ballantine, 1998), p. 72.

31. entire fortunes were spent on forty rare tulips Charles Mackay, *Extraordinary Popular Delusions and the Madness of Crowds* (New York: Harmony, 1980, first published in 1841 and 1852), pp. 90–91.

32. "the *greatest* day in our history" Strachey, *Queen Victoria*, p. 121, quoting *The Letters of Queen Victoria*, vol. 2, pp. 317–318.

33. "strange and neglected races" Journalist Watts Phillips' *The Wild Tribes of London* is quoted in John Marriott, *The Other Empire: Metropolis, India and Progress in the Colonial Imagination* (Manchester and New York: Manchester University Press, 2003), p. 122.

33. "a clash of contest, man against man" Charles Booth, *Charles Booth's London*, Albert Fried and Richard Ellman, eds. (London: Hutchinson, 1968), p. 37.

33. "there is a bitter struggle to live" Charles Booth, *Charles Booth's London*, p. 207.

Chapter 2: Nature Belongs to Man

34. the explorers of this period were seen as self-effacing, duty-driven civil servants Robert A. Stafford, "Scientific Exploration and Empire," in *The Oxford History of the British Empire: The Nineteenth Century*, Andrew Porter, ed., vol. 3 (Oxford: Oxford University Press, 1999), p. 307.

36. "chills the heart, and imparts a feeling of loneliness" MacGregor Laird and R.A.K. Oldfield, *Narrative of an Expedition into the Interior of Africa by the River Niger*, 2nd ed., 2 vols. (London, 1837), vol. 1, p. 181.

36. died at a rate of three hundred to seven hundred per thousand Philip D. Curtin, "The White Man's Grave: Image and Reality, 1780–1850," *Journal of British Studies*, vol. 1 (Nov. 1961), p. 95.

37. during Victoria's reign it was the most powerful antimalarial medicine known to man Quinine is coming back into favor with the recent increase of new strains of malaria resistant to the more widely produced synthetic drugs.

37. Cinchona belongs to the Rubiaceae Cinchona gets its name from the legend of the Countess of Chinchón, which was later shown to be suspect, since her actual stay in Peru did not correspond with the period described in the legend. However, romance often wins out over fact, especially when an angel of mercy is involved. The spelling of her name, Chinchón, and the tree's (cinchona) have been confused ever since, especially by the English, as we shall see. The legend is recounted in Anthony Smith, *Explorers of the Amazon* (New York: Viking, 1990), pp. 260–61; and Charles M. Poser and George W. Bruyn, *An Illustrated History of Malaria* (New York: Parthenon, 1999), p. 78.

37. the mountainous forests of Colombia, Peru, Ecuador, and Bolivia A few members of the species live in the mountainous regions of Panama and Costa Rica, but the alkaloid content of their bark is not as high as in the Andes.

38. "with every necessary for men's wants" T. W. Archer's *Economic Botany*, a highly influential book for its time and one that seemed to echo Kew's vision of its mission, is quoted in Richard Drayton, *Nature's Government: Science, Imperial Britain, and the "Improvement" of the World* (New Haven, CT: Yale University Press, 2000), p. 198.

39. "waltz through life in a dream" Botanist Gustav Mann is quoted in Drayton, *Nature's Government*, p. 233.

39. "the very existence of [tropical] colonies as civilized communities" Ibid., pp. 233–234.

39. "to prevent the utilization of the immense natural resources" Benjamin Kidd is quoted in Drayton, *Nature's Government*, p. 233.

40. "the Mother Country in everything that is useful" William J. Bean, *The Royal Botanic Gardens, Kew: Historical and Descriptive* (London: Cassell, 1908), p. xvii.

40. "the founding of new colonies" Ibid.

40. By 1854, Hooker could boast Drayton, *Nature's Government*, p. 195.

41. "essential to a great commercial country" Ibid.

41. In 1830, Britain imported 211 kilograms Warren Dean, *Brazil and the Struggle for Rubber: A Study in Environmental History* (Cambridge: Cambridge University Press, 1987), p. 220.

41. Brazil was becoming the world center Ibid., p. 169.

41. "in Jamaica and the East Indies" Thomas Hancock is quoted in "On Rubber," *Gardener's Chronicle*, vol. 19 (1855), p. 381.

41. "render any assistance in his power to parties disposed to make the attempt" Ibid.

41. shrink less during transport Water in cured latex tends to dry out during shipment, often at substantial rates. Such shrinkage occurred as much as 15–20 percent in "Pará fine," but up to 42 percent in other species or coarser grades.

42. "gutta-percha" Although *India rubber* and *gutta-percha* were often used as interchangeable terms, thus confusing everyone, the *Scientific American* of January 22, 1860, called gutta-percha "similar, though inferior" to the Amazon breed. The two shared many physical characteristics when vulcanized, and both gums had a chemical composition of seven-eighths carbon to one-eighth hydrogen, but gutta-percha also contained oxygen, while "Pará fine" did not. Given enough moisture, warmth, and time, especially during shipping, unvulcanized gutta-percha would deteriorate rapidly, growing discolored, then brittle, and finally turn to powder.

42. The Dutch were already mounting a campaign to secure and control cinchona In 1853–54, the Dutch in Java moved first: Justus Charles Hasskarl, superintendent of the Buitzenzorg Garden, traveled to South America in disguise to collect seeds.

43. "irritatingly destined for high office" Smith, *Explorers of the Amazon*, p. 264.

43. "My qualifications for the task" Smith, *Explorers of the Amazon*, p. 262; Donovan Williams, "Clements Robert Markham and the Introduction of the Cinchona Tree Into British India, 1861," *Geographical Journal*, vol. 128 (Dec. 1962), p. 433.

43. he showed a keen understanding Donovan Williams, "Clements Robert Markham and the Introduction of the Cinchona Tree Into British India, 1861," p. 433.

43. "double forcing house" Drayton, *Nature's Government*, p. 208.

44. "narrow-minded jealousy" Donovan Williams, "Clements Robert Markham and the Introduction of the Cinchona Tree into British India, 1861," p. 434.

45. "all the clerks in public offices are changed in every revolution" Williams, "Clements Robert Markham and the Introduction of the Cinchona Tree into British India, 1861," p. 435.

46. "Its consistency is that of good cream" Richard Spruce, *Notes of a Botanist on the Amazon and the Andes*, Alfred Russel Wallace, ed., 2 vol., (Cleveland, OH: Arthur H. Clark, 1908), vol. 1, pp. 50–51.

47. "It is a vulgar error that in the tropics the luxuriance of the vegetation" Alfred Russel Wallace's *Travels on the Amazon and Rio Negro* (London, 1853), quoted in Roy Nash, *The Conquest of Brazil* (New York: AMS Press, 1969, originally published 1926), pp. 385–386.

47–48. "How often have I regretted that England" Richard Spruce is quoted in Peter Mason, *Cauchu, the Weeping Wood: A History of Rubber* (Sydney, Australia: Australian Broadcasting Commission, 1979), p. 31.

48. "Die, you English dog" Richard Spruce, *Notes of a Botanist on the Amazon and the Andes*, p. 465.

49. "But during the night . . . they all got gloriously drunk and burst their balls" Spruce, *Notes of a Botanist on the Amazon and the Andes*, p. 196.

50. **"Her Majesty's Secretary of State for India has entrusted the Hon. Richard Spruce"** Anthony Smith, *Explorers of the Amazon*, p. 256.

50. **"very able and painstaking"** Clements Markham's description of Robert Cross is quoted in Smith, *Explorers of the Amazon*, p. 256.

50. **"Matters are in a very unsettled state here"** Spruce is quoted in Victor Wolfgang Von Hagen, *South America Called Them: Explorations of the Great Naturalists* (New York: Knopf, 1945), p. 289.

50. **"I find reason to thank heaven"** Smith, *Explorers of the Amazon*, p. 258.

51. **"that withers every thing it meets"** Edward J. Goodman, *The Explorers of South America* (New York: Macmillan, 1972), p. 291.

51. **"See how that man is laughing at us?"** The tale told to Spruce by the pilgrim is recounted in Von Hagen, *South America Called Them*, p. 288.

Chapter 3: The New World

53. **"I have been wandering by myself"** Charles Darwin is quoted in Anthony Smith, *Explorers of the Amazon* (New York: Viking, 1990), p. 252.

53. **"the good and soft smell of flowers and trees"** Christopher Columbus is quoted in Jose Pedro de Oliveira Costa, "History of the Brazilian Forest: An Inside View," *The Environmentalist* vol 3, no. 5 (1983), p. 50.

53. **"lovely but sinister temptress called Xtabay"** Ronald Wright, *A Short History of Progress* (New York: Carroll & Graf, 2004), p. 81.

54. **"his many drawings with pen and ink"** Lane, "The Life and Work of Sir Henry Wickham: I—Ancestry and Early Years," p. 15.

54. **"traveling artist"** The 1871 census is quoted in Campbell, "Descendants of Benjamin Wickham," pp. 5–6.

54. **"in which every class of society accepts with cheerfulness that lot"** Lord Palmerston is quoted in Briggs, *Victorian People*, p. 98.

54. **"Of all the sources of income, the life of a farmer is the best"** Cicero's *De officiis* is quoted in Stuart B. Schwartz, *Sugar Plantations in the Formation of Brazilian Society* (Cambridge: Cambridge University Press, 1985), p. 264.

55. **"typical of his generation, when the pioneering spirit, fired by a desire"** Lane, "The Life and Work of Sir Henry Wickham: I—Ancestry and Early Years," p. 16.

55. **"if anyone stated that he was six-feet tall"** Ibid., p. 16.

55. **"unbounded energy"** and **"easy-going indolence"** Ibid.

55. **"that neither Mouth, Nose, Eyes"** Nicholas Rogers, "Caribbean Borderland: Empire, Ethnicity, and the Exotic on the Mosquito Coast," *Eighteenth-Century Life*, vol. 26, no. 3 (Fall 2002), pp. 117–138.

56. **British presence in Nicaragua** *A Document of the Mosquito Nation: document signed Feb. 19, 1840, between Robert Charles Frederic, King of the Mosquito Nation, and Great Britain aboard HMS Honduras, with notes.* Introduction by S. L. Canger, Royal Commonwealth Society Collection: GBR/0115/RCMS 240/27.

56. **The British in Latin America brought not only guns and money but such ideas**

Alan Knight, "Britain and Latin America," in *The Oxford History of the British Empire: The Nineteenth Century*, Andrew Porter, ed., vol. 3 (Oxford: Oxford University Press, 1999), p. 125.

56. "prepossessing . . . crowned with umbrella-shaped trees of great size" Henry Alexander Wickham, *Rough notes of a journey through the wilderness from Trinidad to Para, Brazil, by way of the great cataracts of the Orinoco, Atabagao, and Rio Negro* (London: W.H.J. Carter, 1872), p. 144.

56. "handsome butterfly" Ibid.

57. "altogether a very uninteresting place . . . in the most approved fashion" Ibid., p. 146.

57. "numerous and beautiful, varying from the size of a bat" Ibid., p. 145.

57. "a noise like a kettle-drum" Ibid.

57. "perfectly bewildered" Ibid., p. 146.

58. "concoctions of feathers, chopped and tortured into abnormal forms" Asa Briggs, *Victorian People* (London: B. T. Batsford, 1988), p. 271, quoting "Mrs. Hawers," the Victorian fashion critic and author of *The Art of Beauty* (1878), *The Art of Dress* (1879), and *The Art of Decoration* (1891).

59. "Brother mule, I cannot curse you" "Moravian Civic and Community Values," http://moravians.org.

59. "The little chief seemed to take a great fancy to me" Henry Wickham, *Rough Notes of a Journey Through the Wilderness*, p. 149.

59. The first king, known only as Oldman . . . or assassinated by a "Captain Peter Le Shaw" From "Mosquitos, A Brief History," www.4dw.net/royalark/Nicaragua/mosquito2.htm.

61. "Bowing my head, I stepped across the little trench" Henry Wickham, *Rough Notes of a Journey Through the Wilderness*, p. 165.

61. "lay down and rose again with the sun" Ibid., p. 164.

61. "Left alone . . . I soon found" Ibid.

61. "exceedingly difficult . . . the feathers usually fly off in a cloud" Ibid., p. 167.

61. "strong cup of tea . . . never a greater source of enjoyment than on such an occasion" Ibid., p. 165.

62. "It was a long time before I became used" Ibid., pp. 165–66.

62. "knowing that they are easily killed" Ibid., p. 166.

62. "myriads of minute cockroaches" Ibid., p. 172.

62. "was so unendurable" Ibid., pp. 172–73.

62. "much injured, as they lay helplessly in their hammocks" Ibid., p. 172.

62. "peculiar aversion to wet . . . mouthfuls of water at the head of the column" Ibid., pp. 171–72.

62. "filliped it off" Ibid., p. 172. In the 1980s, zoologist Kenneth Miyata was collecting moths in western Ecuador when one of these ants dropped down the neck of his shirt and stung him four times. "Each sting felt as if a red hot spike was being driven in. My field of vision went red and I felt woozy." After an hour of "burning, blinding pain," Miyata endured a sore back and lymph nodes in his armpit so painfully swollen that he was unable to move

his arm for two days. Adrian Forsyth and Ken Miyata, *Tropical Nature: Life and Death in the Rain Forests of Central and South America* (New York: Scribner's, 1984), p. 108.

63. **"I know of nothing so suggestive of reflection"** Henry Wickham, *Rough Notes of a Journey Through the Wilderness,* p. 168.

63. **"Great First Cause of all . . . broke the unusual stillness"** Ibid.

63. **Spaniards from Honduras** Marc Edelman, "A Central American Genocide: Rubber, Slavery, Nationalism, and the Destruction of the Guatusos-Malekus," *Comparative Studies in Society and History* (1998), vol. 40, no. 2, p. 358.

63. *Castilla elastica,* **the main source of latex in Nicaragua** The British naturalist Thomas Belt said that the trees died after tapping because the harlequin beetle (*Acrocinus longimanus*) laid eggs in the cuts, and the grubs bored "great holes through the trunks." Thomas Belt, *The Naturalist in Nicaragua* (London: Edward Bumpus, 1888, first published 1874), p. 34.

64. **"leaving none alive to tell the tale"** Henry Wickham, *Rough Notes,* p. 159.

64. **"their decidedly light apparel"** Ibid., p. 158.

64. **"scrupulous honesty"** Ibid., p. 217.

64. **"remembering probably that I was but a stranger from some distant land of barbarism"** Ibid., p. 202.

64. **"I am sure if some of those who condemn Indians as a lazy race"** Ibid., p. 215.

64. **"Temple himself was nearly black"** Ibid., p. 181.

65. **"poised in psychological uncertainty . . . on the margin of each but a member of neither"** Everett V. Stonequist, *The Marginal Man: A Study in Personality and Culture Conflict* (New York: Russell & Russell, 1961), p. 1.

65. **"a contrast to the quiet of the Indian part of the encampment . . . leaving the Blewfields trader, his son, and myself alone"** Henry Wickham, *Rough Notes,* pp. 181–82.

66. **"being thirsty without hunger"** Ibid., p. 200.

66. **"driving their covered ways"** Ibid., p. 201.

66. **"go with me into the interior"** Ibid., p.198.

67. **"they continued on their journey north"** Ibid., p. 206.

67. **"As it was Christmas week, he went to a dance in the evening"** Ibid.

67. **"disgusting process"** Ibid., p. 189.

68. **"just before dawn . . . I heard the crying of the women"** Ibid., p. 207.

68. **"and I heard the rattle of their paddles while it was yet dark"** Ibid.

68. **"creeping down the steep bank"** Ibid.

68. **"looked quite pale and complained that he felt very sick"** Ibid., p. 208.

69. **"wild, matted tangle of flowering vines"** Ibid., p. 194.

69. **"our oaks, elms, and beeches stand out"** Ibid.

69. **"[A]t all the other places we passed the Indians had fled"** Ibid., pp. 210–211.

69. **"a view of great extent and beauty"** Ibid., p. 224.

70. **"Temple and I saw enough to convince us"** Ibid.

70. **"were cooking at a stove what looked more like beefsteak"** Ibid., p. 226.

70. **"took me to his room . . . where a dinner of beefsteak and bread was already on the table"** Henry Wickham, *Rough Notes,* p. 227. A good overview of the history of the

Cornish miners in Latin America can be found in "The Cornish in Latin America," www.projects.ex.ac.uk/cornishlatin/anewworldorder.htm; Alan Knight, "Britain and Latin America," in *The Oxford History of the British Empire: The Nineteenth Century*, Andrew Porter, ed., vol. 3 (Oxford: Oxford University Press, 1999), p. 127.

71. **"I recognized him at once"** Henry Wickham, *Rough Notes*, p. 229.

72. **"we passed Kissalala"** Ibid., p. 244.

72. **"The missionary standing, book in hand"** Ibid., p. 262.

72. **"the Moskito men were very superior in war"** Ibid.

72. **"I was surprised to meet one day, near Temple's lodge"** Ibid., p. 282.

73. **"[T]he captain loudly deplored the falling off of the warlike spirit"** Ibid., p. 284.

73. **"It was a strange sight"** Ibid., p. 285.

74. **"The mountains behind Porto Bello looked very beautiful"** Ibid.

74. **"racing to and fro"** Ibid.

74. **"I walked along this line one day for some distance"** Ibid., p. 286.

Chapter 4: The Mortal River

75. **"the very rocks are robed in the deepest green"** Henry Wickham, *Rough Notes*, p. 3.

75–76. **Great Show of Singing and Talking Birds** From "Canaries, Singing, and Talking Birds," *Illustrated London News*, Feb. 15, 1868, www.londonancestor.com.

76. **"all that many Creoles enjoyed along its banks"** Alfred Jackson Hanna and Kathryn Abbey Hanna, *Confederate Exiles in Venezuela* (Tuscaloosa, AL: Confederate Publishing Co., 1960), p. 31.

76. **only 12,978 settled in Venezuela** Ibid., p. 32.

77. **"the main purpose of his journey was to study the rubber trade"** Edward V. Lane, "The Life and Work of Sir Henry Wickham: Part II—A Journey Through the Wilderness," *India Rubber Journal*, vol. 125 (Dec. 12, 1953), p. 17.

77. **"a young Englishman who accompanied me"** Henry Wickham, *Rough Notes*, p. 4.

77. **Lane called him a sailor** Lane, "The Life and Work of Sir Henry Wickham: Part II—A Journey Through the Wilderness," p. 16.

78. **"self-respecting British prig"** Howard Wolf and Ralph Wolf, *Rubber: A Story of Glory and Greed* (New York: Covici, Friede, Publishers, 1936).

78. **"What a difference there is in the appearance of the boat's crew from an English man-of-war"** Henry Wickham, *Rough Notes*, p. 3.

78. **"a very fine specimen of a West Indian soldier"** Ibid., p. 4.

78. **"Our little craft, about the size of a Margate lugger, was well manned"** Ibid, p. 9.

78. **"had traversed the Spanish main"** Ibid., p. 10.

78. **"One hardly expects to find such a pitch of education"** Ibid.

78. **"where the greener water of the sea"** Ibid., p. 6.

79. **"paddling as for dear life"** Ibid., p. 7.

79. **"possess the knowledge of an ointment"** Ibid., p. 8.

79. **"rough-paved, but clean streets"** Ibid., p. 16.

80. "and a more villainous-looking collection of different types of men I think I never beheld" Ibid., p. 21.

80. "one of the last of the southern settlers who came two years before" Ibid., p. 19.

80. 8,000–10,000 former Confederates Lawrence F. Hill, "The Confederate Exodus to Latin America, Part I," *Southwestern Historical Quarterly*, vol. 39, no. 2 (October 1935), p. 103.

81. "a gold and diamond country . . . other commercial agricultural products" Charles Willis Simmons, "Racist Americans in a Multi-Racial Society: Confederate Exiles in Brazil," *Journal of Negro History*, vol. 67, no. 1, p. 35.

81. "the fingers of Manifest Destiny pointed southward" Simmons, "Racist Americans in a Multi-Racial Society: Confederate Exiles in Brazil," p. 35.

81. "the sphere of human knowledge . . . merely incidental" John P. Harrison, "Science and Politics: Origins and Objectives of Mid-Nineteenth Century Government Expeditions to Latin America," *Hispanic American Historical Review*, vol. 35, no. 2 (May 1935), pp. 187–192.

81. Price Grant Frank J. Merli, "Alternative to Appomattox: A Virginian's Vision of an Anglo-Confederate Colony on the Amazon, May 1865," *The Virginia Magazine of History and Biography*, 94:2 (April 1986), p. 216; and Alfred Jackson Hanna and Kathryn Abbey Hanna, *Confederate Exiles in Venezuela* (Tuscaloosa, AL: Confederate Publishing Co., 1960), p. 56.

82. "was not blessed with a particularly amiable temper" Henry Wickham, *Rough Notes*, pp. 21–22.

82. "I believe exercise is even more essential" Ibid., p. 16.

82. "A shock from an eel would send a bather" Ibid., p. 20.

83. "I cannot remember ever having received a more terrible shock" Humboldt is quoted in Anthony Smith, *Explorers of the Amazon*, p. 234.

83. "poor Rogers was stuck by a *raya*" Henry Wickham, *Rough Notes*, p. 20.

83. "who was quite motherly to Rogers" Ibid., p. 22.

83–84. "for picking up, and caring for stray chicks of doubtful pedigree" Ibid.

84. "It was most amusing to see what pride they took in being British subjects" Ibid.

84. "fast little native-built lancha" Ibid., p. 24.

84. "I proposed exploring the Caura" Ibid.

85. "the jingling of little bells" Ibid., p. 31.

85. "There does not appear to be much actual fighting" Ibid., p. 34.

85. "a feeling of giddy faintness" Ibid., pp. 37–38.

85. "I began taking doses of quinine and drinking plentifully cream of tartar water" Wallace is quoted in Redmond O'Hanlon, *In Trouble Again: A Journey Between the Orinoco and the Amazon* (New York: Vintage, 1988), p. 3.

86. "to the brink of the river" Henry Wickham, *Rough Notes*, p. 38.

86. "good natured Barbados woman" Ibid., p. 42.

86. "did not fear as to the result" Ibid.

87. "declaring himself unwell . . . I was much disappointed" Ibid., p. 45.

87. "At this, the height of the rainy season, little or no land is to be met with" Ibid., p. 46.

87. "gave forth their peculiar mewing cry" Ibid., p. 47.

87. "[O]nce it has embraced the trunk of a forest tree" Ibid., p. 50.

88. "stands self-supported, a great tree" Ibid.

89. "They were very noisy" Ibid., p. 56.

89. "When one is unwell, it is especially unpleasant. . . . the ball drilled a hole through the body, and continued its way" Ibid., pp. 57–58.

89. "pitchy-dark" Ibid., p. 59.

89. "wanted to take me home with him" Ibid., p. 60.

90. "I managed to control my legs. . . . I did not remember anything until the fever lessened" Ibid., p. 61.

91. "myriads of fireflies sparkled like gems" Ibid., p. 67.

91. "a pair of ferocious moustaches" Ibid., p. 68.

91. "dreaded Guahibos" Ibid., p. 67.

92. "Shortly after the separation of Venezuela from the mother country" Richard Spruce is quoted in Redmond O'Hanlon, *In Trouble Again: A Journey Between the Orinoco and the Amazon* (New York: Vintage, 1988), p. 32.

92. "an unpleasant air of mortality" Henry Wickham, *Rough Notes*, p. 69.

92. "their faces being covered with black spots" Ibid.

92. "Whilst I gazed into the tomb" Ibid., p. 72.

93. "It is singular . . . that these people endeavour" Ibid., p. 77.

93. "I was sorry to see Castro bend his bright toledo" Ibid., pp. 76–77.

93. "with a stupid grin" Ibid., pp. 79–80.

93. "good-looking matron" Ibid., p. 78.

94. "It is a wonder that these simple people do not even more seclude themselves" Ibid.

95. "The air was heavy with the odour of the flowers" Ibid., p. 83.

95–96. "We found that the whole of the inhabitants had been seized by a kind of mania" Ibid., p. 88.

96. "decidedly stupid" Ibid., p. 93.

97. What exactly had he done? There are many descriptions of the work of the rubber tapper, from Richard Spruce, Henry Bates, and Alfred Wallace to contemporary commentators, but the most painstaking in his observation seems to have been Algot Lange, *The Lower Amazon: A Narrative of Exploration in the Little Known Regions of the State of Para, on the Lower Amazon, etc.* (New York: Putnam's, 1914), pp. 51–60. Like Wickham, Lange collected rubber himself, so one can imagine his boredom as he turned the paddle dipped in latex and began to count the number of times he moved his shoulder and rotated his wrist as he turned the heavy rubber-spit over the smoking fire.

99. "shut out from the rest of the world" Henry Wickham, *Rough Notes*, p. 91.

99. "watch the cold shadows of night gradually creep up from the water" Ibid.

99. Psychologists have suggested that *place*, as a force Laura M. Fredrickson and

Dorothy H. Anderson, "A Qualitative Exploration of the Wilderness Experience as a Source of Spiritual Inspiration," *Journal of Experimental Psychology* 19 (1999), p. 22.

99. "elvish little *ti-ti* monkeys." Henry Wickham, *Rough Notes*, p. 94.

Chapter 5: Instruments of the Elastic God

100. "The constant irritation . . . caused my hands and feet to swell" Henry Wickham, *Rough Notes*, p. 92.

100. "not so painful as I had anticipated" Ibid., p. 101.

100. "the pangs of thirst he will suffer after such a gorge of salt fish" Ibid.

101. "The first time I felt them, I could not imagine what on earth" Ibid., p. 98.

101. "Edible. . . . Good to eat, and wholesome to digest" Ambrose Bierce, *The Devil's Dictionary* (Toronto: Coles, 1978, first published 1881), p. 34.

102. "[E]ach time the fit of nausea returned, I became quite powerless" Henry Wickham, *Rough Notes*, p. 103.

102. "but the sun was too powerful . . . the remainder of my strength fast failing" Ibid., p. 104.

102. "I remember little of what passed" Ibid.

102–3. "For five days I was delirious . . . and cups of creamy coffee" Algot Lange, *In the Amazon Jungle: Adventures in Remote Parts of the Upper Amazon River, Including a Sojourn Among Cannibal Indians* (New York: Putnams, 1912), p. 283.

103. "the voice of the forest . . . the murmuring crowd of a large city" Ibid., pp. 283–284.

103. "I saw myself engulfed . . . a place of terror and death" Ibid., pp. 289–290.

103. "It is almost something unbelievable to those who do not know the jungle" Michael Taussig, "Culture of Terror-space of Death: Roger Casement's Putumayo Report and the Explanation of Terror," *Comparative Studies in Society and History* 26:3 (1984), p. 483, quoting P. Francisco de Vilanova, introduction to P. Francisco de Iqualada, *Indios Amazonicas: Colección Misiones Capuchinas*, vol. VI (Barcelona, 1948).

104. "he found several vultures calmly awaiting his death" Edward V. Lane, "The Life and Work of Sir Henry Wickham: Part II—A Journey Through the Wilderness," p. 18.

104. "I recollect one afternoon. . . . I think I never felt so grateful for anything" Henry Wickham, *Rough Notes*, p. 104.

105. "the little pale man of the forest" Ibid.

105. "the sure precursor of evil" Ibid. The tale of the vengeful spirit is also related in Claude Lévi-Strauss, *The Raw and the Cooked: Introduction to a Science of Mythology*, vol. I., trans. John and Doreen Weightman (New York: Harper Colophon, 1975, first published in France as *Le Cru et le Cuit*, 1964, Librairie Plon), p. 264.

105. The Prayer of the Dry Toad Claude Lévi-Strauss, *Tristes Tropiques* (New York: Atheneum, 1981, first published 1951), trans. John and Doreen Weightman, p. 363.

105. "In so remote a situation . . . there must be a mine of gold in that direction" Henry Wickham, *Rough Notes*, p. 109.

106. "unable to work for some time past" Ibid., p. 111.

106. "helplessly sick" Ibid., p. 115.

106. "on the cool and limpid water of the Black River" Ibid., p. 123.

106. "very suggestive of a return to civilization" Ibid., p. 138.

107. "Nothing has been discovered which would even be a substitute" Thomas Hancock, *Personal Narrative of the Origin and Progress of the Caoutchouc or India-Rubber Manufacture in England* (London: Longman, 1857), p. iii.

107. "the ultimate hard currency of exchange" "Blood for Oil?" *London Review of Books*, April 21, 2005, p. 12.

108. "The introduction of the invaluable *cinchonas* into India" James Collins, "On India Rubber, Its History, Commerce and Supply," *Journal of the Society of Arts*, vol. 18 (Dec. 17, 1869), p. 91.

109. in some cases, as high as 13.7 percent Drayton, *Nature's Government*, p. 210.

109. "it was necessary to do for the india-rubber and caoutchouc-yielding trees" Clements Markham is quoted in Dean, *Brazil and the Struggle for Rubber*, p. 12.

109. "When it is considered that every steam vessel afloat" Clements Markham is quoted in John Loadman, *Tears of the Tree: The Story of Rubber—A Modern Marvel* (Oxford: Oxford University Press, 2005), p. 83.

110. except for some woolen stockings and an "antiglare eyeshade" Susan Orlean, *The Orchid Thief* (New York: Ballantine Books, 1998), p. 59.

111. "It will be long before I cease to hear her voice in the garden" Hooker is quoted in "Jos. D. Hooker: Hooker's Biography: 4. A Botanical Career." www.jdhooker .org.uk. A good overview of Joseph Hooker's life is also included in Faubion O. Bower, "Sir Joseph Dalton Hooker,1817–1911," in *Makers of British Botany: A Collection of Biographies by Living Botanists*, ed. F. W. Oliver (Cambridge: Cambridge University Press, 1913), p. 303.

111. "nervous and high-strung . . . impulsive and somewhat peppery in temper" "Jos. D. Hooker: Hooker's Biography: 4. A Botanical Career." www.jdhooker.org.uk.

111. "not recreational . . . rude romping and games" Ibid.

112. "in a very dilapidated condition" William Scully's *Brazil; Its Provinces and Chief Cities* (London, 1866), p. 358, and Franz Keller's *The Amazon and Madeira Rivers* (Philadelphia, 1840), p. 40, are both quoted in E. Bradford Burns, "Manaus 1910: Portrait of a Boom Town," *Journal of Inter-American Studies*, vol. 7, no. 3 (1965), p. 412.

113. "What has happened? . . . Rubber has happened!" Anthony Smith, *Explorers of the Amazon*, p. 269.

113. In the province of Pará alone Lucille H. Brockway, *Science and Colonial Expansion: The Role of the British Royal Botanic Gardens* (New York: Academic Press, 1979), p. 147.

115. "of every variety, from silks and satins to stuff gowns" Professor and Mrs. Louis Agassiz, *A Journey in Brazil* (Boston: Ticknor and Fields, 1868), p. 280.

115. Power reigned in the warehouses Henry C. Pearson, *The Rubber Country of the Amazon; A Detailed Description of the Great Rubber Industry of the Amazon Valley, etc.* (New York: India Rubber World, 1911), pp. 94–95.

115. "Get rich, get rich!" they cried Richard Collier, *The River that God Forgot: The Story of the Amazon Rubber Boom* (New York: E. P. Dutton, 1968), p. 20.

Chapter 6: The Return of the Planter

119. Crisóstavo wrenched a rubber empire from the forest Michael Edward Stanfield, *Red Rubber, Bleeding Trees: Violence, Slavery and Empire in Northwest Amazonia, 1850–1933* (Albuquerque, NM: University of New Mexico Press, 1998), pp. 26–28.

120. The São Paulo Railway Company, Ltd., operated a railway J. Fred Rippy, "A Century and a Quarter of British Investment in Brazil," *Inter-American Economic Affairs*, 6:1 (Summer 1952), pp. 87–88.

121. Englishmen bought stock in Brazilian mines, banks Ibid., p. 83.

121. but in Latin America especially he often found himself thrust D. C. M. Platt, *The Cinderella Service: British Consuls since 1825* (Hamden, CT.: Archon Books, 1971), p. 16.

121. "lower in dignity" Ibid., p. 1.

122. One consul in Siam Ibid., p. 19.

123. "The labour of extracting rubber is so small" James Drummond-Hay's report is included in Henry Wickham's *Rough Notes*, pp. 294–296.

123. "The rubber-bearing country is so vast" Ibid., p. 296.

123–24. It was one of the few excursions he'd made off the boat Violet Wickham, "Lady Wickham's Diary," p. 2. Wolverhampton Archives and Local Studies, Wolverhampton, England, in the records of the Goodyear Tyre and Rubber Company (Great Britain) Ltd., ref. DB-20/G/6. The only mention of Henry's stop is in Violet's diary, written decades later. She notes in passing that Henry had met a "few American backwoods people . . . on a previous journey." In Henry's account, there is no mention of this; it is as if Santarém never existed.

124. "I have come to the conclusion" Henry Wickham, *Rough Notes*, p. 138.

124. "in remembrance of the many kindnesses" Ibid., frontispiece. Interestingly and ironically, the connection doesn't end there. The Wickhams and the Drummond-Hays would be in time distantly related by marriage. Edward Drummond-Hay's brother-in-law was Thomas Gott Livingstone, and in 1880, Livingston's daughter Frances married Henry Wickham's first cousin, the Rev. Alexander G. H. Lendrum. In Anthony Campbell, "The Descendants of Benjamin Wickham," p. 4.

124. Her name was Violet Case Carter Many accounts give her name as Violet Cave, but this is a mistake that seems traceable to Edward Lane's articles of the 1950s and to a group photo in Santarém in 1875. She appears in the 1871 Census as "Violet C. Carter" (age 21, born London), resident at 12 Regent Street with her parents William H. J. Carter (age 55, born London, bookseller) and Patty Carter (age 46, born London). Source: 1871 Census of London, published online by Ancestry.com.uk, citing the *1871 Census of London*, National Archives, Kew, RG10/133, ED 2, folio 30, page 17.

125. set up his shop at 12 Regent Street Today 12 Regent Street is the site of the Economist Bookstore.

126. Family lore suggests that Carter also subsidized many of Henry's future adventures Edward V. Lane, "The Life and Work of Sir Henry Wickham—III: Santarem," *India Rubber Journal*, vol. 125 (Dec. 26, 1953), p. 18.

126. **"Born within the sound of Bow Bells"** Violet Wickham, "Lady Wickham's Diary," p. 1.

126. **"To be married is, with perhaps the majority of women, the entrance into life"** W.G. Hamley, "Old Maids," *Blackwood's Edinburgh Magazine*, 112 (July 1872), p. 95, quoted in Pat Jalland and John Hooper, eds. *Women from Birth to Death: The Female Life Cycle in Britain, 1830–1914* (Atlantic Highlands, NJ: Humanities Press International, Inc., 1986), p. 126.

126. **"The general aim of English wives"** E. J. Tilt, *Elements of Health and Principles of Female Hygiene* (London, 1852), pp. 258–261, quoted in Jalland and Hooper, *Women from Birth to Death*, p. 124.

126. the **"good wife" of Proverbs** Proverbs 31:23.

127. **"The sense of national honour, . . . pride of blood"** Catherine Hall, "Of Gender and Empire: Reflections on the Nineteenth Century," in *Gender and Empire*, Philippa Levine, ed. (Oxford: Oxford University Press, 2004), p. 46. Hall quotes Herman Merivale, *Lectures on Colonization and Colonies* (London, 1861), p. 675.

127. **A wife embodied the moral standards** Beverly Gartrell, "Colonial Wives: Villains or Victims?" in *The Incorporated Wife*, Hilary Callan and Shirley Ardener, eds. (London: Croom Helm, 1984), pp. 165–185; Deborah Kirkwood, "Settler Wives in Southern Rhodesia: A Case Study," in *The Incorporated Wife*, pp. 143–164.

127. **"tenderly preserve them, as the plantation of mankind"** Samuel Solomon, *A Guide to Health*, 66th edition (1817), p. 131, in Jalland and Hooper, eds. *Women from Birth to Death: The Female Life Cycle in Britain, 1830–1914*, p. 32.

128. **"of independent means"** The Census of 1871, quoted in Anthony Campbell, "Descendants of Benjamin Wickham, a Genealogy."

129. **"was very like being dropped into deep water never having learned to swim"** Violet Wickham, "Lady Wickham's Diary," p. 1.

129. **"a singular winged parasitical insect of a disgusting appearance"** C. Barrington Brown and William Lidstone, *Fifteen Thousand Miles on the Amazon and its Tributaries* (London: Edward Stanford, 1878), pp. 9–10.

130. **"They should be nearly square . . . soon lulling you off"** Violet Wickham, "Lady Wickham's Diary," p. 1.

130. the **Amazon stretches approximately 4,000–4,200 miles** Michael Goulding, Ronaldo Barthem and Efrem Ferreira, *The Smithsonian Atlas of the Amazon* (Washington, DC: Smithsonian Institution, 2003), p. 23.

131. the **Amazon could supply in two hours all the water used by New York City's 7.5 million residents each year** Ibid., p. 28. The rationale is this: New York City consumes about 1.1 billion gallons daily, or nearly 4 trillion gallons a year.

131. **The valley itself . . . relentless, uncomprehending *thing*** The scope of the Amazon and the Amazon Basin are almost unimaginable, but the descriptions and statistics on it are from the following sources: Harald Sioli, "Tropical Rivers as Expressions of their Terrestrial Environments," in *Tropical Ecological Systems*, Frank Golley and Ernesto Medina (New York: Springer-Verlag, 1975), pp. 275–288; Betty J. Meggers, *Amazonia: Man and Culture in a Counterfeit Paradise* (Chicago: Aldine-Atherton, 1971); John Melby,

"Rubber River: An Account of the Rise and Collapse of the Amazon Boom," *Hispanic American Historical Review*, vol. 22, no. 3 (Aug. 1942), pp. 452–469; P. T. Bauer, *The Rubber Industry, A Study in Competition and Monopoly* (Cambridge: Cambridge University Press, 1948); Goulding, Barthem, and Ferreira, *The Smithsonian Atlas of the Amazon* (Washington, DC: Smithsonian Institution, 2003).

132. "**We had a few people traveling with us 2nd class**" Violet Wickham, "Lady Wickham's Diary," p. 1.

132. "**had received tidings that beyond the city of Quito**" Von Hagen, *South America Called Them*, p. 5.

133. "**It was here that they informed us of the existence of the Amazons**" Gaspar de Caraval, *The Discovery of the Amazon: According to the Account of Friar Gaspar de Carvajal and Other Documents*, introduction by José Toribio Medina, trans. Bertram T. Lee (New York: American Geographical Society, 1934), p. 177.

133. **what the Indians themselves called the *Paranáquausú*, or "Great River"** William L. Schurz, "The Amazon, Father of Waters," *National Geographic* (April 1926), p. 445.

133. **An estimated 332,000 people lived in this region, up from 272,000 a decade ago** Arthur Cesar Ferreira Reis, "Economic History of the Brazilian Amazon," in *Man in the Amazon*, Charles Wagley, ed. (Gainesville, FL: University Presses of Florida, 1974), p. 39.

134. "**left us a north-east despoiled of its very rich forests**" José Pedro de Oliveira Costa, "History of the Brazilian Forest: An Inside View," *Environmentalist*, vol. 3, no. 5 (1983), p. 51.

134. **Between 1500 and 1800, the Americas sent to Europe £300 million in gold** Edward J. Rogers, "Monoproductive Traits in Brazil's Economic Past," *Americas*, vol. 23, no. 2 (October 1966), p. 133.

Chapter 7: The Jungle

137. **Some Santarém trading houses had branches** Herbert H. Smith, *Brazil, the Amazons and the Coast* (New York: 1879), p. 118.

137. "**Sixty thousand bows can be sent forth from these villages alone**" Jesuit Father Mauricio de Heriarta's "Description of the State of Maranhão, Pará, Corupá, and the River of the Amazons" (1660) is quoted in Smith, *Brazil, the Amazons and the Coast*, p. 171.

137. "**that is the cause why they are feared of the other Indians**" Ibid.

138. **They danced before the doors of the principal citizens** Henry Walter Bates, *The Naturalist on the River Amazon*, ch. 8 (New York, 1864), www.worldwideschool.org/library/books/sci/earthscience/TheNaturalistontheRiverAmazon.

138. "**All children were born free . . . a child sitting on their hip on the other side**" Violet Wickham, "Lady Wickham's Diary," p. 2.

139. "**all made their house keeping (*sic*) money by sending out their slaves**" Ibid.

139. "**no cases of gross cruelty tho' you could often hear the *palmatore* going**" Ibid., p. 1.

139. "**a little Indian boy who had been given to H to bring up**" Ibid., p. 5.

140. "**of an English pleasure ground**" Richard Spruce, *Notes of a Botanist on the Ama-*

zon and the Andes, Alfred Russel Wallace, ed., 2 vols. (Cleveland, OH.: Arthur H. Clark, 1908), p. 66.

140. "I had not gone far when my English saddle turned around" Violet Wickham, "Lady Wickham's Diary," p. 2.

141. "beset with hard spines" Henry Walter Bates, *The Naturalist on the River Amazon,* ch. 8 (New York, 1864), www.worldwideschool.org/library/books/sci/earthscience/TheNaturalistontheRiverAmazon.

141. "undertook the most toilsome journeys on foot to gather a basketful" Ibid.

142. "a world of eternal verdure and perennial spring" Eugene C. Harter, *The Lost Colony of the Confederacy* (Jackson, MI.: University Press of Mississippi, 1985), p. 26.

142. "lean, hard men with their wives" Roy Nash, *The Conquest of Brazil* (New York: AMS Press, 1969, originally published 1926), p. 152; Mark Jefferson, "An American Colony in Brazil," *Geographical Review,* vol. 18, no. 2 (April 1928), p. 228.

143. where diamonds had been found years earlier David Afton Riker. *O Último Confederado na Amazônia* (Brazil, 1983), p. 112. David Afton Riker is the son of the original David Riker (see page 167). He placed at the end of his own story his father's handwritten memoirs of his days as a pioneering *confederado.*

143. "we would never grow quite as they were" Violet Wickham, "Lady Wickham's Diary," p. 2.

143. "white stockings and legs . . . a damper on my ideas of finery" Ibid.

143. "Alas . . . I have grown as utterly careless" Ibid.

143. Until 1997 . . . the plateau was as it had been in Henry's day Steven Alexander, owner of Bosque Santa Lucia, an educational and botanical preserve that contains the area of the old *confederado* site, said that he has counted 200 species of tree in the preserve's 270 acres. All information about Piquiá-tuba comes from an interview with Alexander on his preserve on October 9, 2005.

144. Why *were* there so many trees? Peter Campbell, "Get Planting" (a review of *The Secret Life of Trees: How They Live and Why They Matter,* by Colin Tudge), *London Review of Books,* Dec. 1, 2005, p. 32.

144–45. "the deafening clamour of frogs" Violet Wickham, "Lady Wickham's Diary," p. 4.

145. "not only weakening them . . . their combs as white as the rest of their bodies" Ibid., p. 5.

145. "bathed in blood" Ibid.

145. "other spots of electricity" Ibid., p. 4.

145. the jaguar and the three men Spruce, *Notes of a Botanist on the Amazon and the Andes,* pp. 122–123.

146. Here on the Tapajós . . . and her breasts beneath her arms Algot Lange, *The Lower Amazon: A Narrative of Exploration in the Little Known Regions of the State of Para, on the Lower Amazon, etc.* (New York: Putnam's, 1914), pp. 427–428.

146. Henry Bates had come to know one named Cecilia Bates, *The Naturalist on the River Amazon,* ch. 8.

147. "fertile field" Lange, *The Lower Amazon,* p. 361.

147. **"till it looks like a gigantic green fringe"** Violet Wickham, "Lady Wickham's Diary," p. 2.

147. **"While green it was pretty"** Ibid.

147. **"started off early in the morning"** Ibid., p. 3.

148. **"I get it burning, put on sauce pan"** Ibid.

148. **"as tired, hot, and unrefreshed as before"** Ibid.

148. **"attack the legs of bathers near the shore"** Bates, *The Naturalist on the River Amazon*, ch. 9.

148. **" 'temporary' went on extending"** Violet Wickham, "Lady Wickham's Diary," p. 3.

149. **"to save time with coming and going"** Ibid.

149. **"and soon left us one after another"** Ibid.

149. **an older man, as Violet suggests, though his age is unrecorded** Ibid.

150. **"If she handled the morning 'clinics' and other encounters"** Deborah Kirkwood, "Settler Wives in Southern Rhodesia: A Case Study," in *The Incorporated Wife*, Hilary Callan and Shirley Ardener, eds. (London: Croom Helm, 1984), p. 151.

150. **"[A] readiness to interest herself in the health"** Ibid.

150. **But more than by labor, Henry was defeated by the soil** Betty J. Meggers, *Amazonia: Man and Culture in a Counterfeit Paradise* (Chicago: Aldine-Atherton, 1971), p. 18.

150. **A 1978 study of absorption showed that 99.9 percent of all calcium 45** Carl F. Jordan, "Amazon Rain Forests," *American Scientist*, vol. 70, no. 4 (July–Aug. 1982), pp. 396–397.

151. **Soon afterward, the second, more terrible crisis** Information on the deaths of Henry Wickham's party comes from three sources: Henry Wickham, "Graves in the Confederate Cemetery" (a drawing); Violet Wickham, "Lady Wickham's Diary," p. 3; Anthony Campbell, "Descendants of Benjamin Wickham," p. 3.

152. **"We alone of the original party picked up"** Violet Wickham, "Lady Wickham's Diary," p. 3.

Chapter 8: The Seeds

153. **"a spur just off from the forest covered table highlands"** The March 1872 letter from Wickham to Hooker is quoted in Dean, *Brazil and the Struggle for Rubber*, pp. 13–15.

154. **he followed up** The date of the package to Kew is uncertain. Ibid., p. 13.

155. **"His drawings of the leaf and seeds"** Ibid., p. 15.

156. **"a Mr. Wickham, at Santarem, who may do the job"** Ibid., p. 13.

156. **"with the view of afterwards sending the young plants out to India"** Royal Botanic Gardens–Kew. *Miscellaneous Reports: India Office: Caoutchouc I*, "Letter May 7, 1873, from India Office to Director of Kew, J.D. Hooker, inquiring into possibility of sending Hevea plants to India after raising them from seeds at Kew," file folder 2.

156. **"I have a correspondent at Santarem on the Amazon"** Ibid., "Reply Hooker to India Office, May 15, 1873. Reply to May 7th letter," file folder 4.

156. **But both letters were apparently misplaced** Dean, *Brazil and the Struggle for Rubber*, p. 15.

157. "quite fresh and in a state for planting" Ibid., p. 13.

157. **Since Farris had two thousand seeds, the Empire paid about twenty-seven dollars** W. Gordon Whaley, "Rubber, Heritage of the American Tropics," *Scientific Monthly*, vol. 62, no. 1 (Jan. 1946), p. 23.

157. **"I thought it important to secure them at once"** Royal Botanic Gardens–Kew. *Miscellaneous Reports: India Office: Caoutchouc I*, "Letter from Clements Markham to J.D. Hooker, June 2, 1873, regarding the purchase by James Collins of 2000 Hevea seeds from a 'Mr. Farris' of Brazil," file folder 5.

157. **the U.S. and French consulates had already made a bid** Dean, *Brazil and the Struggle for Rubber*, p. 13.

157. **"Is that all you managed to shoot?"** Farris's tale to Lord Salisbury is recounted in F. W. Sadler, "Seeds That Began the Great Rubber Industry," *Contemporary Review*, vol. 217, no. 1257 (Oct. 1970), pp. 208–209.

157. **"I would like to take this opportunity to place on official record"** Ibid.

157–58. **"a gross attempt to impose . . . an utterly worthless report on Gutta Percha"** Ibid.

158. **"glad to accept your offer to put me into communication"** Ibid.

158. **"The Consul at Para has written to say"** Royal Botanic Gardens–Kew. *Miscellaneous Reports: India Office: Caoutchouc I*, "Letter, Markham to J. D. Hooker. Sept. 23, 1873," file folder 9.

159. **"obtain any quantity that may be necessary at small expense"** Dean, *Brazil and the Struggle for Rubber*, p. 15.

159. **"I have just received a letter from H. Majesty Consul at Para enquiring"** Royal Botanic Gardens–Kew, *Miscellaneous Reports: India Office: Caoutchouc I*, "Letter from Henry Wickham in Santarem to Hooker, Nov. 8, 1873," file folder 10.

160. **He'd absorbed that portion of the Portuguese character known as *saudade*** Stuart B. Schwartz, "The Portuguese Heritage: Adaptability," in G. Hervey Simm, ed., *Brazilian Mosaic: Portraits of a Diverse People and Culture* (Wilmington, DE.: Scholarly Resources, 1995), pp. 31–32.

160. **"In a radiant land"** Paulo Prado, "Essay on Sadness," in Simm, *Brazilian Mosaic*, p. 19.

160. **"The moon is rising, Mother, Mother!"** Prado, "Essay on Sadness," in Simm, *Brazilian Mosaic*, pp. 19–20.

161. **there are two local Curuás from which to choose** If this is not enough, there is a third Curuá, but it would have been inaccessible to Henry and Violet. The Amazon's last major tributary before it meets the sea is the Xingu River; this drops south into the forest where it is fed by the Iriri River, and this third Curuá is the Iriri's main tributary. Since this is hundreds of miles from Santarém, it is impossible that this could be Wickham's Curuá.

162. **"the very head-quarters"** C. Barrington Brown and William Lidstone, *Fifteen Thousand Miles on the Amazon and its Tributaries* (London: Edward Stanford, 1878), p. 249.

162–63. **"every fragment of food . . . and stung with all his might"** Henry Walter Bates, *The Naturalist on the River Amazon* (New York, 1864), ch. 9, p. 14. The book is available in its entirety at www.worldwideschool.org/library/books/sci/earthscience/ TheNaturalistontheRiverAmazon. According to contemporary travel writer Herbert H. Smith, "As late as 1868 the town was still nearly deserted." Herbert H. Smith, *Brazil, The Amazons and the Coast* (New York: 1879), p. 241. Smith quotes the traveler Sr. Penna: "This settlement . . . is a very beautiful and pleasant place, but without inhabitants because of the *formigas de fogo*. A primary school has been created here by law, but no one has profited by it, because no lives there."

163. **"once more made a hole in the primeval forest to put his house in"** Violet Wickham, "Lady Wickham's Diary," p. 3.

163. **"nature's flophouses for the outcasts"** William Sill, "The Anvil of Evolution," *Earthwatch*, Aug. 2001, p. 27.

164. **He was becoming, in effect, a *caboclo*** Emilio F. Moran, "The Adaptive System of the Amazonian Caboclo," in Charles Wagley, ed., *Man in the Amazon* (Gainesville, FL: University Presses of Florida, 1974), pp. 136–159; and Edward C. Higbee, "The River Is the Plow," *Scientific Monthly*, June 1945, pp. 405–416.

165. **"He stripped his shirt up"** Violet Wickham, "Lady Wickham's Diary," p. 4.

165. **"and then I have said all there is to be said for it"** Ibid.

166. **"which measured forty-two feet in length"** Bates, *The Naturalist on the River Amazon*, ch. 9, p. 11.

166. **"The father and his son went . . . to gather wild fruit"** Ibid. **Author's note.** I never saw an anaconda on the Tapajós, but I did see a huge one once on the Rio Napo north of Iquitos in Peru, and after that I'd never discount such tales of attacks on small children. I was in the middle of the stream, trying to paddle one of the flat-bottomed dugouts; since I was used to the more deeply keeled canoes used on North American rivers, my dugout was essentially tracing a big circle in the water using the weight of my body as the fulcrum. I was alone on the river, around a bend from my camp, when suddenly the forest around me grew very quiet and I thought I heard a *plunk!* I looked down and underwater, directly beneath the dugout, an enormous anaconda glided past. I cannot even begin to imagine how big that thing was: I am sure I would exaggerate. I stopped paddling, and sat still, and stopped breathing—and for the first time in my life knew what people mean when they say their "heart stopped in their throat." The snake just kept going and going; it seemed to have no beginning or end. Later I would recognize that otherness, that impossibility of proportion, as the essence of monstrosity, and also realized that out here I counted, at best, as a hearty meal. There are no verified stories I know of that show anacondas attacking and swallowing human adults, but you couldn't have convinced me of that fact at the moment: I slid my paddle from the water without a drip (as I'd learned in Boy Scouts, in case I ever became an Army Ranger—fat chance, there!); I held it in both hands before me, like a club. I had one of those strange and sudden out-of-body experiences where I observed myself from above, as if from a spy satellite: me, holding the flimsy, ineffective paddle; the snake, going on and on a few feet down. All it had to do was rise a couple of inches and brush the canoe, and I'd be over the side and in his element with him. I wasn't a father yet and hadn't really thought about such matters, but it suddenly popped into my head that *This line of Jacksons ends*

right here. I guess I can be proud of the fact that I didn't panic and start thrashing about, which probably *would* have gotten the monster's attention, but other than that I felt like a truly helpless idiot, and part of me hoped that nobody was around to watch when the snake swallowed me whole. Let people wonder what happened to me: at least there'd be some mystery. But nothing happened, of course, and when the snake finally passed, I sat quietly a little longer then gently dipped my paddle back in the water and found I'd discovered the trick to maneuvering the damn canoe after all. Not that it mattered—I never went alone on the river in one of those dugouts again.

166. The house was never finished Lane, "The Life and Work of Sir Henry Wickham—III: Santarem," *India Rubber Journal*, vol. 125 (Dec. 19, 1953), p. 18.

166. "nitrate pulse" Moran, "The Adaptive System of the Amazonian Caboclo," p. 145.

166. "a horticulturist, a rubber collector, a hired hand" Ibid.

167. "My boarded floor had been taken up" Violet Wickham, "Lady Wickham's Diary," p. 4.

167. "Casa-Piririma" Lane, "The Life and Work of Sir Henry Wickham—III: Santarem," p. 19.

167. "as disgusted as the others" Violet Wickham, "Lady Wickham's Diary," p. 4.

167. Mercia Jane Ferrell from West Moors, Dorset We know very little about the ill-fated Mercia Jane Ferrell besides her brief, anonymous mention in Violet's diary; her place of birth and age in the Census of April 3, 1871; and Henry's drawing of the wooden cross above her grave, the only time her name is mentioned in the various annals of Wickham's time in the Amazon.

168. "I was just delited" David Bowman Riker's "Handwritten Narrative" appears in an Appendix in *O Último Confederado na Amazônia* (The Last Confederate in the Amazon), by his son David Afton Riker, (Brazil, 1983), pp. 112 ff. David B. Riker knew the Wickhams, said he was the first person to plant rubber on the Amazon, and planted the first rubber tree for Henry Ford in Forlandia.

168. "Father put Virginia and myself in school" Ibid.

168. "ride out in to the country" Ibid., p. 384.

169. "It is already very late" Royal Botanic Gardens–Kew, *Miscellaneous Reports: India Office: Caoutchouc I,* "Letter dated Oct. 15, 1874, Wickham at Piquiatuba, near Santarem, to Hooker, regarding transport of seeds," file folder 14.

169. "I reopen this in order to add" Ibid.

169. "With reference to Mr. Wickham's proposal to raise young India Rubber plants" Royal Botanic Gardens–Kew, *Miscellaneous Reports: India Office: Caoutchouc I,* "Letter dated October, 1874, Markham to Hooker," file folder 12.

170. "*any amount* of seeds at the same rate—£10 for 1000" Royal Botanic Gardens–Kew, *Miscellaneous Reports: India Office: Caoutchouc I,* "Letter, Markham to Hooker, Dec. 4, 1874," file folder 13.

170. The secretary of state soon authorized Wickham Dean, *Brazil and the Struggle for Rubber,* p. 16.

170. "once more each struggled on alone" Violet Wickham, "Lady Wickham's Diary," p. 5.

171. "She stayed with me till her death" Ibid., p. 4.

171. "I received a few other (seeds) from an up-river trader" Royal Botanic Gardens–Kew, *Miscellaneous Reports: India Office: Caoutchouc I,* "Letter Wickham at Piquiatuba, Santarem, to Hooker, April 18, 1875, regarding the fact that it is too late in the season to collect seed," file folder 15. According to historian Warren Dean, Wickham also sent a package of seeds soon after this that reached the India Office on September 9, 1875; they were "duly paid for," but failed to germinate. There is also no record today in Kew of their arrival. Dean, *Brazil and the Struggle for Rubber,* p. 16.

172. "Should you have opportunity of recommending me" Ibid.

Chapter 9: *The Voyage of the* Amazonas

173. "After supper the family would unite" Baldwin, "David Riker and *Hevea brasiliensis;* the Taking of Rubber Seeds out of the Amazon," p. 384.

173. "I am just about to start for the 'ciringa' district" Royal Botanic Gardens–Kew, *Miscellaneous Reports: India Office: Caoutchouc I,* "Letter from Henry Wickham at Piquiátuba to Hooker, Jan. 29, 1876," file folder 17.

174. "[T]he collectors must be directed" Roy MacLeod, *Nature and Empire: Science and the Colonial Enterprise,* Osiris series, vol. 15 (Chicago: University of Chicago Press, 2001), p. 179.

174. By locating on the Madeira, he'd entered a region that swallowed lives blithely The Acré was a huge chunk of rain forest between Brazil, Peru, and Bolivia. Wars were fought over it, and corruption and slavery occurred in its depths that rivaled anything committed in the name of gold or oil. By 1870, Brazilian rubber prospectors had penetrated its darkness; by 1875, steam navigation had penetrated 1,228 miles up the Madeira. But three miles above the old town of Santo Antonio (now Porto Velho), the Madeira was broken by 19 cataracts and rapids formed by the meeting of the Brazilian Shield and the Amazonian *planície.* For more than 250 miles the river was unnavigable, and trading vessels hauled up and down in backbreaking portages made no more than three round trips a year. By 1872, the American journalist and speculator George Church had persuaded investors that a 225-mile railroad set east of the rapids was the way into this "Garden of the Lord." He raised £1.7 million in bonds backed by the Brazilian government, set up the Madeira-Mamoré Railway Company, and in 1872 sent out his first crew of British engineers. Their boats sank. Caripune Indians attacked. Fever-plagued crews plunged through the forest in an ill-fated effort to flee. Workers died from disease and the heat. The rain forest was so dense that surveyors could only measure a few feet ahead. By 1873, the project was over, and in London the company's stock tumbled from 68 to 18 points on the Exchange. Remaining workers abandoned their job sites when they heard their employer went bankrupt, leaving equipment to rot. British financial assessors reported that "the region is a welter of putrefaction where men die like flies. Even with all the money in the world and half its population, it is impossible to finish this railway." By 1876, however, the Brazilian government had subsidized the company in a desperate bid to link the international rubber port of Pará with the Acré, and Church was back in Philadelphia, recruiting more men.

174. **Markham was not present to oversee the handoff** Donovan Williams, "Clements Robert Markham and the Geographical Department of the India Office, 1867–1877," *Geographical Journal*, vol. 134 (Sept. 1968), p. 349.

174. **"laxity . . . vigorously and systematically suppressed"** Ibid., p. 351.

175. **The same India Office that grew incensed** Dean, *Brazil and the Struggle for Rubber*, p. 16.

175. **"once again we started by boat"** Violet Wickham, "Lady Wickham's Diary," p. 5.

175. **"[H]e decided to collect himself . . . as is very common there"** Ibid.

176. **"a minor who had been unlawfully taken"** Smith, *Brazil, the Amazons and the Coast*, p. 128.

176. **"the country house of an Englishman . . . got more 'kudos' for pluck than I perhaps deserved"** Violet Wickham, "Lady Wickham's Diary," p. 5.

176. **February to March was the period** Edward V. Lane, "Sir Henry Wickham: British Pioneer; a Brief Summary of the Life Story of the British Pioneer," *Rubber Age*, vol. 73 (Aug. 1953), p. 651.

176. **"left me there . . . while he and the boy went off into the woods"** Violet Wickham, "Lady Wickham's Diary," p. 5.

176. **"I am now collecting Indian Rubber seeds in the 'ciringals' "** Royal Botanic Gardens–Kew, *Miscellaneous Reports: India Office: Caoutchouc I*, "Letter from Henry Wickham to Hooker, March 6, 1876," file folder 19.

177. **Cross was a veteran of the cinchona expedition** William Cross, "Chronology: The Years Robert Cross Spent at Home and Abroad," http://scottishdisasters.tripod.com/robertmckenziecrossbotanicalexplorerkewgardens; also, e-mail interview with William Cross on April 3, 2006. William Cross is the descendant of Robert Cross and maintains an exacting and informative website on the often overlooked Kew gardener.

177. **"to cover all expenses and include remuneration"** Royal Botanic Gardens–Kew, *Miscellaneous Reports: India Office: Caoutchouc I*, "Letter from Henry Wickham to Hooker, March 6, 1876," file folder 18.

177. **"their exact place of origin was in 3 degrees of south latitude"** Henry Wickham, "The Introduction and Cultivation of the Hevea in India," *India-Rubber and Gutta-Percha Trades Journal*, vol. 23 (Jan. 20, 1902), p. 81. David Riker would later head to the same area when prospecting for seeds for the Ford plantation. Wickham "made several trips by boat to Boim," Riker said during the Ford era, adding that the seeds for Britain's Asian plantations hailed from there. In the 1960s, environmental historian Warren Dean interviewed Julio David Serique, whose father had been a *patrão* in Boim when Wickham arrived and who confirmed Riker's words. William Schurz, whose 1925 *Rubber Production in the Amazon Valley* influenced Henry Ford's decision to open Fordlandia, interviewed a "Moyses Serique" of the same family. "Boim is the first place on the Tapajoz in which wild *Hevea* is found," Schurz discovered in his own explorations, "and it is not probable that Wickham went further up the river." Not probable, he said, because *estradas* were already being worked in the area and further up the river a "weak" species of *Hevea* that produced an inferior grade of rubber began to predominate. Baldwin, "David Riker and *Hevea brasiliensis*; the Taking of Rubber Seeds out of the Amazon,"

p. 384; Dean, *Brazil and the Struggle for Rubber*, p. 17; William L. Schurz, *Rubber Production in the Amazon Valley*, Department of Commerce: Bureau of Foreign and Domestic Commerce (Washington, DC: Government Printing Office, 1925), p. 133.

178. Boim was more important commercially A general overview of the Jewish migration to the Amazon can be found in several excellent articles: Ambrosio B. Peres, "Judaism in the Amazon Jungle," in *Studies on the History of Portuguese Jews from Their Expulsion in 1497 through Their Dispersion*, Israel J. Katz and M. Mitchell Serels, eds. (New York: Sepher-Hermon Press, 2000), pp. 175–183; Susan Gilson Miller, "Kippur on the Amazon: Jewish Emigration from Northern Morocco in the Late Nineteenth Century," in *Sephardi and Middle Eastern Jewries: History and Culture in the Modern Era*, Harvey E. Goldberg, ed. (Bloomington, IN.: Indiana University Press, 1996), pp. 190–209; "Brazil," *Encyclopedia Judaica*, vol. 4 (Jerusalem: Macmillan, 1971), and "Morocco," ibid., vol. 12; and "Sephardic Genealogy Resources; Indiana Jones Meets Tangier Moshe," www.orthohelp.com/geneal/amazon.htm.

178. Boim's four trading families had come from Tangiers in Morocco Interview, Elisio Eden Cohen, Boim, October 21, 2005.

178. "One must take great care in the jungle on entering" Quoted in Miller, "Kippur on the Amazon: Jewish Emigration from Northern Morocco in the Late Nineteenth Century," p. 201.

179. one of the owners of the "Franco & Sons" cattle ranch C. Barrington Brown and William Lidstone, *Fifteen Thousand Miles on the Amazon and its Tributaries* (London: Edward Stanford, 1878), p. 251.

179. "that the people who annually penetrate into these forests" Henry Wickham, "The Introduction and Cultivation of the Hevea in India," pp. 81–82.

179. the ancient sites covered in a deep, stiff Indian black earth Interview, Elisio Eden Cohen, Boim, Oct. 21, 2005.

179. "a circumference of 10 ft. to 12 ft. in the bole" Henry Wickham, "The Introduction and Cultivation of the Hevea in India," p. 81.

179. "I daily ranged the forest" Henry Wickham, *On the Plantation, Cultivation and Curing of Pará Indian Rubber*, p. 50.

180. "the bark is thickly coated with growths of moss, ferns, and orchids" Henry Wickham, "The Introduction and Cultivation of the Hevea in India," p. 81.

180. "out of seventeen varieties" William Chauncey Geer, *The Reign of Rubber* (New York: Century, 1922), p. 73.

180. The black-bark version was said to yield more latex C. C. Webster and E. C. Paardekooper, "The Botany of the Rubber Tree," in *Rubber*, C. C. Webster & W. J. Baulkwill, eds. (Essex, UK: Longman, 1989), pp. 60–61.

180. Now in his sixties, he'd had the stump of the Mother Tree pointed out to him Interview, Elisio Eden Cohen, Boim, Oct. 21, 2005.

180. as seven men barely stretched their arms around it, standing fingertip to fingertip Was the Mother Tree as big as Cohen said? One rule for measuring girth is that a man's reach is about the same length as his height. Given that the average Amazon Indian or *caboclo* was five feet six inches, seven men circling the tree would give a circumference of 38.5 feet. This gives a radius of 6.13 feet, and doubling that, a diameter of 12.26 feet. This is a big tree, but not unheard of. General Sherman, the giant sequoia in California acknowledged until recently to be the world's largest living tree, has a diameter of 102 feet and a height of 362 feet, or about the size of a thirty-six-story building. The maximum height usually listed today for hevea in the Amazon is 30 meters, but heights of 40 meters, or about 130 feet, were said to exist, if never confirmed, in the virgin forests of Wickham's day.

181. "[D]uring times of rest, I would sit down and look into the leafy arches above" Henry Wickham, *On the Plantation, Cultivation and Curing of Pará Indian Rubber (Hevea brasiliensis) with an Account of Its Introduction from the West to the Eastern Tropics* (London: K. Paul, Trench, Trübner and Co., 1908), pp. 50–51.

181. of the same design as those found along the Amazon today Interview, Elisio Eden Cohen, Boim, Oct. 21, 2005. Also, see the photo of the basket woven by Herica Maria Cohen, fourteen, daughter of Elisio Eden Cohen. It is the same design as those used by Wickham and by Indians and *caboclos* for centuries. The only difference in Wickham's baskets is that they would have been larger.

181. "I got the Tapuyo village maids" Henry Wickham, *On the Plantation, Cultivation and Curing of Pará Indian Rubber*, p. 51.

183. "first of the new line of Inman line steamships" Henry Wickham, *On the Plantation, Cultivation and Curing of Pará Indian Rubber*, pp. 47–48; and John Loadman, *Tears of the Tree: The Story of Rubber—A Modern Marvel* (Oxford: Oxford University Press, 2005), pp. 89–91. Loadman's detective work on the *Amazonas* is the best yet, uncovering sailing records and crew manifests that were previously unknown. Built in 1874 by A. Simey and Co. for the Liverpool & Amazon Royal Mail Steamship Company and registered in 1875, everything about the *Amazonas* was new. She sailed almost immediately under the E. E. Inman flag. Henry said this was its inaugural voyage, but in this he seems mistaken: She'd originally sailed from Liverpool on December 24, 1875, arrived in Pará on January 19, 1876, continued on to Manaus, and was home in Liverpool on March 14.

183. Crew records suggest a complement of thirty-two men Loadman, *Tears of the Tree*, p. 90.

183. "The thing was well-done" Henry Wickham, *On the Plantation, Cultivation and Curing of Pará Indian Rubber*, p. 48.

184. "occurred one of those chances, such as a man has to take at top-tide" Ibid., pp. 48–49.

185. "This suggests that the rapid charter [of the *Amazonas*] was to beat" P. R. Wycherley, "Introduction of Hevea to the Orient," *The Planter, Magazine of the Incorporated Society of Planters* (March 1968), p. 130. According to Wycherley, Wilkens wrote in the September 1940 issue of the *RRI Planters' Bulletin* that Brazilian authorities told Henry he "would not" be able to export the seeds; in the December 1967 issue of the *Planter*, he said they told him he "might not" be able to. By then, Wilkens himself was getting up in years; thus, the truth may be clouded by the fuzzy memories of both Wickham and Wilkens.

186. there is no mention of rubber seeds in the cargo manifest Loadman, *Tears of the Tree*, p. 90.

186. **"When [Henry] had collected and packed"** Violet Wickham, "Lady Wickham's Diary," p. 5.

187. **"What seems most likely . . . is that Wickham managed to persuade"** Dean, *Brazil and the Struggle for Rubber,* p. 19.

187. **John Joseph Wickham, his wife Christine, and son Harry** Interview with Anthony Campbell via e-mail, April 3, 2006, from his own genealogical research.

187. **widower of Henry's sister Harriette Jane** Anthony Campbell, "Descendants of Benjamin Wickham, a Genealogy" (self-published, Jan. 30, 2005).

187. **"slung up fore and aft in their crates"** Henry Wickham, *On the Plantation, Cultivation and Curing of Pará Indian Rubber,* p. 53.

188. **"crabbed and sore . . . so as not much to heed Murray's grumpiness"** Ibid.

188. **"It was perfectly certain in my mind"** Ibid.

188. **"a straight offer to do it; pay to follow result"** Ibid., p. 47.

188. **"a number of Brazilians had been much amused"** Austin Coates, *The Commerce in Rubber: The First 250 Years* (Singapore: Oxford University Press, 1987), p. 67.

188. **"an obstacle of appalling magnitude"** Anthony Smith, *Explorers of the Amazon,* p. 281.

188. **"a friend in court"** Henry Wickham, *On the Plantation, Cultivation and Curing of Pará Indian Rubber,* p. 53.

189. **"quite [entered] into the spirit of the thing"** Ibid., pp. 53–54.

189. **a commoner named Ulrich** Dean, *Brazil and the Struggle for Rubber,* p. 19.

190. **The evening was pleasant and cordial** Coates, *The Commerce in Rubber,* p. 67.

190. **"I could breathe easy"** Henry Wickham, *On the Plantation, Cultivation and Curing of Pará Indian Rubber,* p. 54.

190. **"Products destined for Cabinets of Natural History"** The Brazilian customs regulations are quoted in Dean, *Brazil and the Struggle for Rubber,* p. 19.

191. **"hardly defensible in international law"** Loadman, *Tears of the Tree,* p. 92.

192. **Madagascar never made a dime** "Living Rainforest: Cancer Cured by the Rosy Periwinkle," www.livingrainforest.org/about/economic/rosyperiwinkle.

192. **O. Labroz and V. Cayla of Brazil claimed that authorities** Ibid.

193. **"to appropriate the goods of others"** Ibid., p. 21.

193. **"some higher vision of property"** Ibid., p. 22.

Chapter 10: The Edge of the World

197. **Hooker was an insomniac** Richard Collier, *The River That God Forgot: The Story of the Amazon Rubber Boom* (New York: E. P. Dutton, 1968), pp. 35–36.

198. **he'd leave Liverpool on June 19** Robert Cross listed the dates of his departure from Liverpool and arrival in Pará in his report on the *Investigation and Collecting of Plants and Seeds of the India Rubber Trees of Pará and Ceara and Balsam of Copaiba,* completed in Edinburgh on March 29, 1877. Excerpts from his report are included in William Cross, "Robert McKenzie Cross: Botanical Explorer, Kew Gardens: Chronology: The

Years Robert Cross Spent at Home and Abroad," http://www.scottishdisasters.tripod
.com/robertmckenziecrossbotanicalexplorerkewgardens.

198. **"Not even the wildest imagination could have contemplated"** Coates, *The Commerce in Rubber*, p. 68.

198. **a vast greenhouse called the "seed-pit"** Lane, "The Life and Work of Sir Henry
Wickham: Part IV—Kew" (Dec. 26, 1953), p. 8.

198. **placed in the care of R. Irwin Lynch** "R. Irwin Lynch," *Journal of the Kew Guild*,
vol. 4, no. 32 (1925), p. 341. Lynch, foreman of the tropical department, was an "Old
Kewite," trained by his grandfather, "himself an Old Kewite," and joined the staff as a
student gardener in 1867 at the age of seventeen.

199. **"We knew it was touch and go"** Sir William Thiselton-Dyer quoted in Coates,
The Commerce in Rubber, p. 68.

199. **"70,000 seeds of *Hevea brasiliensis* were received from Mr. H. A. Wickham"**
Royal Botanic Gardens–Kew, *Miscellaneous Reports: India Office: Caoutchouc I,* Unsigned
note, July 7, 1876, file folder 20.

199. **"Many hundreds are now 15 inches long and all are in vigorous health"** Royal
Botanic Gardens–Kew, *Miscellaneous Reports: India Office: Caoutchouc I,* Ibid.

200. **"made some experiments in planting"** Dean, *Brazil and the Struggle for Rubber*, p. 24.

200. **"as if to underscore his ignorance of botany"** Ibid.

200. **On August 20, 1876, The *Evening Herald* ran a short story** Royal Botanic
Gardens–Kew, *Miscellaneous Reports: India Office: Caoutchouc I,* "*Evening Herald,* Aug.
20, 1876, describing growth of seeds at Kew," file folder 43.

200. **"Mr. Wickham seems to have taken very great pains with the seeds"** Dean, *Brazil
and the Struggle for Rubber*, p. 24.

200. **"I have had a long conversation with Mr. Wickham"** Lane, "The Life and Work
of Sir Henry Wickham: Part IV—Kew," p. 6.

201. **"we have no knowledge of his horticultural competence"** Ibid.

201–2. **"I did not mean to suggest my taking entire charge of the plants"** Ibid.

202. **"dishonourable"** Donovan Williams, "Clements Robert Markham and the Geographical Department of the India Office, 1867–1877," *Geographical Journal*, vol. 134
(Sept. 1968), p. 350.

202. **"comply with official rules, or go"** Ibid, p. 351.

202. **"the Malay Peninsula is most likely to combine the climactic conditions required"**
Lane, "The Life and Work of Sir Henry Wickham: Part IV—Kew," p. 7.

202. **"I have known trees, grown in the open"** Henry Wickham, *On the Plantation,
Cultivation and Curing of Pará Indian Rubber*, p. 58.

203. **"What is more . . . its coffin bore the wrong name"** Lane, "The Life and Work
of Sir Henry Wickham: Part IV—Kew," p. 7.

203. **though £740 according to a memo by Hooker** Dean, *Brazil and the Struggle for
Rubber*, p. 24. This letter, dated June 24, 1876 and addressed to Clements Markham,
introduced Wickham "who has been collecting seeds for you. He has brought 74,000

which have all been planted," thus implying that, according to the agreement, he would be paid £740.

203. **"though Kew authorities advocated it"** Violet Wickham, "Lady Wickham's Diary," p. 5.

204. **The P&O Company had caused a revolution in sea transport** Coates, *The Commerce in Rubber*, p. 69.

204. **an enraged Clements Markham expedited the payment back in London** The affair of the *Duke of Devonshire* is a good example of British bureaucracy at its worst and of Clements Markham's frustrations with the India Office's new regime. When H. K. Thwaites, Director of the Peredeniya Gardens, telegraphed for help, Markham was called in. He spent the next week carrying the freight documents from desk to desk, all the way to the detested Louis Mallet himself, provoking some comments about the "new bureaucracy" that probably hastened his departure the following year. Thistelton-Dyer was drawn in, and on September 18, 1876, Markham raged to him in a letter about all the "fatuous processes" needed to gain approval for even the smallest expenses. "Bad as the senseless routine was before," he lamented, "it has become much worse since Sir Louis Mallet and Lord Salisbury have been here." His frustration and disgust were obvious, and he did little to hide them. More instructive, however, was Markham's detailed list of a penny-pinching requisition process that discouraged new initiatives and effectively stifled change. The payment of a simple freight bill, which at best should require an invoice and payment, took ten steps drawn out over thirty days. Markham called it nothing more than "the ordinary circumlocution:

> Aug. 18—I sent down request for sanction to pay freight.
>   22—Sir L. Mallet sends to Lord Salisbury.
>   29—Lord Salisbury sends it to a c'tee of Council.
> Sept. 7—The c'tee sent it back to Sir L. Mallet.
>   9—Sir L. Mallet sent it to the Council.
>   10—The Finance C'tee sent it back.
>   11—Sir L. Mallet sent it to the Council.
>   14—The Council sanctioned the payment.
>   15—It was sent back to me.
>   16—It was paid.

Markham ended the tale by remarking that this letter was "not official, or you would not get it for a month." The released seeds were planted in Colombo, and by 1880, about three hundred of them were still alive. Royal Botanic Gardens–Kew, *Miscellaneous Reports: India Office: Caoutchouc I*, "Letter from Markham to Thiselton-Dyer, Sept. 18, 1876," file folder 59.

204. **while Kew admitted that twelve hevea plants from Cross's mission had managed to survive** As sickly as Cross's hevea specimens were, they became the center of a controversy that still rages among historians. On June 11, 1877, Kew sent twenty-two hevea to Singapore that may or may not have come from Cross's trees.

Taken together, the number of trees sent to India and the Far East totaled several hundred more than the sum of Wickham's and Cross's collections combined. The only explanation, reasoned historian Warren Dean, was that "there had been some propagation through cuttings," the phenomenon whereby detached plant parts can regenerate missing roots, stems, and leaves to form complete new plants. This is good news for centers of economic botany like Kew, for one isn't shackled to seed production, and growing stock can increase geometrically. Yet when Kew sent out shipments, they never kept records of these cuttings. No one knows which came from Wickham's seeds and which came from Cross's young trees. This is important, if only to illustrate the spite growing around Wickham, as well as the fate of those who serve. By the 1920s, when Great Britain controlled the world rubber market, Wickham was honored, Cross forgotten, and experts took sides. People asked, *Who really started the plantation rubber industry?* Was it Wickham, with his more robust seeds from the highlands behind Boim and his greater number of seedlings? Or Cross, whose sickly seedlings from the swamps around Pará may or may not have constituted the June 1877 batch of trees to Singapore, said to form the backbone of the vast Malayan rubber plantations?

The intricacies of the Cross vs. Wickham debate are so arcane that it's tempting to ignore it altogether, yet to do so ignores Cross's contributions and sidesteps one reason historians tend to dislike Wickham. It also ignores the pedigree of the anti-Wickham chorus hailing from Kew, whose records in this respect are so contradictory as to be self-canceling. Some documents state that the Singapore batch came from Wickham's trees, others from Cross's, and the confusion over cuttings just muddies the water. In time, the debate turned into a culture war. While businessmen and planters favored Wickham, Kew's botanists echoed Joseph Hooker's prejudice and bet on Cross, a member of the club. Although the bulk of seeds shipped around the world came from Henry's stock, the debate turned on an unsubstantiated statement by Henry N. Ridley, Hooker's protégée and director of the Singapore Botanical Gardens, that the 1877 Singapore shipment came from Cross's trees. Thiselton-Dyer—also in the Hooker camp—supported Ridley. However, when one checks the numbers, one must admit that by this point the Cross and Wickham stocks were so intermingled by cuttings and loose record keeping as to be inseparable. An observation by early rubber planters that they noticed "an extraordinary variety in their trees" supports such mixing.

The pedigree of Cross's seeds was much different from Wickham's. He did not penetrate the interior, as had Wickham, but stayed close to Pará. He left Liverpool on June 19, 1876, the same day that the first of Henry's seeds began to sprout in Kew's seed pit, and arrived at Pará on July 15. He set to work, replanting, tending, and packing his 1,080 specimens. He collected most of his rubber from the swamps and flood plains surrounding the city, which put his opinions on hevea's natural habitat in direct opposition to Henry's. Consul Green helped him as he had Henry, rendering him "every assistance possible." He embarked on the *Paraense* of the Liverpool Red Cross Line, collected his sixty specimens of Ceará rubber when the ship stopped at the Brazilian port of Fortaleza, and returned to England on November 22, 1876.

The distribution of Wickham's and Cross's plants was pieced together and cross-referenced from six primary sources, each of which leave out some detail but all of which seem to agree chronologically: John Loadman, *Tears of the Tree: The Story of Rubber—A Modern Marvel* (Oxford: Oxford University Press, 2005); Warren Dean, *Brazil and the Struggle for Rubber: A Study in Environmental History* (Cambridge: Cambridge University Press, 1987); Austin Coates, *The Commerce in Rubber: The First 250 Years* (Singapore: Oxford University Press, 1987); William Chauncey Geer, *The Reign of Rubber* (New York: Century, 1922); Edward Valentine Lane, "The Life and Work of Sir Henry Wickham: Part IV—Kew," *India Rubber Journal*, vol. 125 (Dec. 26, 1953), pp. 5–8; and Lane, "Sir Henry Wickham: British Pioneer; a Brief Summary of the Life Story of the British Pioneer," *Rubber Age*, vol. 73 (Aug. 1953), pp. 649–656.

The Cross vs. Wickham debate will probably never be solved. As stated in the text, Ridley and Thiselton-Dyer made statements that would suggest Cross's trees as the source of the British rubber monopoly, but both are suspect: Thiselton-Dyer because he would do little to contradict Hooker, and Ridley because he actively disliked Wickham and made several disparaging statements throughout his life to try to diminish Henry's importance. There also seemed some jealousy at play—Henry would be knighted, but Ridley was not. In 1914, David Prain, who became Kew's director after Thiselton-Dyer's retirement in 1905, questioned "whether a single plant brought back by Cross ever became fit to send" anywhere in Asia. He could not find "any entry in our archives that could be so interpreted." (Dean, 28) Prain's statement is also interesting because he was the first director not related by blood or marriage to the Hooker and William Thiselton-Dyer family circle. Warren Dean seemed to concur with Prain, stating that "evidently, the Wickham selection provided the overwhelming genetic stock for the spread of cultivation in the British colonies" (Dean, 27), but even he was intrigued by the mystery and indulged in speculation. John Loadman, the most recent of the long line of rubber historians, clearly sides with Cross after some meticulous detective work, calling Cross, not Wickham, the "father of the rubber plantation industry"—yet even he admits that "in spite of all the detailed records kept by Kew, one piece of information is missing, and that is the source of those . . . seedlings" (Loadman, 94).

205. **"The flat, low lying, moist tracts, subject to inundation"** Royal Botanic Gardens–Kew, *Miscellaneous Reports: India Office: Caoutchouc I*, Robert Cross, "Report on the Investigation and Collecting of Plants and Seeds of the India-rubber Trees of Para and Ceara and Basalm of Copaiba (March 29, 1877)," file folders 78–93. p. 7.

205. **Robert Cross's opinions** Cross fared little better than his trees. His treatment at the hands of the British government was as bad that of the original hevea prophet, James Collins. By 1881, Cross was forty-seven, stricken like Spruce with debilitating bouts of malaria, and had applied repeatedly for a medical pension. In 1882, he dared to publicize in the *South of India Observer* that the Indian cinchona plantations had lost the empire £2 million when compared to the Dutch plantations, and he revealed for the first time the debacle of Charles Ledger's rejected Bolivian yellow-bark trees. The revelations were not appreciated in the higher levels of government. Joseph Hooker launched an investigation, which revealed nothing about Ledger's offer but discredited the work in India on hybrid varieties of cinchona—the work that Cross supervised.

In a letter to Markham dated July 21, 1882, Cross lamented the fact that his revelations had created so much ill will against him. Some people even blamed him for the depreciation of the Indian cinchona plantations and called for his hide. Although he was eventually exonerated of Hooker's allegations, his career was ruined. Soon afterward, according to the *Nilgiri Express,* he was working at Nilamur overseeing the growth of some new rubber trees when "in reply to some overtures made in his behalf to the Secretary of State, a telegram was received. What the purport of this telegram was we know not, but its contents so disgusted [Cross], that he shook the dust off his feet and departed to seek fresh fields and pastures new." By 1884, Cross had left the service entirely and retired on a £40 annuity. At night, he sweated through malarial dreams in his cottage in Edinburgh, and slept with a gun beneath his pillow that he'd used in Ecuador to fend off snakes and thieves. William Cross, "Robert McKenzie Cross: Botanical Explorer, Kew Gardens: Chronology: The Years Robert Cross Spent at Home and Abroad," p. 6 of 7; Royal Botanic Gardens–Kew, *Miscellaneous Reports 5: Madras-Chinchona, 1860–97,* "Letter from Robert Cross to Clements Markham, July 21, 1882," file folder 131 and 132. The article in the *Nilgiri Express* and an account of Cross's last years in his cottage are carried in William Cross, "Robert McKenzie Cross: Botanical Explorer, Kew Gardens—Last Years at West Cottage Torrance of Campsie," p. 2, www.scottishdisasters.tripod.com/robertmckenziecrossbotanicalexplorerkewgardens.

205. **that nearly doubled production every five, then every three years** Henry Hobhouse, *Seeds of Wealth: Four Plants That Made Men Rich* (London: Macmillan, 2003), p. 134.

206. **More than half of these were Henry Ford's Model Ts** Hobhouse, *Seeds of Wealth,* p. 131.

207. **They pitched it** John R. Millburn and Keith Jarrott, *The Aylesbury Agitator: Edward Richardson: Labourer's Friend and Queensland Agent, 1849–1878* (Aylesbury, Queensland, Australia: Buckingham County Council, 1988), pp. 28–31. Archives of the Institute of Commonwealth Studies, University of London.

207. **the violence in the Australian frontier claimed the lives** "Statistics of Wars, Oppression and Atrocities in the Nineteenth Century," http://users.erol.com/mwhite28/wars19c.htm.

208. **Carl Lumholtz, a Norwegian anthropologist** Carl Lumholtz's *Among Cannibals* (London, 1890), is quoted in G. C. Bolton, *A Thousand Miles Away: A History of North Queens to 1920* (Sydney: Australian National University Press, 1970), p. 95.

208. **"There is nothing extraordinary in it"** The *Queensland Figaro* is quoted in Raymond Evans, "'Kings' in Brass Crescents: Defining Aboriginal Labour Patterns in Colonial Queensland," in *Indentured Labour in the British Empire, 1834–1920,* Kay Saunders, ed. (London: Croom Helm, 1984), p. 196.

208. **"These children are brought in and tied up"** Ibid., p. 196.

208. **"a runaway black child could be hunted and brought back"** Ibid., p. 199.

209. **"in spite of the restrictions on board"** Violet Wickham, "Lady Wickham's Diary," p. 6.

209. his real hope was to cultivate the leaf Lane, "The Life and Work of Sir Henry Wickham: Part V—Pioneering in North Queensland" (Jan. 2, 1954), p. 17.

209. It cost about £110 J. C. R. Camm, "Farm-making Costs in Southern Queensland, 1890–1915," *Australian Geographical Studies*, vol. 12, no. 2 (1974), p. 177.

209. "Dear Land," as they called it Geoffrey Blainey, *The Tyranny of Distance* (London: Macmillan, 1968), pp. 165–167. The Crown Land Act of 1868 and the Homestead Act allowed the selection of 80– to 160–acre homesteads, and records show that about 3 million acres were taken. But about half of this went to 267 people, which translated into vast sugar plantations and, in the north, cattle ranches.

209. "assured him if he could produce that quality" Violet Wickham, "Lady Wickham's Diary," p. 7.

210. "once more there was the old work of cutting down the site for the house" Violet Wickham, "Lady Wickham's Diary," p. 7.

210. "which rather amused his neighbors" Ibid.

210. "When the dew was off I set fire to it" Ibid.

210. "Saddles, flour, etc., might have been saved" Ibid., p. 9.

210. "leaked the whole way" Ibid., p. 10.

211. "Rain does not express it" Ibid., p. 9.

211. "After having burnt out, it seemed necessary for me to try the water cure" Ibid., p. 10.

211. "I fell asleep" Ibid.

212. "cutting off the leaves and young shoots . . . even on the other side of the creek" Ibid., p. 9.

212. farmers discovered that if they could get the victim drunk on rum Charles H. Eden, *My Wife and I in Queensland: An Eight Year's Experience in the Above Colony, with Some Account of Polynesian Labour* (London: Longmans, Green, 1872), pp. 146–147.

212. "You may imagine I slipped out of it as quickly and quietly as I could" Violet Wickham, "Lady Wickham's Diary," p. 9.

213. "The fowls are just having bad dreams" Violet Wickham, "Lady Wickham's Diary," p. 8.

213. "Directions for Tobacco Growing and Curing in North Queensland" Lane, "The Life and Work of Sir Henry Wickham: Part V—Pioneering in North Queensland," p. 17.

213. "I did not need anyone with me" Violet Wickham, "Lady Wickham's Diary," p. 8.

213. "though I expect I considerably scandalized those neighbors" Ibid.

214. a number of government commissions The regulations enacted for Kanaka labor can be found in the following sources: Lane, "The Life and Work of Sir Henry Wickham: Part V—Pioneering in North Queensland," pp. 17–18; G. C. Bolton, *Planters and Pacific Islanders* (Croydon, Victoria, Australia: Longman's, 1967), pp. 22–23; G. C. Bolton, *A Thousand Miles Away: A History of North Queens to 1920* (Sydney: Australian National University Press, 1970), pp. 79–83; "Queensland Sugar Industry," *MacKay Mercury*, Sept. 25, 1878; and the Royal Commonwealth Society Archives, *Cuttings of the Queensland Sugar Industry*, GBR/0115/RCMS 294.

214. **"The Kanaka is at best a savage"** "The Labour Question," *The Queenslander,* May 14, 1881. Royal Commonwealth Society Archives, *Cuttings of the Queensland Sugar Industry,* GBR/0115/RCMS 294.

214. **"A favorite device . . . was to hold up two or three fingers"** Bolton, *A Thousand Miles Away,* p. 79.

214. **"breaking in"** Ibid.

215. **From 1883 to 1885, nearly seven thousand people were kidnapped or duped** Kay Saunders, "The Workers' Paradox: Indentured Labour in the Queensland Sugar Industry to 1920," in *Indentured Labour in the British Empire 1834–1920,* Kay Saunders, ed. (London: Croom Helm, 1984), p. 226.

215. **"The chief Magistrate of the district"** Violet Wickham, "Lady Wickham's Diary," p. 7.

215. **"[W]e found them trustworthy"** Ibid., pp. 7–8.

216. **"I have often wondered . . . whether it was not a plot"** Lane, "The Life and Work of Sir Henry Wickham: Part V—Pioneering in North Queensland," p. 19.

216. **"probably little more than sufficient for the fares home"** Ibid.

Chapter 11: The Talking Cross

217. **"agreed to join a friend in journeying to British Honduras"** Lane, "The Life and Work of Sir Henry Wickham: Part VI—Pioneering in British Honduras," *India Rubber Journal,* vol. 126 (Jan. 9, 1954), p. 17.

217. **"I let him go back some six months in advance"** Violet Wickham, "Lady Wickham's Diary," p. 11.

218. **a world record** The *James Stafford*'s passage across the Pacific was a record for sailing ships that would stand until 1995.

218. **the origin of Belize City** Sir Eric Swayne, "British Honduras," *Geographical Journal,* vol. 50, no. 3 (Sept. 1917), p. 162; John C. Everitt, "The Growth and Development of Belize City," *Journal of Latin American Studies,* vol. 18, no. 1 (May 1986), p. 78.

219. **logwood sold for about £100 a ton** Swayne, "British Honduras," p. 162; David Cordingly, *Under the Black Flag: The Romance and Reality of Life Among the Pirates* (New York: Harcourt, Brace, 1995), p. 150. Cordingly puts the trade in perspective. Logwood was profitable, but not the most profitable trade of the time. During the same period, the colonies of Virginia and Maryland were together exporting seventy thousand hogsheads of tobacco annually worth £300,000 each year. "Log cutting was always a minor industry carried on by a few hundred ex-seamen and pirates in a remote corner of the globe," Cordingly said (p. 150), but in British Honduras at the time, it was the only game in town.

219. **By 1705, the British shipped most of their logwood from the Belize River area** According to a government report, some 4,965 tons of logwood were exported to England from 1713 to 1716 at no less than £60,000 per annum. By 1725, the production had increased to 18,000 tons a year.

219. mahogany exports climbed to twelve thousand tons Robert A. Naylor, *Penny Ante Imperialism: The Mosquito Shore and the Bay of Honduras, 1600–1914, A Case Study in British Informal Empire* (London: Associated University Presses, 1989), p. 103.

220. the white population dropped from 4 percent in 1845 to 1 percent in 1881 Everitt, "The Growth and Development of Belize City," p. 93, file folder 85.

220. "monied cutters" Ibid., p. 90.

221. the colony's treasury, which had a £90,000 surplus Everitt, "The Growth and Development of Belize City," p. 96. Actually, the public works improvements that cast Goldsworthy as the villain had been conceived before he arrived. Known as the Siccama Plan of 1880, named after Baron Siccama, the engineer who devised it, this was a comprehensive and ambitious scheme for municipal improvements that proposed filling in low-lying lots, increasing water-storage facilities, building a pier, and dredging the canals, which had not been cleaned since the 1860s. It is easy to see why the government had a surplus. It had never addressed these major problems, which ran quickly through the surplus when the Siccama Plan was begun. Because of the controversy, the plan was abandoned, and Belize returned to its old pattern of squalor and decay through the rest of the century.

221. "never for one moment ceased to be a friend of the least reputable portion" Wayne M. Clegern, *British Honduras, Colonial Dead End, 1859–1900* (Baton Rouge: Louisiana State University Press, 1967), p. 80, quoting the October 4, 1890 *Colonial Guardian*.

221. The governor was fond of the rubber thief Lane, "The Life and Work of Sir Henry Wickham: Part VI—Pioneering in British Honduras," p. 17.

221. Goldsworthy may have prayed that he'd never have to return When he left the Colonial Offices at Government House, he received a military salute from an honor guard, but as his barge steered past the breakwater, the crowd collected there hissed and jeered. Goldsworthy responded with a "sardonic" gesture, and that unleashed a storm. Some women and children tried to throw stones from the roadside, but the police prevented that, so they rushed into the lagoon, plucked rocks from the bottom and began chucking them at the barge. The crew of the barge pulled for all their worth to escape being pelted; when they reached the steamer, a lighter passed and on it a citizen held up a white banner with the words "Catfish Still Uneaten" in red letters, a reference to Goldsworthy's comment that "he'd make the people eat catfish before he was done with them." The steamer took an hour to get under way: the crowds at the wharf jeered and hooted; the lighter circled and circled, dipping its flag. Clegern, *British Honduras, Colonial Dead End, 1859–1900*, p. 78, quoting the October 23, 1886, *Colonial Guardian*.

221. "Whether it was simply a matter of completing a routine tour of duty" Ibid., p. 80.

222. "A friend and I persuaded him to take a Government post" Violet Wickham, "Lady Wickham's Diary," p. 12.

222. Henry did a little of everything National Archives, Kew, *Honduras Gazette*, 1887,

1888, 1889, 1893. CO 127/6,7, & 8: *Honduras Gazette,* May 7, 1887, p. 78; *Honduras Gazette,* Dec. 17, 1887; *Honduras Gazette,* May 12, 1888, p. 81; *Honduras Gazette,* Dec. 15, 1888, p. 215; *Honduras Gazette,* May 4, 1889, p. 75.

222. **In 1890, he became inspector of forests** Lane, "The Life and Work of Sir Henry Wickham: Part VI—Pioneering in British Honduras," p. 17.

222. **"Being in contact with the Governor I was invited"** Violet Wickham, "Lady Wickham's Diary," p. 11.

223. **"Mr. Wickham is a large-framed idealist, dreamy, sympathetic"** Lane, "The Life and Work of Sir Henry Wickham: Part VI—Pioneering in British Honduras," p. 17.

224. **"I believe [Peck's] errand to be somewhat fanciful"** National Archives, Kew, *Colonial Office: British Honduras, Register of Correspondence, 1883–1888,* CO 348/10. Despatch No. 5: "Mission of Mr. J. B. Peck of New York to British Honduras," Sir Roger Goldsworthy, January 17, 1888. What follows is the text of Goldsworthy's dispatch to the Foreign Office, showing his amusement and naming Wickham as the dig's watchdog:

Sir:

1. I have the honor to inform you that a schooner-rigged yacht the "Maria" arrived at this port in the 28th Ultimo from New York, in charge of a Mr. John Benjamin Peck, said to be a special Treasury agent of the United States.

2. On the day of his arrival the mails from America brought news that Mr. Peck's journey to Belize, though supposed to be in search of hidden treasure, was in reality connected with some filibustering expedition against the neighboring friendly republic of Honduras.

3. Mr. Peck's action, on arrival, in endeavouring to enter into a business agreement with the Belize Estate and Produce Company in relation to his intention to search for treasure on lands belonging to that company and his subsequent steps to secure my approval, subject to such conditions as I might wish to impose, appeared sufficient proof to me, apart from the visit of two Customs House officers on board the "Maria" to ascertain whether he was armed, that Mr. Peck's object was seemingly what he represented it to be, and that there was, at least at present, no hostile intentions to be apprehended.

4. I informed the Consuls of Guatemala and Honduras accordingly, letting them understand that any news I might receive contrary to these convictions should be at once communicated to them, and they have expressed themselves gratified by the courtesy shown to their governments by my action.

5. I enclose a copy of the agreement that I entered into with Mr. Peck in the event of his finding another "Solomon's Mines" and you will observe that it has been drawn up so as to meet the possible case of treasure trove being found on Crown Lands when I would claim the whole on behalf of the Crown under the common law of England, subject to surrender in part

or whole under subsequent arrangement—Mr. Wickham accompanies the expedition on behalf of the Government as a precautionary measure.

6. I enclose herein copy of a despatch which I have addressed to Sir Lionel Sackville, British Minister at Washington.

> I have the honor to be, Sir,
> Your most obedient, humble servant,
> Roger Tuckfield Goldsworthy, Governor

224. eight hundred thousand dollars in gold specie A list of shipwrecks along the Belizean coast is found in "Overview of Belizean History," www.ambergriscaye.com/fieldguide/history2.html.

224. An iron human skeleton Lindsay W. Bristowe, *Handbook of British Honduras for 1891–1892, Comprising Historical, Statistical and General Information Concerning the Colony* (London: William Blackwood, 1891), p. 46.

224–25. "[Henry] believes they really were on the spot" . . . with all hands during a gale Violet Wickham, "Lady Wickham's Diary," p. 18; Lane, "The Life and Work of Sir Henry Wickham: Part VI—Pioneering in British Honduras," p. 18.

225. "[W]ild tales are told of men" Swayne, "British Honduras," p. 167.

225. jumped in the channel and swam home Lane, "The Life and Work of Sir Henry Wickham: Part VI—Pioneering in British Honduras," p. 18. **Author's note:** Lacking Violet's thoughts, I asked my wife, Kathy, what her response would be. "Being gone a night is a lot better than dead," she replied.

225. "Strange as it may seem in a colony so old" J. Bellamy, "Report on the Expedition to the Corkscrew Mountains," *Proceedings of the Royal Geographical Society and Monthly Record of Geography*, vol. 11, no. 9 (Sept. 1889), p. 552. Original in Royal Geographical Society Archives, JMS/5/73.

226. "congested state of the mother country" Ibid.

226. "impossible . . . to recover lost ground" Swayne, "British Honduras," pp. 164–165.

226. "Mr. Wickham continued the ascent" Bellamy, "Report on the Expedition to the Corkscrew Mountains," p. 549.

226–27. "the final ascent became in sensation very like crawling over the edge of a great sponge" Bristowe, *Handbook of British Honduras for 1891–1892*, p. 24.

227. "returning with the good news of his success" Bellamy, "Report on the Expedition to the Corkscrew Mountains," p. 549.

227. "having recovered sufficient breath" Ibid., p. 550.

227. "During the night one of the Carib porters" Ibid.

228. In 1850, the Mayan insurgents were on the brink of defeat Information on the Caste War of the Yucatán, Chan Santa Cruz, and the Talking Cross comes from a variety of sources: Nelson A. Reed, *The Caste War of Yucatan* (Stanford, CA: Stanford University Press, 2001); "Chan Santa Cruz," www.absoluteastronomy.com; "Caste War of the Yucatan: Information from Answers.Com," www.answers.com; "Northern Belize—The Caste War of the Yucatan and Northern Belize," www.northernbelize.com;

"Historic Folk Saints," http://upea.utb.edu/elnino/researcharticles/historicfolksaint
hood.html; J. M. Rosado, "A Refugee of the War of the Castes Makes Belize His Home,"
The Memoirs of J. M. Rosado, ed. Richard Buhler, Occasional Publication No. 2, Belize
Institute for Social Research and Action, (Belize: Berex Press, 1977); Archives of the
Institute of Commonwealth Studies, University of London; Swayne, "British Honduras,"
p. 164; Jennifer L. Dornan, "Document Based Account of the Caste War," www.bol.ucla
.edu/~jdornan/castewar.html; Jeanine Kitchel, "Tales from the Yucatan," in www
.planeta.com/ecotravel/mexico/yucatan/tales; "Statistics of Wars, Oppression and Atrocities in the Nineteenth Century," http://users.erol.com/mwhite28/wars19c.htm,
quoting "Correlates of War Project," www.correlatesofwar.org.

228. **The Talking cross was not God Himself, but Santo Jesucristo, God's intermediary** Spanish "testimonials" from the Cross were written for the Chosen. The most
famous and important promised that the whites would lose and the People of the Cross
would win.

229. **"We are . . . a people living under our own laws"** National Archives, Kew, *Colonial
Office: British Honduras, Register of Correspondence, 1883–1888,* CO 348/10. Despatch No.
11, Enclosure 2: "A Statement by the Santa Cruz Indians," January 8, 1888.

229. **"When I arrived there he had just lost the sight of one eye"** William Miller, "A
Journey from British Honduras to Santa Cruz, Yucatan, with a map," *Proceedings of the
Royal Geographical Society and Monthly Record of Geography,* vol. 11, no. 1 (Jan. 1889),
p. 27. The handwritten copy of this article, with changes and a map, are preserved in the
Archives of the Royal Geographical Society, London, JMS/5/74: received July 1888.

229. **"[T]he governor, fearing a raid by the Santa Cruz Indians"** Lane, "The Life and
Work of Sir Henry Wickham: Part VI—Pioneering in British Honduras," p. 18.

230. **The Cross sat in the center in profound darkness** Reed, *The Caste War of Yucatan,*
p. 266. The description of the Ceremony of the Cross is culled from the accounts of several witnesses over the years.

231. **"Alas . . . back came [Henry's] old longing for plantation life"** Violet Wickham,
"Lady Wickham's Diary," p. 11.

231. **Five houses were located on the Temash River** British Honduras, *Report and
Results of the Census of the Colony of British Honduras, taken April 5th, 1891* (London:
Waterlow & Sons, 1892), p. 11. Archives of the Institute of Colonial Studies, University
of London.

231–32. **"as the man who brought the rubber seeds from the Amazon"** National
Archives, Kew, *Colonial Office and Predecessors: British Honduras, Original Correspondence
1744–1951,* "Mr. H. A. Wickham's Temash Concession, (Pleadings in court case)," 1892.
CO 123/200.

232. **"value to the extent of $10,000 to consist of India Rubber trees"** Lane, "The Life
and Work of Sir Henry Wickham: Part VI—Pioneering in British Honduras," p. 19.

232. **In 1889–90, fever swept through Belize** Bristowe, *Handbook of British Honduras for 1891–1892,* pp. 30–31.

232. **"maliciously fabricating false reports to the detriment of the colony"** Ibid., p. 33.

232. **"as nearly as possible died"** Violet Wickham, "Lady Wickham's Diary," p. 11.

232. **"He lived contentedly enough"** Ibid., p. 12.

233. **the editor was ordered to pay court costs** Bristowe, *Handbook of British Honduras for 1891–1892*, p. 33.

233. **the sale or lease of land to small settlers like Henry** Swayne, "British Honduras," p. 170.

233. **his account books showed a monthly balance between $13 and $47.96** Lane, "The Life and Work of Sir Henry Wickham: Part VI—Pioneering in British Honduras," p. 18.

234. **His replacement, Sir C. Alfred Maloney, made twelve thousand pounds** Bristowe, *Handbook of British Honduras for 1891–1892*, p. 13.

234. **"honest men, as a rule, [kept] aloof"** Clegern, *British Honduras, Colonial Dead End, 1859–1900*, p. 80, quoting the *Colonial Guardian* of October 4, 1890.

235. **"great and rare experience"** National Archives, Kew, *Colonial Office and Predecessors: British Honduras, Original Correspondence 1744–1951*, "Mr. H. A. Wickham's Temash Concession, (Pleadings in court case)."

235. **In 1892, Victoria had been on the throne for fifty-five years** Hector Bolitho, ed., *Further Letters of Queen Victoria: From the Archives of the House of Brandenburg-Prussia*, trans. Mrs. J. Pudney and Lord Sudley (London: Thornton Butterworth, 1971, first printed 1938), pp. 259–261.

235. **Victoria understood perfectly the importance of her colonies** Lytton Strachey, *Queen Victoria* (London: Penguin, 1971, first published 1921), pp. 236–242.

236. **"Let Justice be done. Victoria R. & I"** Lane, "The Life and Work of Sir Henry Wickham: Part VI—Pioneering in British Honduras," p. 19.

Chapter 12: Rubber Madness

237. **on September 7, the courts awarded him $14,500 in damages"** National Archives, Kew, *Colonial Office and Predecessors: British Honduras, Original Correspondence 1744–1951*, "State of Wickham's Case," 1893. CO 123/281.

237. **A hurricane hammered the colony that summer** National Archives, Kew, *Colonial Office and Predecessors: British Honduras, Original Correspondence 1744–1951*, "Damage Caused by Gale," 1893. CO 123/204.

237. **"His keen analytical mind and authoritarian manner"** Lane, "The Life and Work of Sir Henry Wickham: Part VII—The Conflict Islands and New Guinea," *India Rubber Journal*, vol. 126 (Jan. 16, 1954), p. 10.

238. **"appalling roughness . . . I have encountered nowhere such difficulties as in New Guinea"** Henry O. Forbes, "British New Guinea as a Colony," *Blackwoods Magazine*, vol. 152 (July 1892), p. 85.

238. **"great and salubrious 'Treasure Island' "** Ibid., p. 82.

238. **on the island of Samarai** Arthur Watts Allen, "The Occupational Adventures of an Observant Nomad," an unpublished memoir written by Allen and kept in the care of David Harris and Jenepher Allen Harris. The author has not seen the book, which is apparently uncopied and in fragile shape, but the Harrises described its contents in detail in an e-mail message dated December 30, 2006.

239. **"like the Cocos Islands, in the Indian Ocean"** J. Douglas, "Notes on a Recent Cruise through the Louisiade Group of Islands," *Transactions of the Royal Geographical Society of Australia, Victorian Branch*, vol. 5, part 1 (March 1888), p. 55.

239. **The islands varied in size** The description of the islands and central lagoon is found in "The Conflict Islands," www.conflictislands.net.

239. **"business is a dirty one but profitable"** Quoted in Bolton, *A Thousand Miles Away*, p. 141.

240. **Sir William Macgregor** "MacGregor, Sir William," www.electricscotland.com.

240. **According to a tale told to distant relatives** Arthur Watts Allen, "The Occupational Adventures of an Observant Nomad," Chapter 1.

241. **"This gentleman has been making trial of the sponges"** Lane, "The Life and Work of Sir Henry Wickham: Part VII—The Conflict Islands and New Guinea, " p. 8.

241. **"roughly ceiled to make a loft or sleeping place"** Violet Wickham, "Lady Wickham's Diary," p. 12.

242. **First there were sponges** *The Sponging Industry*, A booklet of the Exhibition of Historical Documents held at the Public Records Office (Nassau, Bahamas: Public Records Office, 1974), Archives of the Institute of Commonwealth Studies, University of London.

242. **This was planting coconut palms, from which he sold copra** "Coconuts and Copra," www.msstarship.com/sciencenew.

243. **"Then they turn it on its back"** Violet Wickham, "Lady Wickham's Diary," p. 12.

243. **"not a locality where anyone would, or could, work Mother-of-Pearl"** Lane, "The Life and Work of Sir Henry Wickham: Part VII—The Conflict Islands and New Guinea, " p. 9.

244. **"During the whole time of my sojourn there"** Violet Wickham, "Lady Wickham's Diary," p. 12.

244. **"We expected to be respected, have privileges, be superior"** Colonist Judy Tudor is quoted in James A. Boutilier, "European Women in the Solomon Islands, 1900–1942: Accommodation and Change on the Pacific Frontier," in *Rethinking Women's Roles: Perspectives from the Pacific*, Denise O'Brien and Sharon W. Tiffany, eds. (Berkeley, CA: University of California Press, 1984), p. 181; also, Chilla Bulbeck, *Staying in Line or Getting Out of Place: The Experiences of Expatriate Women in Papua New Guinea 1920–1960: Issues of Race and Gender* (London: Sir Robert Menzies Centre for Australian Studies, Institute of Commonwealth Studies, University of London, 1988), Working Papers in Australian Studies, no. 35.

244. **"shaped like the claws of a crab"** Basil H. Thomson, "New Guinea: Narrative of an Exploring Expedition to the Louisiade and D'Entrecasteaux Islands," *Proceedings of the Royal Geographical Society and Monthly Record of Geography*, vol. 11, no. 9 (Sept. 1889), p. 527.

244. **"I did not come in contact with their family life"** Violet Wickham, "Lady Wickham's Diary," p. 12.

245. **"boy-proof" sleeping rooms, enclosed in heavy chicken wire** Bulbeck, *Staying in Line or Getting Out of Place*, p. 8.

245. **"a few Government weatherboard buildings"** Forbes, "British New Guinea as a Colony," pp. 91–92.

245. "we woke to find our boys had gone off with one of the boats" Violet Wickham, "Lady Wickham's Diary," p. 13.

246. cannibal tales served as an "agenda" Frank Lestringant, *Cannibals: The Discovery and Representation of the Cannibal from Columbus to Jules Verne* (Berkley, CA: University of California Press, 1997), pp. 28–29.

246. As late as 1901, the missionary James Chalmers Diane Langmore, "James Chalmers: Missionary," in *Papua New Guinea Portraits: The Expatriate Experience,* ed. James Griffin (Canberra, Australia: Australian National University Press, 1978), p. 24.

247. North American rubber imports jumped . . . half of all the rubber produced in the world Michael Edward Stanfield, *Red Rubber, Bleeding Trees: Violence, Slavery and Empire in Northwest Amazonia, 1850–1933* (Albuquerque, NM: University of New Mexico Press, 1998), pp. 20–21.

247. £14 million in rubber that came down the Rio Negro Collier, *The River that God Forgot,* p. 18.

248. "I ought to have chosen rubber" Ibid., p. 19.

248. The "trade gun" became notorious Coates, *The Commerce in Rubber,* pp. 139–140.

249. an estimated 131,000–149,000 men were tapping from 21.4 million hevea trees Bradford L. Barham and Oliver T. Coomes. "Wild Rubber: Industrial Organisation and the Microeconomics of Extraction During the Amazon Rubber Boom (1860–1920)," *Journal of Latin American Studies* (Feb. 1994), vol. 26, no. 1, p. 41.

249. the average tapper produced about 1,750 pounds a year Charles H. Townsend, *Report on the Brazilian Rubber Situation* (Belterra, Pará, Brazil, May 17, 1958), p. 3. Other sources estimated that, based on a 100-day season, the average tapper would harvest 550–660 pounds of rubber (Barham and Coomes, "Wild Rubber," p. 45); J. Oakenfull, *Brazil in 1912* (London: Robert Atkinson, 1913), p. 189.

249. A January 1899 report by the U.S. Consul in Pará *U.S. Consular Reports,* vol. 59, no. 220 (Jan. 1899), p. 70.

249. Death rates as high as 50 percent were recorded Barham and Coomes, "Wild Rubber," pp. 10, 36ff.

250. "400 tame Mundurucu Indians" Barbara Weinstein, "The Persistence of Precapitalist Relations of Production in a Tropical Export Economy: The Amazon Rubber Trade, 1850–1920," in Michael Hanagan and Charles Stephenson, ed., *Proletarians and Protest: the Roots of Class Formation in an Industrializing World* (Westport, CT: Greenwood Press, 1986), pp. 2–3.

250. "model, prosperous plantation" Eugene C. Harter, *The Lost Colony of the Confederacy* (Jackson, MS: University Press of Mississippi, 1985), p. 28.

250. "I have made enough to live well on" Ibid.

251. in 1884, he planted rubber, and by 1910, this had grown J. T. Baldwin, "David Riker and *Hevea brasiliensis;* the Taking of Rubber Seeds out of the Amazon," *Economic Botany* 22 (Oct–Dec., 1968), p. 384; Harter, *The Lost Colony of the Confederacy,* pp. 28–29.

251. Tapajós Pará Rubber Forests Ltd. National Archives, Kew, "Articles of Association, Tapajós Pará Rubber Forests Ltd.," BT 31/8165/59032.

251. **an economic expansion so rapid and comprehensive** William Schell Jr., "American Investment in Tropical Mexico: Rubber Plantations, Fraud, and Dollar Diplomacy, 1897–1913," *Business History Review*, vol. 64 (Summer 1990), p. 223.

252. **Journalists were hired to write copy that sold confidence instead of value** Schell, "American Investment in Tropical Mexico," p. 223.

252. **A monthly investment of $5–$150 assured an annual income of $500–$5,000"** Ibid., p. 224, quoting "Why Do You Remain Satisfied?"—an advertisement for the Mexican Development and Construction Co. of Oshkosh, Wisconsin, in *Modern Mexico* (1901).

252. **the Peru Pará Rubber Company, with a reported capital of $3 million** John Melby, "Rubber River: An Account of the Rise and Collapse of the Amazon Boom," *The Hispanic American Historical Review*, vol. 22, no. 3 (Aug. 1942), p. 465.

252. **Lucille Wetherall, who, like thousands of others, lost her life savings** Ibid., p. 227.

252. **enriched the state's treasury by as much as £1.6 million annually** Collier, *The River that God Forgot*, p. 21.

253. **the city's per capita diamond consumption** Collier, *The River that God Forgot*, p. 26.

253. **The leading stores catering to women bore French names** Burns, "Manaus 1910: Portrait of a Boom Town," p. 403.

254. **Every Sunday, the Derby Club held horse races** Robin Ferneaux, *The Amazon: The Story of a Great River* (New York: Putnam's, 1969), pp. 151–155.

254. **Another paid four hundred pounds for a ride in the city's only Mercedes Benz** Collier, *The River that God Forgot*, p. 17.

254. **133 rubber firms and buyers** Burns, "Manaus 1910: Portrait of a Boom Town," p. 415. Although 133 people and firms bought and sold rubber in Manaus, ten of these dominated the market. They are listed here in order of the amount of rubber they exported in 1910:

Dusendchon, Zargas & Co.	3,770,018 kilos
Scholz & Co.	2,509,050
Adelbert H. Alden Ltd.	1,432,907
Gordon & Co.	1,016,398
Anderson Warehouses	224,869
Gunzburger, Lévy, & Co.	130,500
De Lagotellerie & Co.	125,191
J. G. Araujo	106,289
Theodore Levy Camille & Co.	75,090
J. C. Arana y Hermanos	68,557

254. **"clear vision, incomparable energy, and extraordinary activity"** Ibid., p. 416.

255. **"a crusade worthy of this century of progress"** Mason, *Cauchu, the Weeping Wood*, p. 54.

256. **"It is collected by force; the soldiers drive the people into the bush"** Ibid.

256. **"The most rigid injunctions enforcing free trades"** Ibid.

256. **"[I]t is the call to brutality which comes from above"** Ibid, pp. 55–56.

257. **"One, a young man, both of whose hands had been beaten off"** From "Atrocities in the Congo: The Casement Report, 1903," http://web.jjay.cuny.edu/~jobrien.

257. **That equalled one life per every 5 kilograms, a little more than the amount used in one automobile tire** Mason, *Cauchu, the Weeping Wood*, p. 56.

258. **"The insatiable desire to obtain the greatest production in the least time"** Ibid., p. 64.

259. **"He grasped his carbine and machete and began the slaughter"** Walter E. Hardenburg, *The Putumayo, the Devil's Paradise* (London: T. Fisher Unwin, 1912), p. 260.

260. **"advances, bends down, takes the Indian by the hair"** Ibid., p. 236.

260. **"I have seen Indians tied to a tree, their feet about half a yard above the ground"** Ibid.

260. **"designed . . . to just stop short of taking life"** Michael Taussig, "Culture of Terror—Space of Death: Roger Casement's Putumayo Report and the Explanation of Terror," *Comparative Studies in Society and History*, vol. 26, no. 3 (1984), p. 477. Taussig quotes Roger Casement's report, "Correspondence Respecting the Treatment of British Colonial Subjects and Native Indians Employed in the Collection of Rubber in the Putumayo District," *House of Commons Seasonal Papers*, Feb. 14, 1912 to March 7, 1913, vol. 68, p. 35.

260. **mothers were beaten for "just a few strokes" to make them better workers** Ibid, p. 477, quoting Casement's report, p. 17.

260. **"that a man might be a man in Iquitos, but 'you couldn't be a man up there'"** Ibid., p. 478, quoting Casement, p. 55.

261. **"Rubber has taken the blood, the health, and the peace of our people"** Louis Mosch, "Rubber Pirates of the Amazon," *Living Age*, vol. 345 (Nov. 1933), p. 223.

261. **Allen was twenty-three** Allen, "Occupational Adventures."

Chapter 13: The Vindicated Man

263. *maki* Henry Wickham, "The Introduction and Cultivation of the Hevea in India," *India-Rubber and Gutta-Percha Trades Journal*, vol. 23 (Jan. 20, 1902), p. 82.

263. **He invented a machine for smoke-curing latex** Lane, "The Life and Work of Sir Henry Wickham: Part IX—Closing Years," *India Rubber Journal*, vol. 126 (Jan. 30, 1954), p. 6.

263. **He invented a three-bladed tapping knife** Ibid., p. 5.

263. **"most valuable," "quick-growing" tree** Lane, "The Life and Work of Sir Henry Wickham: Part VIII—Piqui-Á and Arghan," *India Rubber Journal*, vol. 126 (Jan. 23, 1954), p. 7.

263. **"of at least equal magnitude to that of Pará rubber"** Ibid.

264. **the Irai Company Ltd.** Ibid., p. 9.

264. **"Its salt-water-resisting qualities are remarkable"** Ibid.

264. **"there would be sufficient demand in Lancashire alone to take up the production"** "Arghan Company, Limited. Commercial Value of the Fibre," *Times* (London), April 4, 1922, p. 20.

264. "**All of us know what a good thing he has done for this country**" Lane, "The Life and Work of Sir Henry Wickham: Part VIII—Piqui-Á and Arghan," p. 9.

266. **He observed the techniques used there for tapping rubber** Dean, *Brazil and the Struggle for Rubber*, pp. 30–31.

267. "**at least eight English thumbs deep**" Simon Winchester, *Krakatoa: The Day the World Exploded, August 27, 1883* (London: Viking, 2003), p. 223.

268. "**little regard for those who did not share his views on botanical matters**" D.J.M. Tate, *The RGA History of the Plantation Industry in the Malay Peninsula* (Kuala Lumpur: Oxford University Press, 1996), p. 201.

268. "**Never mind your body, man, plant these instead!**" O. D. Gallagher, "Rubber Pioneer in his Hundredth Year," *Observer* (London), June 20, 1955; also, Henry N. Ridley, "Evolution of the Rubber Industry," *Proceedings of the Institution of the Rubber Industry*, vol. 2, no. 5 (Oct. 1955), p. 117.

268. "**I looked on him as a 'failed' planter**" Lane, "The Life and Work of Sir Henry Wickham: Part IX—Closing Years," p. 7.

268. "**like junket . . . toffee in vacuum driers**" Wolf and Wolf, *Rubber: A Story of Glory and Greed*, p. 162.

268–69. "**sometime commissioner for the introduction of the Pará (*Hevea*) Indian Rubber Tree**" Ibid., p. 161.

269. "**almost with the air of looking down paternally on his 'children'**" Lane, "The Life and Work of Sir Henry Wickham: Part IX—Closing Years," p. 7.

269. "**Turning over what you said to me in the path-ways yesterday**" Royal Botanic Gardens–Kew, *Miscellaneous Reports: India Office: Caoutchouc I*, "Letter from Wickham to Sir Thistleton-Dyer at Kew, Sept. 4, 1901," file folder 131.

270. **In 1905, Ceylon was still the world leader** In 1905, Ceylon had 40,000 acres planted with hevea, while Malaya had 38,000; by 1907, Ceylon had 150,000 while Malaya had 179,227. Ceylon's plantings stalled after that, totaling 188,000 acres in 1910, while Malaya's acreage totaled 400,000. Herbert Wright, *Hevea Brasiliensis, or Para Rubber: Its Botany, Cultivation, Chemistry and Diseases* (London: Maclaren & Sons, 1912), p. 79. Wright traces the race between the two countries in the following table:

Malaya Takes Premier Position

It will be instructive to compare the planted acreages, under Hevea, in the two leading countries—Ceylon and Malaya:

Year	Ceylon (Middle of each year) Acres	Malay Peninsula (End of each year) Acres
1897	650	350
1902	4,500	7,500
1903	7,500	—

1904	25,000	—
1905	40,000	38,000 (est.)
1906	100,000	99,230
1907	150,000	179,227
1908	170,000	241,138
1909	174,000	292,035
1910	188,000	400,000

271. By 1905, the Straits Settlements were known as the "melting pot" of Asia Frederick Simpich, "Singapore, Crossroads of the East: The World's Greatest Mart for Rubber and Tin was in Recent Times a Pirate-Haunted, Tiger-Infested Jungle Isle," *National Geographic*, March 1926, p. 241.

271. They strolled around Singapore in their golf brogues W. Arthur Wilson. "Malaya—Mostly Gay: All About Rubber: A Guide for Griffins," *British Malaya*, February 1928, pp. 264.

271. Between 1844 and 1910, some 250,000 indentured Indian laborers Ravindra K. Jain, "South Indian Labour in Malaya, 1840–1920: Asylum, Stability and Involution," in *Indentured Labour in the British Empire 1834–1920*, Kay Saunders, ed. (London: Croom Helm, 1984), p.162.

271. "Coolies lines, each room 12 ft. by 12 ft." Ibid., p. 164, quoting the *Selangar Journal* of 1894.

272. "Tamils are . . . cheap and easily managed" T. L. Gilmour, "Life on a Malayan Rubber Plantation," the *Field;* found in "Cuttings from the *Field*," Royal Commonwealth Society Collection, GBR/0115/RCMS 322/11: Malaya.

272. "is a pleasant one" Ibid.

272. plantation rubber was beginning to catch the eye Randolph Resor, "Rubber in Brazil: Dominance and Collapse, 1876–1945," *Business History Review*, vol. 51, no. 3 (Autumn, 1977), p. 349.

272. "Mr. Wickham is no longer a young man" Lane, "Sir Henry Wickham: British Pioneer; a Brief Summary of the Life Story of the British Pioneer," *Rubber Age*, vol. 73 (Aug. 1953), p. 653.

273. "Wickham never sermonized; he just talked" "Palia Dorai." "The Early Days of Rubber: Memories of Henry Wickham," *British Malaya*, vol. 14, no. 12 (April 1940), p. 243.

273. "Having learned your address from the present director of Kew" Royal Botanic Gardens–Kew, *J. D. Hooker Correspondence*, vol. 21, "Wickham to Hooker, Aug. 10th, 1906, London," file folder 120.

273. "My Fellow Planters and Foresters" Henry Wickham, *On the Plantation, Cultivation and Curing of Pará Indian Rubber*, frontispiece.

273. "I was at that time, as one before my time—as one crying in the wilderness" Ibid., pp. 54–57.

274. In April 1910, it reached its peak at $3.06, and the world's rubber consumers let out a howl In 1910, the Amazon accounted for over half of the world's 83,000 tons of wild rubber, and almost all of it high-quality hevea. Africa and Mexico accounted for the rest, with much lower grades of rubber, while the British plantations of the East accounted for a mere 11,000 tons. Although the United States bought 30 percent of the Amazon's rubber, Britain was still the best customer. In 1905–1909, the empire's imports of Brazilian rubber far surpassed its other imports from that country; during that time, it shipped £32 million in rubber out of a total £45 million in imports from Brazil.

274. **"The Rubber Market continued to astonish"** Quoted in Mason, *Cauchu, the Weeping Wood*, p. 58.

274. **"It is a maddening revel of speculation"** Ibid., pp. 58–59.

274. **"New companies continue to be floated"** Ibid., p. 59.

274. **Banks in Pará** Coates, *The Commerce in Rubber*, p. 159.

275. **In the face of such need, it seemed . . . that rubber profits could only soar** In 1910, for example, the rubber barons in Manaus were finishing up their biggest decade of export ever. They'd shipped some 345,079 tons of rubber abroad, 100,000 more than they'd shipped in the previous decade. In 1910 alone, 38,000 tons went to New York, Liverpool, Le Havre, Hamburg, and Antwerp, the world's principal markets. Sixty percent of the rubber sold in New York was Brazilian, and by 1915 the U.S. would buy six times more rubber than Great Britain, eight times more than France or Russia, and twelve times more than Italy or Germany. The price per pound reached its 1910 zenith in April when rubber sold for $2.90 a pound. The price dropped after that, but not enough to depress the market—the average price for 1910 would be $2.01 per pound, compared to the $1.60 average in 1909 and $1.18 in 1908. Bradford L. Barham and Oliver T. Coomes, *Prosperity's Promise: The Amazon Rubber Boom and Distorted Economic Development*, Dellplain Latin American Studies, no. 34, David J. Robinson, ed. (Boulder, CO: Westview Press, 1996), p. 32. Prices quoted from "India Rubber World and Electrical Trades Review" (Nov. 15, 1890; Nov. 1, 1900; Nov. 1, 1905; Nov. 1, 1910).

275. **By 1912, the world had 1.085 million acres planted in rubber** Herbert Wright, *Hevea Brasiliensis, or Para Rubber: Its Botany, Cultivation, Chemistry and Diseases* (London: Maclaren & Sons, 1912), p. 45. The following table gives a breakdown of where plantation rubber was grown in 1912:

Country	Acres
Malaya	420,000
Ceylon	238,000
Dutch East Indies, Borneo, & Pacific Islands	240,000
South India & Burma	42,000
German Colonies	45,000
Mexico, Brazil, Africa, West Indies	100,000
Total	1,085,000

Hevea from Wickham's seeds was cultivated in East Asia. Elsewhere in the world, planters produced rubber from *Castilloa, Ficus, Manihot, Landolphia,* and *Funtimia,* but the quality of these types of rubber was never considered as high.

276. **spent it all on grandiose palaces and payoffs to politicians** Dean, *Brazil and the Struggle for Rubber,* p. 47.

276. **"inexhaustible natural supplies" and the "unrivaled quality" of "Pará fine"** Ibid.

276. **"Yankee speculators"** Ibid.

277. **"English firm of gilt-edged bond folk"** J. T. Baldwin, "David Riker and *Hevea brasiliensis,*" p. 383.

277. **Sometimes those who remained tried to recapture the ostentatious dream** The sources for this account of the Bust include:

Economic reasons for the Bust: Coates, *The Commerce in Rubber: The First 250 Years,* pp. 154–167; Dean, *Brazil and the Struggle for Rubber,* pp. 36–52; James Cooper Lawrence, *The World's Struggle with Rubber* (New York: Harper, 1931), pp. 12–18; W. C. Holmes, "The Tragedy of the Amazon," *Rubber Age,* vol. 9, no. 1 (April 10, 1921), pp. 11–16; John Melby, "Rubber River: An Account of the Rise and Collapse of the Amazon Boom," pp. 452–469; Mason, *Cauchu, the Weeping Wood,* pp. 58–59.

Santarém and Boim: David Bowman Riker, "Handwritten Narrative," in David Afton Riker's *O Último Confederado na Amazônia* (Brazil, 1983), pp. 111–129; Interview with Elisio Eden Cohen, postmaster and historian of Boim, Oct. 21, 2005.

Manaus: E. Bradford Burns, "Manaus 1910: Portrait of a Boom Town," pp. 400–421; Lucile H. Brockway, *Science and Colonial Expansion: The Role of the British Royal Botanic Gardens* (New York: Academic Press, 1979), pp. 151–156.

Iquitos: Harry L. Foster, "Ghost Cities of the Jungle," *New York Herald Tribune,* Sunday magazine section, March 20, 1932.

Obidos: Eric B. Ross, "The Evolution of the Amazon Peasantry," *Journal of Latin American Studies,* vol. 10, no. 2 (1978), p. 215.

278. **nations do not generally raise more than perfunctory complaints** Bernard Porter, *Britain, Europe, and the World, 1850–1986: Delusions of Grandeur* (London: Allen & Unwin, 1987), p. 60.

278. **"The fight for raw materials plays the most important part in world politics"** Jacob Viner, "National Monopolies of Raw Materials," *Foreign Affairs,* vol. 4 (July 1926), p. 585. Viner quotes Dr. Schacht, President of the Reichsbank, in an interview in the March 26, 1926 *New York Times.*

279. **By the end of the war, American factories would churn out 3.9 million gas masks** William Chauncey Geer, *The Reign of Rubber* (New York: Century, 1922), p. 309.

279. **The face piece** Ibid., p. 303.

279. **Observers in balloons watched overhead** Ibid., p. 312.

280. **the United States imported 333.8 million pounds of rubber** Harvey Samuel Firestone, *America Should Produce Its Own Rubber* (Akron, OH: Harvey S. Firestone, 1923), p. 5. Firestone quotes U.S. Department of Commerce records and the *London Financier,* respectively.

280. **This meant that in two months alone, Ford required 78,800 sets of tires** *Ford Times*, vol. 8, no. 10 (July 1914), p. 474.

281. **by introducing the Stevenson Rubber Restriction Plan** Voon Phin-keong, *American Rubber Planting Enterprise in the Philippines, 1900–1930* (London: University of London, Department of Geography, 1977), p. 22.

281. **"I am going to fight this law with all the strength that is in me"** Coates, *The Commerce in Rubber: The First 250 Years*, p. 232.

282. **That meant a $150 million increase "to the crude rubber bill to the United States for 1923"** Allan Nevins and Frank Ernest Hill, *Ford: Expansion and Challenge, 1915–1933* (New York: Scribner's, 1957), p. 231.

282. **"in the future Americans can produce their own rubber"** Coates, *The Commerce in Rubber: The First 250 Years*, p. 233.

282. **"the play an Olympian cast of characters"** Ibid.

282. **the United States alone had imported 2.7 billion pounds of rubber for $1.16 billion** Firestone, *America Should Produce Its Own Rubber*, p. 5. Firestone is quoting U.S. Department of Commerce figures for 1922.

283. **"for services in connexion with the rubber plantation industry in the Far East"** "The King's Birthday. First List of Honours. No Ministerial Dinner," *Times* (London), June 3, 1920, p. 18. Also, *London Gazette*, June 4, 1920, second supplement, p. 6315.

283. **"were loaded by stealth in a small steamer under the nose of a gunboat"** "Death of Sir H. Wickham," *Planter*, vol. 9, no 3 (1928), p. 85.

284. **In 1876, the Brazilian Navy had seventy vessels of war** *The Empire of Brazil at the Universal Exhibition of 1876 in Philadelphia* (Rio de Janeiro: Typographia e Lithographia do Imperial Insitituto Artistico, 1876), p. 144.

285. **Edgar Byrum Davis was odd even by American standards** The tale of Edgar B. Davis can be found in several sources: Robert Gaston, "Edgar B. Davis and the Discovery of the Luling Oilfield," URL: www4.drillinginfo.com; "Handbook of Texas Online: Luling Oilfield," www.tsha.utexas.edu/handbook; "Handbook of Texas Online: Edgar Byrum Davis," www.tsha.utexas.edu/handbook; "Money from God," *Time* (Sept. 2, 1935), *www.TIME.com*; Henry C. Dethloff, "*Edgar B. Davis and Sequences in Business Capitalism: From Shoes to Rubber to Oil*, a Review," *The Journal of Southern History*, vol. 60, no. 4 (Nov. 1994), pp. 829–830; James Cooper Lawrence, *The World's Struggle with Rubber* (New York: Harper, 1931), pp. 30–32; Frank Robert Chalk, *The United States and the International Struggle for Rubber, 1914–1941* (Dissertation, Department of History, University of Wisconsin, 1970), pp. 6–8.

286. **"made a killing in oil"** Quincy Tucker, "A Commentary on the Biography of Sir Henry Wickham," sidebar to Edward Valentine Lane, "Sir Henry Wickham: British Pioneer; a Brief Summary of the Life Story of the British Pioneer," *Rubber Age*, vol. 73 (Aug. 1953), p. 653.

287. **"America is paying the bills"** Chalk, *The United States and the International Struggle for Rubber, 1914–1941*, pp. 7–8.

287. **"If you men think you are doing the best for the industry"** James Cooper Lawrence, *The World's Struggle with Rubber*, p. 31.

287. **"of senile decay"** Lane, "The Life and Work of Sir Henry Wickham: Part IX— Closing Years," p. 7.

288. **the belief was probably mistaken, his wish was fulfilled** Lane, "Sir Henry Wickham: British Pioneer; a Brief Summary of the Life Story of the British Pioneer," p. 656.

288. **"Sir Henry Alexander Wickham . . . was the man who, in the face of extraordinary difficulties"** "Sir Henry Wickham. The Plantation Rubber Industry." *Times* (London), Sept. 28, 1928, p. 19.

288. **"She showed her heroism . . . by remaining on the job"** Tucker, "A Commentary on the Biography of Sir Henry Wickham," p. 653.

289. **And when she sold the rubber shares, they were worthless too** Ibid.

Epilogue: *The Monument of Need*

291. **abundance would change the world** Douglas Brinkley, *Wheels for the World: Henry Ford, His Company, and a Century of Progress, 1903–2003* (New York: Penguin, 2003), p. 135, paraphrasing Arthur M. Schlesinger on "Fordism."

292. **Three of Riker's sons** Sources on Riker and the *confederado* descendants still remaining when Henry Ford arrived include: Harter, *The Lost Colony of the Confederacy,* pp. 107–113; J. T. Baldwin, "David Riker and *Hevea brasiliensis*," pp. 383–384; James E. Edmonds, "They've Gone—Back Home!" *Saturday Evening Post,* Jan. 4, 1941, pp. 30–47.

294. **Ford's vision of a protective, paternal industry** The early years of Fordlandia are chronicled in the following sources: John Galey, "Industrialist in the Wilderness: Henry Ford's Amazon Venture," *Journal of Interamerican Studies and World Affairs* vol. 21, no. 2 (May 1979), pp. 261–276; Nevins and Hill, *Ford: Expansion and Challenge, 1915–1933,* pp. 230–238; Joseph A. Russell, "Fordlandia and Belterra: Rubber Plantations on the Tapajos River, Brazil," *Economic Geography,* vol. 18, no. 2 (April 1942), pp. 125–145; Mary A. Dempsey, "Fordlandia," *Michigan History Magazine,* Jan. 24, 2006, www.michigan historymagazine.com/extra/fordlandia/fordlandia.html; Dean, *Brazil and the Struggle for Rubber,* pp. 67–86; Coates, *The Commerce in Rubber,* pp. 232–235.

295. **"A great business is really too big to be human"** Quoted in Galey, "Industrialist in the Wilderness: Henry Ford's Amazon Venture," p. 276.

295. **"They tried to do to these Brazilians what Northerners had always wanted to do to the South"** Harter, *The Lost Colony of the Confederacy,* p. 111.

296. **while corn flakes made them gag** Sources describing the first Fordlandia riot include: Interview, Doña Olinda Pereira Branco, Fordlandia, Oct. 21, 2005; Galey, "Industrialist in the Wilderness: Henry Ford's Amazon Venture," p. 277; Harter, *The Lost Colony of the Confederacy,* pp. 111–112; Dempsey, "Fordlandia," p. 5 of 9.

296. **"In one night . . . the officials of the Ford Motor Company learned more sociology"** Vianna Moog is quoted in Harter, *The Lost Colony of the Confederacy,* p. 112.

296–97. **What happened to the immigrants after their dismissal remains a mystery** Dempsey, "Fordlandia," p. 5 of 9.

297. "I was really scared of them" Interview, Doña Olinda Pereira Branco, Fordlandia, Oct. 21, 2005.

298. "Practically all the branches of the trees throughout the estate . . . terminate in naked stems" Dean, *Brazil and the Struggle for Rubber*, p. 77; also, Rubber Research Institute of Malaya, "Memorandum on South American Leaf Disease of Rubber" (Kuala Lumpur: Rubber Research Institute of Malaya, May 1948).

298. "Henry Ford has never yet seen one of his big plans fail" Roger D. Stone, *Dreams of Amazonia* (New York: Viking, 1985), p. 85.

298. "threatens not only the sane progress of the world" Quoted in Viner, "National Monopolies of Raw Materials," p. 586.

300. "take from us our flora and fauna" Michael Astor, "Fears of Biopiracy Hampering Research in Brazilian Amazon," *Americas' Intelligence Wire*, Oct. 20, 2005.

300. "Brazil has lost the capacity to control its own resources" Ibid.

300. A few aging veterans from the Ford era still remain in the dusty town Interviews, Doña Olinda Pereira Branco and Biamor de Sousa Pessoa, Fordlandia, Oct. 21, 2005.

303. "The brown current ran swiftly out of the heart of darkness" Conrad, "Heart of Darkness," p. 67.

BIBLIOGRAPHY

I. Printed Sources

Adalbert, Prince of Prussia. *Travels of His Royal Highness Prince Adalbert of Prussia in the South of Europe and in Brazil, with a Voyage up the Amazon and the Xingu*, trans. Sir Robert H. Schomburgk and John Edward Taylor, vol. 2 (London: David Bogue, Fleet Street, 1849).

Agassiz, Professor, and Mrs. Louis. *A Journey in Brazil* (Boston: Ticknor and Fields, 1868).

Akers, C. E. *The Rubber Industry in Brazil and the Orient* (London: Methuen, 1914).

Allen, Arthur Watts. "The Occupational Adventures of an Observant Nomad," an unpublished memoir written by Allen and kept in the care of David Harris and Jenepher Allen Harris.

Allen, P. W. *Natural Rubber and Synthetics* (New York: Wiley, 1972).

"Amazon Bubble: U.S. Rubber Program Fizzles; Brazil Will Handle Production," *Newsweek*, Feb. 21, 1944, p. 66.

"Amazon Rubber Atrocities: The Brazilian System of Indian Protection," *Times* (London), November 27, 1913, p. 7.

Amazon Steam Navigation Company, *The Great River, Notes on the Amazon and Its Tributaries and the Steamer Service* (London: Simpkin, Marshall, Hamilton, Kent, 1904).

"Arghan Company, Limited: Commercial Value of the Fibre," *Times* (London), April 4, 1922, p. 20.

Arnold, David. "Cholera and Colonialism in British India," *Past and Present* 113 (Nov. 1986), pp. 118–151.

Asimont, W.F.C. *Para Rubber in the Malay Peninsula* (London: L. Upcott Gill, 1908).

Astor, Michael. "Fears of Biopiracy Hampering Research in Brazilian Amazon," *Americas' Intelligence Wire*, Oct. 20, 2005.

"The Atrocities in the Putumayo: British Consul's Report," *Times* (London), April 5, 1913, p. 7.

"Automobiles and Rubber: How the Automobile, and Especially Ford Cars, Has Revolutionized the Rubber Industry," *Ford Times*, July 1914, vol. 7, no. 10, pp. 473–476.

Baldwin, J. T. "David Riker and *Hevea brasiliensis;* the Taking of Rubber Seeds out of the Amazon," *Economic Botany* 22 (Oct–Dec., 1968), pp. 383–384.

Baldwin, Neil. *Henry Ford and the Jews: The Mass Production of Hate* (New York: PublicAffairs Books, 2001).

Barham, Bradford L., and Oliver T. Coomes. "Wild Rubber: Industrial Organisation and the Microeconomics of Extraction During the Amazon Rubber Boom (1860–1920)," *Journal of Latin American Studies* (Feb. 1994), vol. 26, no. 1, pp. 37–72.

———. *Prosperity's Promise: The Amazon Rubber Boom and Distorted Economic Development*, Dellplain Latin American Studies, no. 34, David J. Robinson, ed. (Boulder, CO: Westview Press, 1996).

———. "Reinterpreting the Amazon Rubber Boom: Investment, the State, and Dutch Disease," *Latin American Research Review* 29:2 (1994), pp. 73–109.

Bates, Henry Walter. *The Naturalist on the River Amazon* (New York, 1864), worldwide school.org/library/books/sci/earthscience/TheNaturalistontheRiverAmazon.

Bauer, P. T. *The Rubber Industry, A Study in Competition and Monopoly* (Cambridge: Cambridge University Press, 1948).

Baulkwill, W. J. "The History of Natural Rubber Production," in C. C. Webster and W. J. Baulkwill, eds., *Rubber* (Essex, UK: Longman, Scientific & Technical, 1989), pp. 1–56.

Bayly, Susan. "The Evolution of Colonial Cultures: Nineteenth-Century Asia," in *The Oxford History of the British Empire: The Nineteenth Century*, Andrew Porter, ed., (Oxford: Oxford University Press, 1999), vol. 3, pp. 447–469.

Bean, William J. *The Royal Botanic Gardens, Kew: Historical and Descriptive* (London: Cassell, 1908).

"Beginnings of Plantation Rubber," *Rubber Age*, vol. 8, no. 7, (Jan. 10, 1921), pp. 273–274.

Bekkedahl, Norman. "Brazil's Research for Increased Rubber Production," *Scientific Monthly*, 61:3 (September 1945), pp. 199–209.

Bellamy, J. "Report on the Expedition to the Corkscrew Mountains," *Proceedings of the Royal Geographical Society and Monthly Record of Geography*, vol. 11, no. 9 (September, 1889), pp. 542–552. Original in Royal Geographical Society Archives, JMS/5/73.

Belt, Thomas. *The Naturalist in Nicaragua* (London: Edward Bumpus, 1888, first published 1874).

Bentham, George. "On the North Brazilian Euphorbiaceae in the Collections of Mr. Spruce," *Hooker's Journal of Botany* 6 (Dec., 1854), pp. 363–377.

Bevan, Theodore F. "Further Exploration in the Regions Bordering upon the Papuan Gulf," *Proceedings of the Royal Geographical Society and Monthly Record of Geography*, vol. 11, no. 2 (February 1889), pp. 82–90.

———. "The Gold Rush to British New Guinea," *Transactions of the Royal Geographical Society of Australia, Victorian Branch*, vol. 15, 1898, pp. 16–23.

Bierce, Ambrose. *The Devil's Dictionary* (Toronto: Coles, 1978, first published 1881).

"Big Excursion to Mexico." *New York Times*, Jan. 4, 1893, p. 3.

Blainey, Geoffrey. *The Tyranny of Distance* (Macmillan: London, 1968).

"Blood for Oil?" *London Review of Books*, April 21, 2005, pp. 12–16.

Bois, Sir Stanley. "The Importance of Rubber in Economic and Social Progress," *Journal of the Royal Society of Arts*, vol. 75 (Nov. 19, 1926), pp. 28–39.

Bolitho, Hector, ed. *Further Letters of Queen Victoria: From the Archives of the House of Brandenburg-Prussia*, trans. Mrs. J. Pudney & Lord Sudley (London: Thornton Butterworth, 1971, first printed 1938).

Bolton, G. C. *Planters and Pacific Islanders* (Croydon, Victoria, Australia: Longman's, 1967).

———. *A Thousand Miles Away: A History of North Queens to 1920* (Sydney: Australian National University Press, 1970).

Bonner, James, and Arthur W. Galston. "The Physiology and Biochemistry of Rubber Formation in Plants," *Botanical Review* vol. 13, no. 10 (December 1947), pp. 543–596.

Bonnerjea, Biren. *A Dictionary of Superstitions and Mythology* (London: Folk Press, 1927).

Booth, Charles. *Charles Booth's London: A Portrait of the Poor at the Turn of the Century, Drawn from his "Life and Labour of the People in London,"* Albert Fried and Richard M. Elman, eds. (London: Hutchinson, 1969).

Boutilier, James A. "European Women in the Solomon Islands, 1900–1942: Accommodation and Change on the Pacific Frontier," in *Rethinking Women's Roles: Perspectives from the Pacific,* Denise O'Brien and Sharon W. Tiffany, eds., (Berkeley: University of California Press, 1984), pp. 173–200.

Bower, Faubion O. "Sir Joseph Dalton Hooker, 1817–1911," in *Makers of British Botany: A Collection of Biographies by Living Botanists,* F. W. Oliver, ed. (Cambridge: Cambridge University Press, 1913), pp. 302–323.

Bowman, Isaiah. "Geographical Aspects of the New Madeira–Mamoré Railroad," *Bulletin of the American Geographical Society,* vol. 45 (1913), pp. 275–281.

Briggs, Asa. *Victorian People* (London: B. T. Batsford, 1988).

Brinkley, Douglas. *Wheels for the World: Henry Ford, His Company, and a Century of Progress, 1903–2003* (New York: Penguin Books, 2003).

British Honduras. *Report and Results of the Census of the Colony of British Honduras, taken April 5th, 1891* (London: Waterlow, 1892). Archives of the Institute of Colonial Studies, University of London.

———. *Report on the Toledo District for the Year Ending 31st December 1956.* Archives of the Institute of Colonial Studies, University of London.

Bristowe, Lindsay W. *Handbook of British Honduras for 1891–1892, Comprising Historical, Statistical and General Information Concerning the Colony* (London: William Blackwood, 1891).

Brockway, Lucile H. *Science and Colonial Expansion: The Role of the British Royal Botanic Gardens* (New York: Academic Press, 1979).

———. "Science and Colonial Expansion: The Role of the British Royal Botanic Gardens," *American Ethnologist,* vol. 6, no. 3, Interdisciplinary Anthropology (August 1979), pp. 449–465.

Brooks, Collin. *Something in the City: Men and Markets in London* (London: Country Life, 1931).

Brown, C. Barrington. *Canoe and Camp Life in British Guiana* (London: Edward Stanford, 1876).

Brown, C. Barrington, and William Lidstone, *Fifteen Thousand Miles on the Amazon and its Tributaries* (London: Edward Stanford, 1878).

Brownfoot, Janice N. "Memsahibs in Colonial Malaya: A Study of European Wives in a British Colony Protectorate, 1900–1940," in *The Incorporated Wife*, Hilary Callan and Shirley Ardener, eds. (London: Croom Helm, 1984), pp. 186–210.

Buenzle, Alphonse, Asst. Field Technician. Letter to B. V. Worth, Senior Field Technician, Rubber Development Corp., Rio de Janeiro, Nov. 2, 1944. Warren Dean Collection, New York Botanical Gardens archives.

Bulbeck, Chilla. *Staying in Line or Getting Out of Place: The Experiences of Expatriate Women in Papua New Guinea 1920–1960: Issues of Race and Gender* (London: Sir Robert Menzies Centre for Australian Studies, Institute of Commonwealth Studies, University of London, 1988). Working Papers in Australian Studies, no. 35.

Burn, W. L. *The Age of Equipoise: A Study of the Mid-Victorian Generation* (New York: Norton, 1965).

Burns, E. Bradford. "Manaus 1910: Portrait of a Boom Town," *Journal of Inter-American Studies* vol. 7, no. 3, (1965), pp. 400–421.

Byrd, Richard Evelyn. *Little America* (New York: Putnam's, 1930).

Cain, P. J. "Economics and Empire: The Metropolitan Context," in *The Oxford History of the British Empire: The Nineteenth Century*, Andrew Porter, ed., vol. 3. (Oxford: Oxford University Press, 1999), pp. 31–52.

Callan, Hilary, and Shirley Ardener, eds. *The Incorporated Wife* (London: Croom Helm, 1984).

Campbell, Anthony. "Descendants of Benjamin Wickham, a Genealogy" (self-published, Jan. 30, 2005).

Campbell, Peter. "Get Planting," a review of *The Secret Life of Trees: How They Live and Why They Matter*, by Colin Tudge, *London Review of Books* (December 1, 2005), pp. 32–33.

Camm, J.C.R. "Farm-making Costs in Southern Queensland, 1890–1915," *Australian Geographical Studies* vol. 12 (1974), no. 2: pp. 173–189.

———. "The Queensland Agricultural Land Purchase Act of 1894 and Rural Settlement: A Case Study of Jimbour," *Australian Geographer* (Sidney), vol. 10 (September 1967), no. 4: pp. 263–274.

Caraval, Gaspar de. *The Discovery of the Amazon: According to the Account of Friar Gaspar de Carvajal and Other Documents*, introduction by José Toribio Medina, Bertram T. Lee, trans. (New York: American Geographical Society, 1934).

Casement, Roger. "The Putumayo Indians," *Contemporary Review* 102 (1912), pp. 317–328.

Chalk, Frank Robert. *The United States and the International Struggle for Rubber, 1914–1941*. Doctoral dissertation for the Department of History, University of Wisconsin, 1970.

"Cholera," *Times* (London), Dec. 21, 1854, p. 7.

Clarke, Nell Ray. "The Haunts of the Caribbean Corsairs: The West Indies a Geographic Background for the Most Adventurous Episodes in the History of the Western Hemisphere," *National Geographic Magazine*, vol. 41, no. 2, (February 1922), pp. 147–188.

Clegern, Wayne M. *British Honduras, Colonial Dead End, 1859–1900* (Baton Rouge: Louisiana State University Press, 1967).

Coates, Austin. *The Commerce in Rubber: The First 250 Years* (Singapore: Oxford University Press, 1987).

Cohen, Elisio Eden. *Vila de Boim: 1690–1986*, 2nd ed. (Boim, Brazil: self-published).

Collier, Richard. *The River that God Forgot: The Story of the Amazon Rubber Boom* (New York: E. P. Dutton, 1968).

Collins, James. "On the Commercial Kinds of India Rubber or Caoutchouc," *Journal of Botany British and Foreign*, vol. 6 (January 1868), pp. 2–22.

———. "On India Rubber, Its History, Commerce and Supply," *Journal of the Society of Arts*, vol. 18 (December 17, 1869), pp. 81–93.

Companhia Goodyear do Brasil. *A List of English Words and Portuguese Equivalents of Terms Relating to the Tire and Rubber Industry* (Sao Paulo, Brazil: Companhia Goodyear do Brasil, 1940).

Conrad, Joseph. "Conrad in the Congo," in *Heart of Darkness: An Authoritative Text, Backgrounds and Souces, Criticism*, 3rd ed., Robert Kimbrough, ed. (New York: Norton, 1988), pp. 142–194.

———. "Heart of Darkness," in *Heart of Darkness: An Authoritative Text, Backgrounds and Souces, Criticism*, 3rd ed. Robert Kimbrough, ed. (New York: Norton, 1988), pp. 7–76.

———"The Planter of Malata," first published as a novella in the collection *Within the Tides* (London, 1915), reprinted on the World Wide Web at URL: www.readbookonline .net.

Cook, O. F. "Beginnings of Rubber Culture," *The Journal of Heredity*, vol. 19, no. 5, (May 1928), pp. 204–215.

Coomes, Oliver T., and Bradford L. Barham. "The Amazon Rubber Boom: Labor Control, Resistance, and Failed Plantation Development Revisited," *Hispanic American Historical Review*, vol. 74, no. 2 (May 1994), pp. 231–257.

Cordingly, David. *Under the Black Flag: The Romance and Reality of Life Among the Pirates* (New York: Harcourt, Brace, 1995).

Costa, José Pedro de Oliveira. "History of the Brazilian Forest: An Inside View," *Environmentalist*, vol. 3, no. 5 (1983), pp. 50–56.

Cunningham, Frank. "The Lost Colony of the Confederacy," *American Mercury*, vol. 93 (July 1961), pp. 33–38.

Curtin, Philip D. "The White Man's Grave: Image and Reality, 1780–1850," *Journal of British Studies* vol. 1 (Nov. 1961), pp. 94–110.

D'Auzac, Jean, Jean-Louis Jacob, Hervé Chrestin, *Physiology of Rubber Tree Latex: The Lactiferous Cell and Latex—A Model of Cytoplasm* (Boca Raton, FL: CRC Press, 1989).

Dawsey, Cyrus B., and James M. Dawsey, eds. *The Confederados: Old South Immigrants in Brazil* (Tuscaloosa, AL: University of Alabama Press, 1995).

Dean, Warren. *Brazil and the Struggle for Rubber: A Study in Environmental History* (Cambridge: Cambridge University Press, 1987).

———. *Rio Claro: A Brazilian Plantation System, 1820–1920* (Stanford, CA: Stanford University Press, 1976).

"Death of Sir H. Wickham," *Planter*, vol. 9, no. 3 (1928), pp. 84–85.

De Courtais, Georgine. *Women's Headdress and Hairstyles: In England from AD 600 to the Present Day* (London: B. T. Batsford, 1973).

De la Hamba, Louis. "Belize, the Awakening Land," *National Geographic Magazine*, vol. 141, no. 1, (January 1972), pp. 124–146.

De Lange, Nicholas. *Atlas of the Jewish World* (New York: Facts on File Publications, 1984).

De Léry, Jean. *History of a Voyage to the Land of Brazil, Otherwise Called America: Containing the Navigation and the Remarkable Things Seen on the Sea by the Author; the Behavior of Villegagnon in That Country; the Customs and Strange Ways of Life of the American Savages; Together with the Description of Various Animals, Trees, Plants, and Other Singular Things Completely Unknown Over Here*, Janet Whatley, trans. (Berkeley, CA: University of California Press, 1990, translated from the original of 1578).

Denoon, Donald, with Marivic Wyndham. "Australia and the Western Pacific," in *The Oxford History of the British Empire: The Nineteenth Century*, Andrew Porter, ed., vol. 3. (Oxford: Oxford University Press, 1999), pp. 546–572.

Dethloff, Henry C. *"Edgar B. Davis and Sequences in Business Capitalism: From Shoes to Rubber to Oil*, a Review," *Journal of Southern History*, vol. 60, no. 4 (Nov. 1994), pp. 829–830.

Docker, Edward Wybergh. *The Blackbirders: The Recruiting of South Seas Labour for Queensland, 1863–1907* (Sydney: Angus & Robertson, 1970).

"Dorai, Palia." "The Early Days of Rubber: Memories of Henry Wickham," *British Malaya*, vol. 14, no. 12 (April 1940), pp. 241–243.

Douglas, J. "Notes on a Recent Cruise through the Louisiade Group of Islands," *Transactions of the Royal Geographical Society of Australia, Victorian Branch*, vol. 5, part 1 (March 1888), pp. 46–59.

Drayton, Richard. *Nature's Government: Science, Imperial Britain, and the "Improvement" of the World* (New Haven, CT: Yale University Press, 2000).

Dunn, Frederick L. "On the Antiquity of Malaria in the Western Hemisphere," *Human Biology* vol. 37, no. 4 (Dec. 1965), pp. 383–393.

Eakin, Marshall C. "Business Imperialism and British Enterprise in Brazil: The St. John d'el Rey Mining Company, Limited, 1830–1960," *Hispanic American Historical Review* vol. 66, no. 4 (1986), pp. 697–742.

Edelman, Marc. "A Central American Genocide: Rubber, Slavery, Nationalism, and the Destruction of the Guatusos-Malekus," *Comparative Studies in Society and History* vol. 40, no. 2 (1998), pp. 356–390.

Eden, Charles H. *My Wife and I in Queensland: An Eight Year's Experience in the Above Colony, with Some Account of Polynesian Labour* (London: Longmans, Green, 1872). Cambridge University Library: Royal Commonwealth Society Library.

Edmonds, James E. "They've Gone—Back Home!" *Saturday Evening Post*, Jan. 4, 1941, pp. 30–47.

Edwards, William H. *A Voyage up the River Amazon, Including a Residence at Para* (London: John Murray, 1861).

Eidt, Robert C. "The Climatology of South America," in *Biogeography and Ecology in South America*, E. J. Fittkau, H. Klinge, and H. Sioloi, eds. (The Hague: 1968).

"Emigration to Brazil," *New York Times*, Nov. 25, 1866, p. 1.

The Empire of Brazil at the Universal Exhibition of 1876 in Philadelphia (Rio de Janeiro: Typographia e Lithographia do Imperial Insitituto Artistico, 1876).

Encyclopedia Judiaca (Jerusalem: Macmillan, 1971), vol. 4, "Brazil," and vol. 12, "Morocco."

Encyclopedia of World Mythology (Hong Kong: BPC Publishing, 1975).

Evans, Raymond. " 'Kings' in Brass Crescents: Defining Aboriginal Labour Patterns in Colonial Queensland," in *Indentured Labour in the British Empire 1834–1920*, Kay Saunders, ed. (London: Croom Helm, 1984), pp. 183–212.

Everitt, John C. "The Growth and Development of Belize City," *Journal of Latin American Studies* vol. 18, no. 1, (May 1986), pp. 75–112.

Fairchild, David. "Dr. Ridley of Singapore and the Beginnings of the Rubber Industry," *Journal of Heredity*, vol. 19, no. 5, (May 1928), pp. 193–203.

Ferguson, J., ed. *All About Rubber and Gutta-Percha: The India-Rubber Planter's Manual, with the Latest Statistics and Information, More Particulalrly in Regard to Cultivation and Scientific Experiments in Trinidad and Ceylon* (London: J. Hadden, 1899).

Ferneaux, Robin. *The Amazon: The Story of a Great River* (New York: Putnam's, 1969).

Firestone, Harvey Samuel. *America Should Produce Its Own Rubber* (Akron, OH: Harvey S. Firestone, 1923).

Firestone, Harvey Samuel and Samuel Crowther. *Men and Rubber: The Story of Business* (Garden City, NY: Doubleday, Page, 1926).

Forbes, Henry O. "British New Guinea as a Colony," *Blackwoods Magazine*, vol. 152 (July 1892), pp. 82–100.

Ford Times, July 1914, vol. 8, no. 10.

Forsyth, Adrian, and Ken Miyata. *Tropical Nature: Life and Death in the Rain Forests of Central and South America* (New York: Scribner's, 1984).

Foster, Harry L. "Ghost Cities of the Jungle," *New York Herald Tribune*, Sunday magazine section, March 20, 1932.

Fredrickson, Laura M., and Dorothy H. Anderson. "A Qualitative Exploration of the Wilderness Experience as a Source of Spiritual Inspiration," *Journal of Experimental Psychology*, vol. 19, (1999), pp. 21–39.

Frenchman, Michael. "Unique Link with Amazon," *Times* (London), May 3, 1976, p. 2.

Galey, John. "Industrialist in the Wilderness: Henry Ford's Amazon Venture," *Journal of Interamerican Studies and World Affairs* vol. 21, no. 2, (May 1979), pp. 261–289.

Gallagher, John, and Ronald Robinson. "The Imperialism of Free Trade," *Economic History Review* vol. 6, no. 1, (2nd series), 1953, pp. 1–15.

Gallagher, O. D. "Rubber Pioneer in His Hundredth Year," *Observer* (London), June 20, 1955.

Gardiner, John. *The Victorians: An Age in Retrospect* (London: Hambledon and London, 2002).

Gartrell, Beverley. "Colonial Wives: Villains or Victims?" in *The Incorporated Wife*, Hilary Callan and Shirley Ardener, eds. (London: Croom Helm, 1984), pp. 165–185.

Gaskell, G. A. *Dictionary of All Scriptures and Myths* (New York: Julian Press, 1960).

Geer, William Chauncey. *The Reign of Rubber* (New York: Century, 1922).

Gilbert, Pamela K. *Mapping the Victorian Social Body* (Albany, NY: State University of New York Press, 2004).

———. "A Sinful and Suffering Nation: Cholera and the Evolution of Medical and Religious Authority in Britain, 1832–1866," *Nineteenth Century Prose* vol. 25, no. 1, (Spring 1998), pp. 26–45.

Gilpen, Robert. "Economic Interdependence and National Security in Historical Perspective," in *Economic Issues and National Security*, Klaus Knorr and Frank N. Trager, eds. (Lawrence, KS: Allen Press, 1977), pp. 19–66.

Glave, E. J. "Cruelty in the Congo Free State," *Century Magazine*, vol. 54 (Sept. 1897), pp. 691–715.

———. "New Conditions in Central Africa," *Century Magazine*, vol. 53 (April 1897), pp. 900–915.

Goldberg, Harvey E., ed. *Sephardi and Middle Eastern Jewries: History and Culture in the Modern Era* (Bloomington, IN: Indiana University Press, 1996).

Golley, Frank B., and Ernesto Medina, eds. *Tropical Ecological Systems: Trends in Terrestrial and Aquatic Research* (New York: Springer-Verlag, 1975).

Gomez, J. B. *Anatomy of Hevea and Its Influence on Latex Production* (Kuala Lumpur: Malaysian Rubber Research and Development Board, 1982).

Goodman, Edward J. *The Explorers of South America* (New York: Macmillan, 1972).

Goulding, Michael, Ronaldo Barthem, and Efrem Ferreira. *The Smithsonian Atlas of the Amazon* (Washington, DC: Smithsonian Institution, 2003).

Graham, Gerald S. *A Concise History of the British Empire.* (New York: Thames and Hudson, 1978).

Graham, Richard. "Sepoys and Imperialists: Techniques of British Power in Nineteenth-Century Brazil," *Inter-American Economic Affairs* vol. 23, no. 2, (Autumn 1969), pp. 23–37.

"Great Demand for Rubber," *New York Times*, Oct. 12, 1903, p. 7.

Greely, General A. W. "Rubber Forests of Nicaragua and Sierra Leone," *National Geographic*, March 1897, pp. 83–88.

Griggs, Willliam Clark. *The Elusive Eden: Frank McMullan's Confederate Colony in Brazil* (Austin, TX: University of Texas Press, 1987).

Hahner, June E. "Women and Work in Brazil, 1850–1920: A Preliminary Investigation," in *Essays Concerning the Socioeconomic History of Brazil and Portuguese India*, Davril Aldens and Warren Dean, eds. (Gainesville, FL: University Press of Florida, 1977), pp. 87–117.

Hall, Catherine. "Of Gender and Empire: Reflections on the Nineteenth Century," in *Gender and Empire*, Philippa Levine, ed. (Oxford: Oxford University Press, 2004), pp. 46–76.

Hammerton, A. James. "Gender and Migration," in *Gender and Empire*, Philippa Levine, ed. (Oxford: Oxford University Press, 2004), pp. 156–180.

Hancock, Thomas. *Personal Narrative of the Origin and Progress of the Caoutchouc or India-Rubber Manufacture in England* (London: Longman, 1857).

Hanna, Alfred Jackson, and Kathryn Abbey Hanna. *Confederate Exiles in Venezuela* (Tuscaloosa, AL: Confederate Publishing, 1960).

Hanson, Earl Parker, ed. *South from the Spanish Main: South America Seen Through the Eyes of its Discoverers* (New York: Delacorte Press, 1967).

Hardenburg, Walter E. *The Putumayo, the Devil's Paradise* (London: T. Fisher Unwin, 1912).

Harrison, John P. "Science and Politics: Origins and Objectives of Mid-nineteenth Century Government Expeditions to Latin America," *Hispanic American Historical Review*, vol. 35, no. 2 (May 1935), pp. 175–202.

Harter, Eugene C. *The Lost Colony of the Confederacy* (Jackson, MS: University Press of Mississippi, 1985).

Hartt, Charles Frederick. "Preliminary Report of the Morgan Expeditions, 1870–1871: Report of a Reconnaissance of the Lower Tapajos," *Bulletin of the Cornell University (Science)*, vol. 1, no. 1 (Ithaca, NY: University Press, 1874), pp. 11–37.

Hecht, Susanna, and Alexander Cockburn. *The Fate of the Forest: Developers, Destroyers and Defenders of the Amazon* (New York: HarperPerennial, 1990).

Herndon, William Lewis, and Lardner Gibbon, *Exploration of the Valley of the Amazon*, 2 vols., (Washington, DC, 1853), vol. 1, pp. 285, 330–331 (House Document no. 53, Thirty-third Congress, Washington, DC, 1854).

Hepper, F. Nigel, ed. *Royal Botanic Gardens Kew: Gardens for Science and Pleasure* (Owings Mill, MD: Stemmer House, 1982).

Hessel, M. S., W. J. Murphy, and F. A. Hessel. *Strategic Materials in Hemisphere Defense* (New York: Hastings House, 1942).

Heuman, Gad. "The British West Indies," in *The Oxford History of the British Empire: The Nineteenth Century*, Andrew Porter, ed., vol. 3 (Oxford: Oxford University Press, 1999), pp. 470–493.

Higbee, Edward C. "Of Man and the Amazon," *Geographical Review* vol. 41, no. 3 (July 1951), pp. 401–420.

———. "The River Is the Plow," *Scientific Monthly* (June 1945), pp. 405–416.

Hill, Jonathan D., ed. *Rethinking History and Myth: Indigenous South American Perspectives of the Past* (Urbana, IL: University of Illinois Press, 1988).

Hill, Lawrence F. "Confederate Exiles to Brazil," *Hispanic American Historical Review*, vol. 7, no. 2 (May 1927), pp. 192–210.

———. "The Confederate Exodus to Latin America, Part I," *Southwestern Historical Quarterly*, vol. 39, no. 2 (October 1935), pp. 100–134; "Part II," vol. 39, no. 3 (January 1936), pp. 161–199; and "Part III," vol. 39, no. 4 (April 1936), pp. 309–326.

Hobhouse, Henry. *Seeds of Wealth: Four Plants That Made Men Rich* (London: Macmillan, 2003).

Hochschild, Adam. *King Leopold's Ghost: A Story of Greed, Terror, and Heroism in Colonial Africa* (Boston: Houghton Mifflin Co., 1998).

Hoffer, Eric. *The True Believer: Thoughts on the Nature of Mass Movements* (New York: Harper & Row, 1966, first published 1951).

Holdridge, Desmond. "Native Returns to the Amazon," *Living Age*, vol. 360 (April 1941), pp. 153–158.

Holmes, Harry N. *Strategic Materials and National Strength.* (New York: MacMillan, 1942).

Holmes, W. C. "The Tragedy of the Amazon," *Rubber Age*, vol. 9, no. 1 (April 10, 1921), pp. 11–16.

Holthouse, Hector. *Cannibal Cargoes* (Adelaide, Australia: Rigby, 1969).

Hopper, Janice H., comp. *Indians of Brazil in the Twentieth Century* (Washington, DC: Institute for Cross-Cultural Research, 1967).

"India-Rubber, Gutta-Percha, and Other Insulating Materials Compared," *India-Rubber and Gutta-Percha and Electrical Trades Journal*, vol. 4, no. 2 (Sept. 8, 1887), pp. 25–28.

Ireland, Emilienne. "Cerebral Savage: The Whiteman as Symbol of Cleverness and Savagery in Waurá Myth," *Rethinking History and Myth: Indigenous South American Perspectives on the Past*, Jonathan D. Hill, ed. (Urbana, IL: University of Illinois Press, 1988), pp. 157–173.

Jackson, Derrick Z. "U.S. Takes the Lead in Trashing Planet," *Virginian-Pilot*, April 16, 2005, editorial page.

Jain, Ravindra K. "South Indian Labour in Malaya, 1840–1920: Asylum, Stability and Involution," in *Indentured Labour in the British Empire 1834–1920*, Kay Saunders, ed. (London: Croom Helm, 1984), pp. 158–182.

Jalland, Pat, and John Hooper, eds. *Women from Birth to Death: The Female Life Cycle in Britain, 1830–1914* (Atlantic Highlands, NJ: Humanities Press International, 1986).

Jayarathnam, K. "Pests," in M. R. Sethuraj and N. M. Mathew, eds. *Natural Rubber: Biology, Cultivation, and Technology* (Amsterdam: Elsevier, 1992), pp. 360–69.

Jefferson, Mark. "An American Colony in Brazil," *Geographical Review* vol. 18, no. 2 (April 1928), pp. 226–231.

Johnston, W. Ross. *A Documentary History of Queensland.* (St. Lucia, Queensland, Australia: University of Queensland Press, 1988).

Jordan, Carl F. "Amazon Rain Forests," *American Scientist* vol. 70, no. 4 (July–Aug 1982), pp. 394–401.

Kandell, Jonathan. *Passage Through El Dorado: Traveling the World's Last Great Wilderness* (New York: William Morrow, 1984).

Katz, Israel J., and M. Mitchell Serels, eds. *Studies on the History of Portuguese Jews from Their Expulsion in 1497 through Their Dispersion* (New York: Sepher-Hermon Press, 2000).

Keket, Susan, and Watorea Ivara. *A Checklist of the Islands of Papua New Guinea* (Boroko, Papua New Guinea: National Library Service of Papua New Guinea, 1982).

Kimbrough, Robert, ed. "Conrad in the Congo," in *Heart of Darkness: An Authoritative*

Text, Backgrounds and Souces, Criticism, 3rd ed., Robert Kimbrough, ed. (New York: Norton, 1988). pp. 142–194.

"The King's Birthday, First List of Honours, No Ministerial Dinner," *Times* (London), June 3, 1920, p. 18.

Kirkwood, Deborah. "Settler Wives in Southern Rhodesia: A Case Study," in *The Incorporated Wife*, Hilary Callan and Shirley Ardener, eds. (London: Croom Helm, 1984), pp. 143–164.

———. "The Suitable Wife: Preparation for Marriage in London and Rhodesia/ Zimbabwe," in *The Incorporated Wife*, Hilary Callan and Shirley Ardener, eds. (London: Croom Helm, 1984), pp. 106–119.

Knight, Alan. "Britain and Latin America," in *The Oxford History of the British Empire: The Nineteenth Century*, Andrew Porter, ed., vol. 3 (Oxford: Oxford University Press, 1999), pp. 122–145.

Knorr, Klaus, and Frank N. Trager, eds. *Economic Issues and National Security* (Lawrence, KS: Allen Press, 1977).

Kramer, Paul J., and Theodore T. Kozlowski. *Physiology of Trees* (New York: McGraw-Hill, 1960).

Kreier, Julius P., ed. *Malaria: Epidemiology, Chemotherapy, Morphology, and Metabolism* (New York: Academic Press, 1980).

Laird, MacGregor, and R. A. K. Oldfield. *Narrative of an Expedition into the Interior of Africa by the River Niger*, 2nd ed., 2 vols. (London, 1837).

Lane, Edward V. "The Life and Work of Sir Henry Wickham, Parts I–IX," *India Rubber Journal* vols. 125 (Dec. 5, 12, 19, 26, 1953) and 126 (January 2, 9, 19, 23, 30, 1954). These include: "Part I—Ancestry and Early Years" (Dec. 5, 1953), pp. 14–17; "Part II— A Journey Through the Wilderness" (Dec. 12, 1953), pp. 16–18; "Part III—Santarem" (Dec. 19, 1953), pp. 18–23; "Part IV—Kew" (Dec. 26, 1953), pp. 5–8; "Part V—Pioneering in North Queensland" (Jan. 2, 1954), pp. 17–19; "Part VI—Pioneering in British Honduras" (Jan. 9, 1954), pp. 17–23; "Part VII—The Conflict Islands and New Guinea" (Jan. 16, 1954), pp. 7–10; "Part VIII—Piqui-Á and Arghan" (Jan. 23, 1954), pp. 7–10; and "Part IX—Closing Years" (Jan. 30, 1954), pp. 5–8.

———. "Sir Henry Wickham: British Pioneer; a Brief Summary of the Life Story of the British Pioneer," *Rubber Age* 73 (Aug. 1953), pp. 649–656.

Lange, Algot. *In the Amazon Jungle: Adventures in Remote Parts of the Upper Amazon River, Including a Sojourn Among Cannibal Indians* (New York: Putnam's, 1912).

———. *The Lower Amazon: A Narrative of Exploration in the Little Known Regions of the State of Para, on the Lower Amazon, etc.* (New York: Putnam's, 1914).

———. "The Rubber Workers of the Amazon," *Bulletin of the American Geographical Society*, vol. 43, no. 1 (1911), pp. 33–36.

Langmore, Diane. "James Chalmers: Missionary," in *Papua New Guinea Portraits: The Expatriate Experience*, James Griffin, ed. (Canberra, Australia: Australian National University Press, 1978), pp. 1–27.

Lawrence, James Cooper. *The World's Struggle with Rubber* (New York: Harper, 1931).

Lebergott, Stanley. "The Returns to U.S. Imperialism, 1890–1929," *Journal of Economic History* vol. 40, no. 2 (June 1980), pp. 229–252.

Leff, Nathaniel. "Tropical Trade and Development in the Nineteenth Century: The Brazilian Experience," *Journal of Political Economy* vol. 81, no. 3 (1973), pp. 678–696.

Lestringant, Frank. *Cannibals: The Discovery and Representation of the Cannibal from Columbus to Jules Verne* (Berkley, CA: University of California Press, 1997).

Lévi-Strauss, Claude. *The Raw and the Cooked: Introduction to a Science of Mythology,* vol. 1, John and Doreen Weightman, trans. (New York: Harper Colophon, 1975, first published in France as *Le Cru et le Cuit,* 1964, Librairie Plon).

———. *Tristes Tropiques,* John and Doreen Weightman, trans. (New York: Atheneum, 1981, first published 1951).

Levine, Philippa, ed. *Gender and Empire* (Oxford: Oxford University Press, 2004).

———. "Sexuality, Gender, and Empire," in *Gender and Empire,* Philippa Levine, ed. (Oxford: Oxford University Press, 2004), pp. 134–155.

Lloyd, Francis E. "Plantation Rubber: Its Source and Acquisition," *Scientific Monthly,* vol. 23, no. 3 (September 1926), pp. 268–278.

Loadman, John. *Tears of the Tree: The Story of Rubber—A Modern Marvel* (Oxford: Oxford University Press, 2005).

Loh Fook Seng, Philip. *The Malay States, 1877–1895: Political Change and Social Policy* (Singapore: Oxford University Press, 1969).

London Gazette, June 4, 1920, second supplement, pp. 6313–6315. "King's Birthday Honours."

———. November 20, 1928, p. 7604. "Sir Henry Alexander Wickham, Deceased."

London, Jack. *The People of the Abyss* (London: Pluto Press, 2001, first published in Great Britain in 1903).

"R. Irwin Lynch," *Journal of the Kew Guild* vol. 4, no. 32 (1925), pp. 341–342.

Lynn, Martin. "British Policy, Trade, and Informal Empire in the Mid-Nineteenth Century," in *The Oxford History of the British Empire: The Nineteenth Century,* Andrew Porter, ed., vol. 3 (Oxford: Oxford University Press, 1999), pp. 101–121.

Macdonald, A. C. "Notes on the Panama and Nicaragua Canal Schemes," *Transactions of the Royal Geographical Society of Australia, Victorian Branch,* vol. 16 (1899), pp. 46–57.

Mackay, Charles. *Extraordinary Popular Delusions and the Madness of Crowds* (New York: Harmony Books, 1980, first published in 1841 and 1852).

MacLeod, Roy. *Nature and Empire: Science and the Colonial Enterprise,* Osiris series, vol. 15 (Chicago: University of Chicago Press, 2001).

Markham, Clements R. "The Cultivation of Caoutchouc-Yielding Trees in British India," *Journal of the Society of Arts,* vol. 24, no. 1220 (April 7, 1876), pp. 475–481.

———. *Peruvian Bark: A Popular Account of the Introduction of Chinchona Cultivation into British India, 1860–1880* (London, John Murray, 1880).

Marriott, John. *The Other Empire: Metropolis, India and Progress in the Colonial Imagination* (Manchester and New York: Manchester University Press, 2003).

Mason, Peter. *Cauchu, the Weeping Wood: A History of Rubber* (Sydney, Australia: The Australian Broadcasting Commission, 1979).

Mathew, N. M. "Physical and Technological Properties of Natural Rubber," in M. R. Sethuraj and N. M. Mathew, eds. *Natural Rubber: Biology, Cultivation, and Technology* (Amsterdam: Elsevier, 1992), pp. 399–425.

May, George O. "Rubber: The Inquiry and the Facts," *Atlantic Monthly*, 127 (June 1926), pp. 805–812.

Maynard, Margaret. "'A Great Deal Too Good for the Bush': Women and the Experience of Dress in Queensland," in *On the Edge: Womens' Experiences in Queensland,* Gail Reekie, ed. (St. Lucia, Queensland, Australia: University of Queensland Press, 1994), pp. 51–65.

McClain, Michael E., Reynaldo L. Victoria, and Jeffrey E. Richey, eds. *The Biogeochemistry of the Amazon Basin* (New York: Oxford University Press, 2001).

McNeill, William H. *Plagues and Peoples* (Garden City, NY: Anchor Books, 1976).

Meggers, Betty J. *Amazonia: Man and Culture in a Counterfeit Paradise* (Chicago: Aldine-Atherton, 1971).

———. "Environment and Culture in Amazonia," in *Man in the Amazon*, Charles Wagley, ed. (Gainesville: University of Florida Press, 1974), pp. 91–110.

———. "The Indigenous Peoples of Amazonia, Their Cultures, Land Use Patterns and Effects on the Landscape and Biota," in Harald Sioli, ed. *The Amazon: Limnology and Landscape Ecology of a Mighty Tropical River and Its Basin* (The Hague: Der W. Junk, 1984), pp. 627–648.

Melby, John. "Rubber River: An Account of the Rise and Collapse of the Amazon Boom," *Hispanic American Historical Review*, vol. 22, no. 3 (Aug. 1942), pp. 452–469.

"Memorandum on South American Leaf Disease of Rubber" (Kuala Lumpur: Rubber Research Institute of Malaya, May 1948). Warren Dean collection, New York Botanical Gardens archives.

Merli, Frank J. "Alternative to Appomattox: A Virginian's Vision of an Anglo-Confederate Colony on the Amazon, May 1865," *Virginia Magazine of History and Biography*, vol. 94, no. 2 (April 1986), pp. 210–219.

Millburn, John R., and Keith Jarrott. *The Aylesbury Agitator: Edward Richardson: Labourer's Friend and Queensland Agent, 1849–1878* (Aylesbury, Queensland, Australia: Buckingham County Council, 1988). Archives of the Institute of Commonwealth Studies, University of London.

Miller, Susan Gilson. "Kippur on the Amazon: Jewish Emigration from Northern Morocco in the Late Nineteenth Century," in *Sephardi and Middle Eastern Jewries: History and Culture in the Modern Era,* Harvey E. Goldberg, ed. (Bloomington, IN: Indiana University Press, 1996), pp. 190–209.

Miller, William. "A Journey from British Honduras to Santa Cruz, Yucatan, with a map," *Proceedings of the Royal Geographical Society and Monthly Record of Geography*, vol. 11, no. 1 (Jan. 1889), pp. 23–28. The handwritten copy of this article, with changes and a map, received July 1888, are preserved in the Archives of the Royal Geographical Society, London, JMS/5/74.

Montagnini, Florencia. "Nutrient Considerations in the Use of Silviculture for Land Development and Rehabilitation in the Amazon," in Michael E. McClain, Reynaldo

L. Victoria, and Jeffrey E. Richey, eds. *The Biogeochemistry of the Amazon Basin* (New York: Oxford University Press, 2001), pp. 106–121.

Moran, Emilio F. "The Adaptive System of the Amazonian Caboclo," in Charles Wagley, ed., *Man in the Amazon* (Gainesville, FL: University Presses of Florida, 1974), pp. 136–159.

Morel, E. D. *Red Rubber: The Story of the Rubber Slave Trade Flourishing on the Congo in the Year of Grace 1906* (New York: Negro Universities Press, 1906).

Morris, D. *The Colony of British Honduras, Its Resources and Prospects; with Particular Reference to Its Indigineous Plants and Economic Productions* (London: Edward Stanford, 1883). Archives of the Institute of Commonwealth Studies, University of London.

Morrison, Tony, Ann Brown, and Ann Row, comps. *A Victorian Lady's Amazon Adventure.* (London: BBC, 1985).

Mosch, Louis. "Rubber Pirates of the Amazon," *Living Age*, vol. 345 (Nov. 1933), pp. 219–224.

"Mosquito Day Luncheon: The Importance of Tropical Medicine in Empire Development: Speeches by the Prime Minister and Mr. Chester Beatty," *British Malaya*, vol. 14, no. 2 (June 1939), pp. 43–44.

Murphy, Robert F., and Julian H. Steward. "Tappers and Trappers: Parallel Process in Acculturation," *Economic Development and Cultural Change*, vol. 4, no. 4 (July 1956), pp. 335–355.

Nash, Roy. *The Conquest of Brazil* (New York: AMS Press, 1969, originally published 1926).

Naylor, Robert A. *Penny Ante Imperialism: The Mosquito Shore and the Bay of Honduras, 1600–1914, A Case Study in British Informal Empire* (London: Associated University Presses, 1989).

Nead, Lynda. *Victorian Babylon: People, Streets and Images in Nineteenth-Century London* (New Haven: Yale University Press, 2000).

Nevill, Barry St.-John, ed. *Life at the Court of Queen Victoria: With Selections from the Journals of Queen Victoria* (Exeter, England: Webb & Bower, 1984).

Nevins, Allan, and Frank Ernest Hill. *Ford: Expansion and Challenge, 1915–1933* (New York: Scribner's, 1957).

New York Tribune, July 17 and 20, 1865.

Oakenfull, J. *Brazil in 1912* (London: Robert Atkinson, 1913).

O'Brien, Denise, and Sharon W. Tiffany, eds. *Rethinking Women's Roles: Perspectives from the Pacific* (Berkeley: University of California Press, 1984).

"Obituary: Sir Henry Wickham, The Plantation Rubber Industry," *Times* (London), September 28, 1928, p. 19.

O'Hanlon, Redmond. *In Trouble Again: A Journey Between the Orinoco and the Amazon* (New York: Vintage Books, 1988).

Olivier, Sydney. "British Honduras," in *British America: The British Empire Series*, vol. 3, (London: Kegan Paul, Trench, Trubner, 1900), pp. 476–496.

"On Rubber," *Gardener's Chronicle*, vol. 19 (1855), p. 381.

Orlean, Susan. *The Orchid Thief* (New York: Ballantine Books, 1998).

Paneth, Nigel, Peter Vinten-Johansen, Howard Brody, and Michael Rip. "A Rivalry of Foulness: Unofficial Investigations of the London Cholera Epidemic of 1854," *American Journal of Public Health*, vol. 88, no. 10 (Oct. 1998) pp. 1545–1553.

"Para Rubber, Additional Series," *Kew Bulletin of Miscellaneous Information*, vol. 7 (1906), pp. 241–242.

Park, Mungo. *Travels into the Interior of Africa* (London: Eland Press, 1983).

Parmer, J. Norman. *Colonial Labor Policy and Administration: A History of Labor in the Rubber Plantation Industry in Malaya, 1910–1941* (Locust Valley, NY: J. J. Augustin, 1960).

Peachey, S. J. "The New Process for Vulcanizing Rubber," *Rubber Age* vol. 8, no. 4 (Nov. 25, 1920), pp. 141–142.

Pearson, Henry C. *The Rubber Country of the Amazon; a Detailed Description of the Great Rubber Industry of the Amazon Valley, etc.* (New York: India Rubber World, 1911).

Peres, Ambrosio B. "Judaism in the Amazon Jungle," in *Studies on the History of Portuguese Jews from Their Expulsion in 1497 through Their Dispersion*, Israel J. Katz and M. Mitchell Serels, eds. (New York: Sepher-Hermon Press, 2000), pp. 175–183.

Phin-keong, Voon. *American Rubber Planting Enterprise in the Philippines, 1900–1930* (London: University of London, Department of Geography, 1977).

———. *Western Rubber Planting Enterprise in Southeast Asia, 1876–1921* (Kuala Lumpur: Penerbit Universiti Malaya, 1976).

Pike, Fredrick B. *The United States and Latin America: Myths and Stereotypes of Civilization and Nature* (Austin, TX: University of Texas Press, 1992).

Platt, D. C. M. *The Cinderella Service: British Consuls Since 1825* (Hamden, CT: Archon Books, 1971).

Porritt, B. D. *The Chemistry of Rubber* (London: Gurney & Jackson, 1913).

Porter, Andrew, ed. *The Oxford History of the British Empire: The Nineteenth Century*, vol. 3 (Oxford: Oxford University Press, 1999).

Porter, Bernard. *Britain, Europe, and the World, 1850–1986: Delusions of Grandeur* (London: Allen & Unwin, 1987).

Poser, Charles M., and George W. Bruyn. *An Illustrated History of Malaria* (New York: Parthenon Publishing Group, 1999).

Prado, Paulo. "Essay on Sadness," in G. Hervey Simm, ed., *Brazilian Mosaic: Portraits of a Diverse People and Culture* (Wilmington, DE: Scholarly Resources, 1995), pp. 18–20.

Purseglove, J. W. "Ridley, Malaya's Greatest Naturalist," *Malayan Nature Journal*, vol. 10 (Dec. 1955), pp. 43–55.

Queensland. *Queensland Census of 1871, Taken on the 1st day of September, Being the Fourth Taken in the Colony* (Brisbane: James C. Beal, Government Printer, 1872). Archives of the Institute of Colonial Studies, University of London.

Radin, Paul. *Indians of South America* (Garden City, NY: Doubleday, Doran, 1942), American Museum of Natural History, Science Series.

"Raw Rubber and Romance," *British Malaya: The Magazine of the Association of British Malaya*, vol. 2, no. 8 (Dec. 1927), p. 191.

Reed, Nelson A. *The Caste War of Yucatan* (Stanford, CA: Stanford University Press, 2001).

Reeve, Mary-Elizabeth. "Caucha Uras: Lowland Quicha Histories of the Amazon Rubber Boom," *Rethinking History and Myth: Indigenous South American Perspectives on the Past*, Jonathan D. Hill, ed. (Urbana, IL: University of Illinois Press, 1988), pp. 19–34.

Reis, Arthur Cesar Ferreira. "Economic History of the Brazilian Amazon," in *Man in the Amazon*, Charles Wagley, ed. (Gainesville, FL: University Presses of Florida, 1974), pp. 33–44.

Resor, Randolph. "Rubber in Brazil: Dominance and Collapse, 1876–1945," *Business History Review* vol. 51, no. 3 (Autumn, 1977), pp. 341–366.

Ridley, H. N. "Evolution of the Rubber Industry," *Proceedings of the Institution of the Rubber Industry*, vol. 2, no. 5 (October 1955), pp. 114–122.

———. "History of the Evolution of the Cultivated Rubber Industry," *Bulletin of the Rubber Growers Association*, vol. 10, no. 1 (January 1928), pp. 45–49.

Riker, David Afton. *O Último Confederado na Amazônia* (Brazil, 1983).

Riker, David Bowman. "Handwritten Narrative," in David Afton Riker's *O Último Confederado na Amazônia* (Brazil, 1983), pp. 111–129.

Rippy, J. Fred. "British Investments in Latin America, End of 1913," *Inter-American Economic Affairs*, vol. 5, no. 2 (Autumn 1951), pp. 90–100.

———. "British Investments in Latin America: A Sample of Profitable Enterprises," *Inter-American Economic Affairs*, vol. 6, no. 4 (Spring 1953), pp. 3–17.

———. "A Century and a Quarter of British Investment in Brazil," *Inter-American Economic Affairs*, vol. 6, no. 1 (Summer 1952), pp. 83–92.

Rogers, Edward J. "Monoproductive Traits in Brazil's Economic Past," *Americas*, vol. 23, no. 2 (October 1966), pp. 130–141.

Rogers, Nicholas. "Caribbean Borderland: Empire, Ethnicity, and the Exotic on the Mosquito Coast," *Eighteenth-Century Life*, vol. 26, no. 3 (Fall 2002), pp. 117–138.

"The Romance of Rubber." *New York Times*, Sept. 23, 1906, p. SM3.

Rosado, J. M. "A Refugee of the War of the Castes Makes Belize His Home," *The Memoirs of J. M. Rosado*, ed. Richard Buhler. Occasional Publication, No. 2, Belize Institute for Social Research and Action (Belize: Berex Press, 1977). Archives of the Institute of Commonwealth Studies, University of London.

Ross, Eric B. "The Evolution of the Amazon Peasantry," *Journal of Latin American Studies* vol. 10, no. 2 (1978), 193–218.

Roosevelt, Theodore. *Through the Brazilian Wilderness* (New York: Scribner's, 1914).

"Rubber Becoming Scarce." *New York Times*, Dec. 10, 1902, p. 1.

"Rubber from Papua," *Bulletin of the Imperial Institute*, vol. 10 (1912), pp. 386–388.

Rubber Growers' Association £10,000 Prize Scheme, *A Handbook on Rubber Uses and Their Development* (London: Rubber Growers Association, 1923).

"Rubber in its Home: Conditions in the Amazon Valley," *Times* (London), July 7, 1911, p. 24.

"Rubber Plantations in Mexico and Central America," *National Geographic*, November 1903, pp. 409–414.

Rubber Research Institute of Malaya. "Memorandum on South American Leaf Disease of Rubber" (Kuala Lumpur: Rubber Research Institute of Malaya, May 1948).

Rusk, Rev. John. *The Beautiful Life and Illustrious Reign of Queen Victoria: A Memorial Volume* (Chicago: Monarch Book Company, 1901).

Russell, Joseph A. "Fordlandia and Belterra: Rubber Plantations on the Tapajos River, Brazil," *Economic Geography,* vol. 18, no. 2 (April 1942), pp. 125–145.

Sadler, F. W. "Seeds That Began the Great Rubber Industry," *Contemporary Review,* vol. 217, no. 1257 (Oct. 1970), pp. 208–210.

St. Aubyn, Giles. *Queen Victoria: A Portrait* (New York: Atheneum, 1992).

Saunders, Kay, ed. *Indentured Labour in the British Empire 1834–1920* (London: Croom Helm, 1984).

———. "The Workers' Paradox: Indentured Labour in the Queensland Sugar Industry to 1920," in *Indentured Labour in the British Empire 1834–1920,* Kay Saunders, ed. (London: Croom Helm, 1984), pp. 213–259.

Schell, William Jr. "American Investment in Tropical Mexico: Rubber Plantations, Fraud, and Dollar Diplomacy, 1897–1913," *Business History Review,* vol. 64 (Summer 1990), pp. 217–254.

Schneider, Franz Jr. "The World's Rubber Supply," *International Relations and Problems,* vol. 12, no. 1 (July 1926), pp. 153–158.

Schultes, Richard Evans. "The History of Taxonomic Studies in Hevea," in *Essays in Biohistory and Other Contributions,* P. Smit and R.J.Ch.V. Ter Laage, eds. (Utrecht, Netherlands: International Association for Plant Taxonomy, 1970).

———. "The Odyssey of the Cultivated Rubber Tree," *Endeavor,* n.s., vol. 1, nos. 3–4 (1977), pp. 133–137.

Schurz, William L. "The Amazon, Father of Waters," *National Geographic,* April 1926, pp. 444–463.

———. *Rubber Production in the Amazon Valley,* Department of Commerce: Bureau of Foreign and Domestic Commerce (Washington, DC: Government Printing Office, 1925).

Schuyler, George S. "Uncle Sam's Black Step-Child," *American Mercury,* June 1933, pp. 147–156.

Schwartz, Stuart B. "The Portuguese Heritage: Adaptability," in G. Hervey Simm, ed., *Brazilian Mosaic: Portraits of a Diverse People and Culture* (Wilmington, DE: Scholarly Resources, 1995), pp. 29–35.

———. *Sugar Plantations in the Formation of Brazilian Society* (Cambridge: Cambridge University Press, 1985).

Sethuraj, M. R., and N. M. Mathew, eds. *Natural Rubber: Biology, Cultivation, and Technology* (Amsterdam: Elsevier, 1992).

Shapin, Steven. "Sick City," *New Yorker,* Nov. 6, 2006, pp. 110–115, a review of Steven Johnson, *The Ghost Map: The Story of London's Most Terrifying Epidemic and How It Changed Science, Cities, and the Modern World* (New York: Riverhead, 2006).

Shoumatoff, Alex. *The Rivers Amazon* (San Francisco: Sierra Club Books, 1978).

Shtier, Ann B. *Cultivating Women, Cultivating Science: Flora's Daughters and Botany in England, 1760 to 1860* (Baltimore, MD: Johns Hopkins University Press, 1996).

Sill, William. "The Anvil of Evolution," *Earthwatch,* August 2001, pp. 24–31.

Simmons, Charles Willis. "Racist Americans in a Multi-Racial Society: Confederate Exiles in Brazil," *Journal of Negro History*, vol. 67, no. 1, pp. 34–39.

Simon, John. *Report on the Cholera Epidemic of 1854, as It Prevailed in the City of London* (London: C. Dawson, 1854).

Simon, Sir John. *Report on the Last Two Cholera Epidemics of London, as Affected by the Consumption of Impure Water; addressed to the Rt. Hon. the President of the General Board of Health* (London: George E. Eyre and William Spottiswoode, 1856).

Simpich, Frederick. "Singapore, Crossroads of the East: The World's Greatest Mart for Rubber and Tin was in Recent Times a Pirate-Haunted, Tiger-Infested Jungle Isle," *National Geographic*, March 1926, pp. 235–269.

Sioli, Harald, ed. *The Amazon: Limnology and Landscape Ecology of a Mighty Tropical River and Its Basin* (The Hague: Der W. Junk, 1984).

———. "Tropical Rivers as Expressions of their Terrestrial Environments," in *Tropical Ecological Systems*, Frank Golley and Ernesto Medina (New York: Springer-Verlag, 1975).

———. "Unifying Principles of Amazonian Landscape Ecology and Their Implications," in Harald Sioli, ed. *The Amazon: Limnology and Landscape Ecology of a Mighty Tropical River and Its Basin* (The Hague: Der W. Junk, 1984), pp. 615–625.

"Sir H. Wickham and the Rubber Industry: Gift from the Straits Settlement," *Times* (London), October 12, 1926, p. 14.

"Sir Henry Wickham: The Plantation Rubber Industry." *Times* (London), Sept. 28, 1928, p. 19.

Smith, Anthony. *Explorers of the Amazon* (New York: Viking, 1990).

Smith, Herbert H. *Brazil, the Amazons and the Coast* (New York: 1879).

Spearritt, Katie. "The Sexual Economics of Colonial Marriage," in *On the Edge: Women's Experiences of Queensland*, Gail Reekie, ed. (St. Lucia, Queensland, Australia: University of Queensland Press, 1994), pp. 66–79.

The Sponging Industry. A booklet of the Exhibition of Historical Documents held at the Public Records Office (Nassau, Bahamas: Public Records Office, 1974). Archives of the Institute of Commonwealth Studies, University of London.

Spruce, Richard. *Notes of a Botanist on the Amazon and the Andes*, Alfred Russel Wallace, ed., 2 vols. (Cleveland, OH: Arthur H. Clark, 1908).

———. "Note on the India Rubber of the Amazon," *Hooker's Journal of Botany*, vol. 7 (July 1855), pp. 193–196.

Stafford, Robert A. "Scientific Exploration and Empire," in *The Oxford History of the British Empire: The Nineteenth Century*, Andrew Porter, ed., vol. 3. (Oxford: Oxford University Press, 1999), pp. 294–319.

Stanfield, Michael Edward. *Red Rubber, Bleeding Trees: Violence, Slavery and Empire in Northwest Amazonia, 1850–1933* (Albuquerque, NM: University of New Mexico Press, 1998).

Steward, Julian H., and Louis C. Faron. *Native Peoples of South America* (New York: McGraw-Hill, 1959).

Stockwell, A. J. "British Expansion and Rule in South-East Asia," in *The Oxford History of the British Empire: The Nineteenth Century*, Andrew Porter, ed., vol. 3 (Oxford: Oxford University Press, 1999), pp. 371–394.

Stone, Irving. "British Direct and Portfolio Investment in Latin America Before 1914," *Journal of Economic History*, vol. 37, no. 3 (Sept. 1977), pp. 690–722.

Stone, Roger D. *Dreams of Amazonia* (New York: Viking, 1985).

Stonequist, Everett V. *The Marginal Man: A Study in Personality and Culture Conflict* (New York: Russell & Russell, 1961).

Storr, Anthony. *Churchill's Black Dog, Kafka's Mice: And Other Phenomena of the Human Mind* (New York: Ballantine, 1988, first published 1965).

Strachey, Lytton. *Queen Victoria* (London: Penguin, 1971, first published 1921).

Sturt, Charles. *Narrative of an Expedition into Central Australia*, 2nd ed., 2 vols. (Adelaide, Australia: 1849).

Summ, G. Harvey, ed. *Brazilian Mosaic: Portraits of a Diverse People and Culture* (Wilmington, DE: Scholarly Resources, 1995).

Swayne, Sir Eric. "British Honduras," *Geographical Journal*, vol. 50, no. 3 (Sept. 1917), pp. 161–175.

Tate, D.J.M. *The RGA History of the Plantation Industry in the Malay Peninsula* (Kuala Lumpur: Oxford University Press, 1996).

Taussig, Michael. "Culture of Terror-Space of Death: Roger Casement's Putumayo Report and the Explanation of Terror," *Comparative Studies in Society and History*, vol. 26, no. 3 (1984), pp. 467–497.

Thiselton-Dyer, W. T. "Historical Account of Kew to 1841," *Kew Bulletin of Miscellaneous Information*, vol. 60 (1890), pp. 279–327.

Thomson, Basil H. "New Guinea: Narrative of an Exploring Expedition to the Louisiade and D'Entrecasteaux Islands," *Proceedings of the Royal Geographical Society and Monthly Record of Geography*, vol. 11, no. 9 (Sept. 1889), pp. 525–542.

Tomlinson, B. R. "Economics and Empire: The Periphery and the Imperial Economy," in *The Oxford History of the British Empire: The Nineteenth Century*, Andrew Porter, ed., vol. 3 (Oxford: Oxford University Press, 1999), pp. 53–74.

Townsend, Charles H. T. *Report on the Brazilian Rubber Situation* (Belterra, Para, Brazil: May 17, 1958). Warren Dean collection, New York Botanical Gardens archives.

Tucker, Quincy. "A Commentary on the Biography of Sir Henry Wickham," sidebar to Edward Valentine Lane, "Sir Henry Wickham: British Pioneer; a Brief Summary of the Life Story of the British Pioneer," *Rubber Age*, vol. 73 (Aug. 1953), p. 653.

Tuttle, John B. "The Action of Heat and Light on Vulcanized Rubber," *Rubber Age*, vol. 8, no. 7 (Jan. 10, 1921), pp. 271–272.

U.S. Consular Reports, vol. 59, no. 220 (Jan. 1899).

Uyangoda, Jayadeva. *Life Under Milk Wood: Women Workers in Rubber Plantations, an Overview* (Colombo, Sri Lanka: Women's Education and Research Centre, 1995).

Van der Giessen, C., and F. W. Ostendorf, "The Oldest Hevea Trees in Java," *Chronica Naturae*, vol. 104 (July 1948), pp. 197–200.

Vietmeyer, Noel D. "Rediscovering America's Forgotten Crops," *National Geographic*, vol. 159, no. 5 (May 1981), pp. 702–712.

Viner, Jacob. "National Monopolies of Raw Materials," *Foreign Affairs*, vol. 4 (July 1926), pp. 585–600.

Von Hagen, Victor Wolfgang. *South America Called Them: Explorations of the Great Naturalists*. (New York: Knopf, 1945).

Wagley, Charles. *Amazon Town*, rev. ed. (New York: Knopf, 1964).

———. ed. *Man in the Amazon* (Gainesville, FL: University Presses of Florida, 1974).

Walker, Jay, American Consul to Brazil. Letter to the U.S. Department of State, Aug. 5, 1941, Para, Brazil. "Ford Rubber Plantations in Northern Brazil." Warren Dean collection, New York Botanical Gardens archives.

Wallace, Alfred Russel. *A Narrative of Travels on the Amazon and Rio Negro, with an Account of the Native Tribes, and Observations on the Climate, Geology, and Natural History of the Amazon Valley* (New York: Greenwood Press, 1969, originally published 1895).

Wallace, Benjamin Bruce, and Lynn Ramsay Edminster. *International Control of Raw Materials* (New York: Garland, 1983).

Wallace, Robert. "The Para Rubber Tree," *Times* (London), March 2, 1911, p. 22.

Weaver, Blanche Henry Clark. "Confederate Emigration to Brazil," *Journal of Southern History*, vol. 27, no. 1 (Feb. 1961), pp. 33–53.

———. "Confederate Immigrants and Evangelical Churches in Brazil," *Journal of Southern History*, vol. 18, no. 4 (Nov. 1952), pp. 446–468.

Webster, C. C., and E. C. Paardekooper. "The Botany of the Rubber Tree," in *Rubber*, C. C. Webster and W. J. Baulkwill, eds. (Essex, UK: Longman, Scientific & Technical, 1989), p. 57ff.

Weinstein, Barbara. *The Amazon Rubber Boom, 1850–1920* (Stanford, CA: Stanford University Press, 1983).

———. "The Persistence of Precapitalist Relations of Production in a Tropical Export Economy: The Amazon Rubber Trade, 1850–1920," in Michael Hanagan and Charles Stephenson, *Proletarians and Protest: The Roots of Class Formation in an Industrializing World* (Westport, CT: Greenwood Press, 1986).

Werner, Louis. "The Mad Mary, All Aboard to Nowhere," *Américas*, vol. 42, no. 4 (1990), pp. 6–17.

"Westminster Abbey Fund, Further Donations," *Times* (London), July 5, 1920, p. 9.

Whaley, W. Gordon. "Rubber, Heritage of the American Tropics," *Scientific Monthly*, vol. 62, no. 1 (Jan. 1946), pp. 21–31.

Wickham, Henry Alexander. "The Introduction and Cultivation of the Hevea in India," *India-Rubber and Gutta-Percha Trades Journal*, vol. 23 (Jan. 20, 1902), pp. 81–82.

———. *On the Plantation, Cultivation and Curing of Pará Indian Rubber (Hevea brasiliensis) with an Account of Its Introduction from the West to the Eastern Tropics* (London: K. Paul, Trench, Trübner, 1908).

———. *Rough Notes of a Journey Through the Wilderness from Trinidad to Para, Brazil, by Way of the Great Cataracts of the Orinoco, Atabagao, and Rio Negro* (London: W. H. J. Carter, 1872).

Wickham, Violet. "Lady Wickham's Diary," photocopy and typescript. Wolverhampton Archives and Local Studies, Wolverhampton, England. In the records of the Goodyear Tyre and Rubber Company (Great Britain), ref. DB-20/G/6.

Williams, Donovan. "Clements Robert Markham and the Geographical Department of the India Office, 1867–1877," *Geographical Journal*, vol. 134 (Sept. 1968), pp. 343–352.

———. Clements Robert Markham and the Introduction of the Cinchona Tree into British India, 1861," *Geographical Journal*, vol. 128 (Dec. 1962), pp. 431–442.

Wilson, Colin. *The History of Murder* (New York: Carroll & Graf, 1990).

Wilson, W. Arthur. "Malaya—Mostly Gay: All About Rubber: A Guide for Griffins," *British Malaya*, February 1928, pp. 261–265.

Winchester, Simon. *Krakatoa: The Day the World Exploded, August 27, 1883* (London: Viking, 2003).

Wolf, Howard, and Ralph Wolf. *Rubber: A Story of Glory and Greed* (New York: Covici, Friede, 1936).

Woodroffe, Joseph F. *The Upper Reaches of the Amazon* (London: Methuen, 1914).

Woshner, Mike. *India-Rubber and Gutta-Percha in the Civil War Era: An Illustrated History of Rubber and Pre-Plastic Antiques and Militaria* (Alexandria, VA: O'Donnell Publications, 1999).

"Wrecked Emigrants," *New York Times*, March 28, 1867, p. 2.

Wright, Herbert. *Hevea Brasiliensis, or Para Rubber: Its Botany, Cultivation, Chemistry and Diseases* (London: Maclaren, 1912).

———. *Rubber Cultivation in the British Empire: A Lecture Delivered Before the Society of Arts* (London: Maclaren, 1907).

Wright, Ronald. *A Short History of Progress* (New York: Carroll & Graf, 2004).

Wycherley, P. R. "Introduction of Hevea to the Orient," *Planter*, March 1968, pp. 127–137.

Young, H. E. "South American Leaf Blight of Rubber," Report to the Rubber Research Institute of Ceylon, January 4, 1954. Warren Dean collection, New York Botanical Gardens archives.

Yungjohann, John C. *White Gold: the Diary of a Rubber Cutter in the Amazon, 1906–1916* (Oracle, AZ: Synergetic Press, 1989).

Zahl, Paul A. "Seeking the Truth About the Feared Piranha," *National Geographic*, vol. 138, no. 5 (Nov. 1970), pp. 715–732.

Zilles, John A. "Development of Disease Resistant High Yielding Clones in Brazil Through Work of Ford and Goodyear Plantations," Apr. 25, 1983. Warren Dean collection, New York Botanical Gardens archives.

Ziman, John. *The Force of Knowledge: The Scientific Dimension of Society* (Cambridge: Cambridge University Press, 1976).

II. Archival Sourcebooks

National Archives, Kew. "Articles of Association, Tapajos Para Rubber Forests Ltd.," BT 31/8165/59032.

National Archives, Kew. Colonial Office: British Honduras, Register of Correspondence, 1883–1888, CO 348/10. Despatch No. 5: "Mission of Mr. J. B. Peck of New York to British Honduras," Sir Roger Goldsworthy, Jan. 17, 1888.

————. Despatch No. 11, Enclosure 2: "A Statement by the Santa Cruz Indians," Jan. 8, 1888.

National Archives, Kew. Colonial Office and Predecessors: British Honduras, Original Correspondence 1744–1951. CO 123/200. "Mr. H. A. Wickham's Temash Concession, (Pleadings in court case)," 1892.

————. CO 123/204. "Damage Caused by Gale," 1893.

————. CO 123/281 "State of Wickham's Case," 1893.

National Archives, Kew. *Honduras Gazette*, 1887, 1888, 1889, 1893. CO 127/6, 7, 8.

————. *Honduras Gazette*, May 7, 1887, p. 78.

————. *Honduras Gazette*, Dec. 17, 1887.

————. *Honduras Gazette*, May 12, 1888, p. 81.

————. *Honduras Gazette*, Dec. 15, 1888, p. 215.

————. *Honduras Gazette*, May 4, 1889, p. 75.

————. *Honduras Gazette*, May 6, 1893, p. 154. Notice of the forfeit of Wickham's lease of 2,500 acres along the Temash River.

National Archives, Kew. "Map Showing Portion of the Republic of Yucatan Shewing the Occupation of the Chan Santa Cruz Indians." FO 925/1288.

Royal Botanic Gardens–Kew. *J. D. Hooker Correspondence*, vol. 21. file folder 120: Wickham to Hooker, Aug. 10, 1906, London.

Royal Botanic Gardens–Kew. *Miscellaneous Reports: Brazil: Jequié Maniçoba and General, 1879–1913:* file folder 3: Memos in the Inwards and Outwards Book, 1873, regarding the purchase of hevea seeds from James Collins in 1873.

————. file folder 4: D. Prain to Ivor Etherington, editor of "The Tropical Agriculturist," re: Collins's seeds, May 17, 1907.

————. file folder 30–32: "The Rubber Growers' Association Third Annual Report of the Council: Transactions of the year, June 1910–June 1911."

————. file folder 388–389: Wickham to D. Prain, director of Kew, on the cultivation of the Piquia tree, Aug. 11, 1906.

————. file folder 389: Wickham to D. Prain, notes on location of "the forest of the Piquia," Aug. 21, 1906.

————. file folder 390: D. Prain to Wickham, notes on location of the Piquia nut, Aug. 14, 1906.

————. file folder 391: Reply by the British vice-consul in Manaus to earlier letter by the Director of Kew, Sept. 18, 1906. Note that the Piquía trees in Manaus are different than those in Para and Maranhão, so he passes Kew's request to the Consul at Pará.

————. file folder 392: Wickham to Director of Kew, Feb. 7, 1907. Types of Piquiá tree and estimation of economic importance to the British Empire.

————. file folder 393: Letter from Wickham to Director of Kew, March 14, 1907. Request to send Piquiá seeds to A. W. Copeland, director of the Malacca Rubber Plantations, and a description of the fruit of the Piquiá.

————. file folder 398: Letter from Wickham to Director of Kew, June 8, 1911. Urges introduction of Piquiá tree to plantations in Malaya.

Royal Botanic Gardens–Kew. *Miscellaneous Reports: Congo Region Miscellaneous, 1883–1928:* file folder 25: from "Congo Rubber," *The India-Rubber and Gutta-Percha Journal, and Electrical Trades Journal,* Sept. 8, 1893, p. 46.

Royal Botanic Gardens–Kew. *Miscellaneous Reports: India Office: Caoutchouc I.* file folder 1: "Introduction of Hevea brasilensis into India," a memo calculating costs of the Wickham transfer.

———. file folder 2: Letter, May 7, 1873, from India Office to Director of Kew, J. D. Hooker, inquiring into possibility of sending hevea plants to India after raising them from seeds at Kew.

———. file folder 4: Reply, Hooker to India Office, May 15, 1873. Reply to May 7th letter.

———. file folder 5: Letter from Clements Markham to "my dear Hooker," June 2, 1873, regarding the purchase by James Collins of 2,000 hevea seeds from a "Mr. Ferris" of Brazil.

———. file folder 9: Markham to "my dear Hooker," Sept. 23, 1873, regarding Wickham's proposal to start a nursery in Santarém for rubber trees.

———. file folder 10: Letter from Henry Wickham in Santarém to Hooker, Nov. 8, 1873, regarding the collection and shipment of rubber seeds.

———. file folder 12: Markham to Hooker, Oct. 1874, regarding Lord Salisbury's paltriness concerning costs of obtaining and transporting rubber seeds.

———. file folder 13: Markham to Hooker, Dec. 4, 1874.

———. file folder 14: Letter dated Oct. 15, 1874, Wickham at Piquiá-tuba, near Santarém, to Hooker, regarding transport of seeds.

———. file folder 15: Letter Wickham at Piquiá-tuba, Santarém, to Hooker, April 18, 1875, regarding the fact that it is too late in the season to collect seeds.

———. file folder 17: Wickham at Piquiá-tuba, to Hooker, Jan. 29, 1876. "I am just about to start for the 'ciringa' district in order to get you as large a supply of the fresh India Rubber seeds as possible."

———. file folder 18: Markham to Hooker, April 1, 1876. Robert Cross is sent to the Amazon to collect hevea seeds.

———. file folder 19: Wickham to Hooker, March 6, 1876 , informing Hooker that he is "now collecting Indian Rubber seeds."

———. file folder 20: Unsigned note, dated July 7, 1876: "70,000 seeds of Hevea brasiliensis were received from Mr. H. A. Wickham on June 14th."

———. file folder 21: "Expenses to send 2,000 seeds from Kew to Ceylon."

———. file folder 43: *Evening Herald,* Aug. 20, 1876, describing growth of seeds at Kew.

———. file folder 59: Letter from Markham to "My dear Mr. Dyer," Sept. 18, 1876, complaining about treatment of seeds after being shipped to the East.

———. file folder 78–93: Robert Cross, "Report on the Investigation and Collecting of Plants and Seeds of the India-rubber Trees of Para and Ceara and Basalm of Copaiba," (March 29, 1877).

———. file folder 131: Letter from Wickham to Sir Thistleton-Dyer at Kew, Sept. 4, 1901.

Royal Botanic Gardens–Kew. *Miscellaneous Reports 5: Madras-Chinchona, 1860–97.* file
 folder 131 & 132: "Letter from Robert Cross to Clements Markham, July 21, 1882."
Royal Commonwealth Society Archives, University of Cambridge Library. Cuttings from
 The Field. GBR/0115/RCMS 322.
———. "Life on a Malayan Rubber Plantation." RCMS 322/11: Malaya.
———. "Sporting and Other Notes of the Amazon." RCMS 322/1: 1888–1894.
———. "Travel and Colonisation," May 13, 1905, p. 10. RCMS 322/3: 1905–1907.
Royal Commonwealth Society Archives, University of Cambridge Library. Cuttings of
 the Queensland Sugar Industry. GBR/0115/RCMS 294.
———. "Britons in the South Seas," *European Mail,* Sept. 24, 1880.
———. "The Esperanze Tragedy in the South Seas," *European Mail,* Sept. 24, 1880.
———. "Idiosyncracies," *MacKay Mercury,* Dec. 1878.
———. "Kanaka Labor," *Queenslander,* Nov. 16, 1879.
———. "Kidnapping in the South Sea Islands," *Sydney Mail,* Apr. 23, 1881.
———. "The Labour Question," *Queenslander,* May 14, 1881.
———. "Letters to the Editor," *Queenslander,* Aug. 30, 1870.
———. "Queensland Sugar Industry," *MacKay Mercury,* Sept. 25, 1878.
Royal Commonwealth Society Archives, University of Cambridge Library. *A Document of
 the Mosquito Nation:* document signed Feb. 19, 1840, between Robert Charles Frederic,
 King of the Mosquito Nation, and Great Britain Aboard HMS *Honduras,* with notes.
 GBR/0115/RCMS 240/27. Introduction by S. L. Canger.
Royal Commonwealth Society Archives, University of Cambridge Library. News cut-
 tings on Colonial Relations with England, March 11, 1869–July 23, 1870.
 GBR/0115/RCMS 18.
———. "Problems in the Colonies," *Evening Standard,* Aug. 31, 1869.
Royal Commonwealth Society Archives, University of Cambridge Library. *Sir Henry
 Wickham, sketch for chart of Conflict Group.* GBR/0115/RCMS 278/58.

III. *World Wide Web*

1871 Census of London, in www.Ancestry.com.uk, citing the National Archives, Kew,
 RG10/133, ed. 2, folio 30.
Adams, Patricia. "Probe International: Odious Debts," Part 1, Chapter 2, www
 .probeinternational.org/probeint/OdiousDebts/OdiousDebts/chapter2.html.
Annual Register for 1928, http://historyonline.chadwyck.co.uk.
"Antipodes," *Wikipedia,* http://en.wikipedia.org.
"Atrocities in the Congo: The Casement Report, 1903," http://web.jjay.cuny.edu/~jobrien.
"Canaries, Singing, and Talking Birds," *Illustrated London News,* Feb. 15, 1868, www
 .londonancestor.com.
"Caste War of the Yucatán," www.answers.com.
"Chan Santa Cruz," www.absoluteastronomy.com.
"Cockneys," *Wikipedia,* http://en.wikipedia.org.

"Coconuts and Copra," www.msstarship.com/sciencenew.

"The Conflict Islands," www.conflictislands.com.

Conrad, Joseph. "The Planter of Malta," www.readbookonline.net.

"The Cornish in Latin America," www.projects.ex.ac.uk/cornishlatin/anewworldorder
.htm.

Cross, William. "Robert McKenzie Cross: Botanical Explorer, Kew Gardens," http://
scottishdisasters.tripod.com/robertmckenziecrossbotanicalexplorerkewgardens/.

Cruikshank, Dan. "The Wages of Sin," in BBC Online-History, www.bbc.co.uk/history/
programmes/zone/georgiansex3.html.

"Danger of Street Crossings in London," *Illustrated London News*, Jan. 3, 1857. www
.londonancestor.com.

Dempsey, Mary A. "Fordlandia," *Michigan History Magazine*, Jan. 24, 2006, www
.michiganhistorymagazine.com/extra/fordlandia/fordlandia.html.

Dornan, Jennifer L. "Document Based Account of the Caste War," www.bol.ucla.edu/
~jdornan/castewar.html.

"Finchley Road and Haverstock Hill," www.gardenvisit.com/travel/london/
finchleyroadhaverstockhill.html.

Gaston, Robert. "Edgar B. Davis and the Discovery of the Luling Oilfield," www4
.drillinginfo.com.

"Genuki: Hampstead History," http://homepages.gold.ac.uk/genuki/MDX/Hampstead/
HampsteadHistory.html.

"Genuki: Hampstead History, Description and History from 1868 *Gazetteer*," http://
homepages.gold.ac.uk/genuki.

"Genuki: Middlesex, Hampstead," http://homepages.gold.ac.uk/genuki/MDX/
Hampstead/index.html.

"Genuki: St. Marylebone History," http://homepages.gold.ac.uk/genuki/MDX/
StMarylebone/StMaryleboneHistory.html.

Grieve, Mrs. M. "Violet, Sweet," in Botanical.com, A Modern Herbal, www.botanical
.com.

"Hampstead: Chalcots/British History Online," www.british-history.ac.uk.

"Hampstead—MDX ENG," http://privatewww.essex.ac.uk/~alan/family/G-Hampstead
.html.

"Hampstead: Social and Cultural Activities/British History Online," www.british-history
.ac.uk.

"Handbook of Texas Online: Edgar Byrum Davis," www.tsha.utexas.edu/handbook.

"Handbook of Texas Online: Luling Oilfield," www.tsha.utexas.edu/handbook.

"Historic Folk Saints," http://upea.utb.edu/elnino/researcharticles/historicfolksainthood
.html.

Iacocca, Lee. "Henry Ford," in *Time 100: Henry Ford*, www.time.com/time/time100/
builder/profile/ford.html.

"Jos. D. Hooker: Hooker's Biography: 4. A Botanical Career." www.jdhooker.org.uk.

Kipling, Rudyard. "The Explorer" (1898). Every one of Kipling's poems can be found,
alphabetically arranged, at www.poetryloverspage.com/poets/kipling.

Kitchel, Jeanine. "Tales from the Yucatan," www.planeta.com/ecotravel/mexico/yucatan/tales.

"Living Rainforest: Cancer Cured by the Rosy Periwinkle," www.livingrainforest.org/about/economic/rosyperiwinkle.

Lovejoy, Paul E. "Gustavas Vassa, alias Olaudah Equiano, on the Mosquito Shore: Plantation Overseer *cum* Abolitionist," www.lsc.edu/collections/economicHistory/seminars/Lovejoy.pdf.

"MacGregor, Sir William," www.electricscotland.com.

Marx, Karl. "Marx and Engels in Manchester, 2 December, 1856," www.marxists.org/archive/marx/works/1856/letters/56_12_02.htm.

"The Milliner," www.colonialwilliamsburg.com/almanack/life/trades/trademln.cfm.

"Money from God," *Time* (Sept. 2, 1935), www.TIME.com.

"Moravian Civic and Community Values," http://moravians.org.

"Mosquitos, A Brief History,' www.4dw.net/royalark/Nicaragua/mosquito2.htm.

"My Name is Norval," Antiquarian Booksellers' Association of America, www.abaa.org.

"Northern Belize—The Caste War of the Yucatan and Northern Belize," www.northernbelize.com.

"Overview of Belizean History," www.ambergriscaye.com/fieldguide/history2.html.

"Patron Saints Index: Saint Ignatius of Loyola," www.catholic-forum.com/saints.

Reismiller, John. "Violet Traditions: Footnotes to the Violet," in *Chamomile Times and Herbal News*, www.chamomiletimes.com/articles/violettraditions.htm.

"Sackville Street: British History Online," www.british-history.ac.uk.

"Sanitation, Not Vaccination the True Protection Against Small-Pox," www.whale.to/vaccine/tcbbl.html.

"Sephardic Genealogy Resources; Indiana Jones Meets Tangier Moshe," www.orthohelp.com/geneal/amazon.htm.

Snow, John, M.D. "On the Origin of the Recent Outbreak of Cholera at West Ham," www.epi.msu.edu/johnsnow.

"Statistics of Wars, Oppression and Atrocities in the Nineteenth Century," http://users.erol.com/mwhite28/wars19c.htm.

"*Vibrio cholerae* and Asiatic Cholera," *Todar's Online Textbook of Bacteriology*, www.textbookofbacteriology.net/cholera.html.

Wojtczak, Helena. "Female Occupations C19th Victorian Social History—Women of Victorian Sussex," www.fashion-era.com/victorian_occupations_wojtczak.htm.

IV. Interviews

Alexander, Steven, owner of "Bosque Santa Lucia," present site of Piquiá-tuba, Oct. 9, 2005.

Campbell, Anthony (descendant of the Wickham family), via Internet, April 3, 2006.

Cohen, Elisio Eden, postmaster and historian of Boim, Oct. 21, 2005.

Cross, William, (descendant of Robert Cross), via Internet, April 2, 2006.

Pereira Branco, Doña Olinda, Fordlandia, Oct. 21, 2005.

Sena, Cristovao, regional rubber historian, Santarém, Pará, Brazil. Oct. 18, 2005.

Serique, Gil, guide, Santarém, Pará, Brazil. Dec. 15, 2005.

V. *Films*

Kautschuk, dir. Eduard von Borsody, with René Deltgen, Gustav Diessl, and Roman Bahn. Ufa Studios, Germany, 1938.

ACKNOWLEDGMENTS

The life of a writer would be impossible if not for the kindness of sympathetic strangers. For a project such as this, one can't help but be humbled by the patience and generosity of those fascinated by the tale of a complicated, globe-trotting dreamer (and his long-suffering wife), whose ambitions changed the world. But where does one start? Henry Wickham was a true nomad, setting off from London to Nicaragua, Venezuela, Brazil, Queensland, Belize, and Papua New Guinea in a tortured quest for acceptance and respectability. His wife, Violet, added Bermuda to the list; other branches of the family drifted to the Caribbean and Texas.

Yet the heart of Wickham's fame, or infamy, rests in the Amazon Valley of Brazil, where the muddy "Father of Waters" meets the deep blue Tapajós River at Santarém. So we begin our acknowledgments there. This book wouldn't be what it is without the help of Gil Serique, for my money the best guide on the Amazon: He anticipated my research, knew Wickham's story like a scholar, and ushered me around as if the saga were his own. Although he is primarily a nature guide, he switched gears to history and exploration as if he were born to the trade; as we ranged up the river in an open boat, our stops at villages and towns along the route gave me greater insight into the realities of life on the Tapajós and Amazon.

But there were many others. Gil's sister Maxie was the best shipboard cook anyone could hope for. Expatriate Steven Alexander, owner of the Bosque Santa Lucia nature preserve at Piquiá-tuba, where the Wickhams first settled, was invaluable for his vast knowledge of rain-forest flora and fauna. Regional historian Cristovão Sena revealed the hidden history of Fordlandia. Eric Jennings, descendant of *confederado* Elizabeth Vaughan, was a font of knowledge concerning his wayward American ancestors. Fellow explorers are always appreciated when setting off into the unknown. In this case, Alyhana Hamad, Deyna Cavacánta, and José Eduardo Siqueiria (better known as Ze) braved the logging trails of the forest to find the hidden route to Taperinha. And I would be remiss if I left out hotelier James Murray, owner of the Amazon Shamrock Inn, who connected me with Gil Serique and Steven Alexander in the early stages of my research.

In Fordlandia, I'm grateful to Doña America Labita, Doña Olinda Pereira Branco, and Biamor Adolfo de Sousa Passoa for opening their homes and explaining the realities of life as Henry Ford tried to tame the rain forest. In Belterra, Divaldo Alves Marques ushered us through the twilight of Ford's ambition, while rubber tapper Raimondo Mirando Lopez, eighty-three years old, graciously provided a crash course in the art of the *seringueiro*. In Boim, regional historian and author Elisio Eden Cohen unveiled the hidden history of Wickham's seed theft, probably doing more to clear up this first modern act of biopiracy than the legions of other commentators I'd read. And Cohen's daughter, fourteen-year-old Herica Maria, was kind enough to show a nosy stranger the basket she'd woven—a basket whose design had been passed down through the generations and was the same as those used to carry seventy thousand stolen rubber seeds from the heart of the Amazon to the greenhouses of Kew.

To comprehend such travels, one needs a framework of theory and history. At times like this, archivists and librarians are a writer's best friend. Christopher Laursen, Science and Technology Librarian at the University of Akron, started the ball rolling for me with the depth of his knowledge amidst what is probably the best collection of rubber-related material in the United States. Elaine Donnelly, a fellow traveler from my teenage years and now an archivist at the National Geographic Society in Washington, D.C., provided sources on the far-flung lands of Wickham's exile. David Steere, senior reference librarian at the Smithsonian National Museum of Natural History, opened the door to a wealth of research. Steve Sinon, head archivist at the New York Botanical Garden's LuEsther T. Mertz Library, provided access to the Warren Dean papers and to the botanical lore of Amazon rubber. In England, Michelle Losse at the archives of the Royal Botanical Gardens–Kew and Rachel Rowe, Smuts Librarian in South Asian and Commonwealth Studies at the University of Cambridge, were gracious and patient guides through the vast botanical and colonial collections of both libraries. Especially helpful was David Clover at the Institute of Commonwealth Studies at the University of London: Thanks to him, I was able to relocate Violet Wickham's memoir of life with her difficult husband, an unpublished manuscript that had been buried in the files of the B.F. Goodrich Company for half a century.

Over the years, I've been fortunate to develop a kind of personal brain trust of friends and professors at Virginia Wesleyan College in Virginia Beach, Virginia, close to where I live. They've helped me in past projects, and this time was no different. Paul Resslar, Batten Professor of Biology and world traveler, actually handed me the subject of this volume on a silver platter when we were talking over dinner about biopiracy and the Amazon. "Why don't you write about Henry Wickham, who smuggled seventy thousand rubber seeds out of the Amazon and killed their economy?" he asked as I passed the shrimp risotto. Friends should be careful with their casual remarks: This one sent me into the heat and mad hornets of the Amazon Valley, the subway bombings in London, and the automated Metro nightmare of Washington, D.C. Susan Wansink, Professor of German and French, guided me through the B-movie translation of *Kautschuk,* a forgotten German potboiler of the 1930s. This was matinee fare, like Jungle Jim or Sheena, Queen of the Jungle, but what it lacked in art was made up by transparent national aspiration, an important point I wouldn't have seen without Susan's help.

John Loadman deserves special mention of his own. A former analytical chemist with the Natural Rubber Producers' Research Association, the research arm of the Malaysian Rubber Board, he turned himself into a world-class expert on the history of natural rubber. Today he maintains the Web site www.bouncingballs.com. a starting point for everything connected to natural and synthetic rubber, and in 2005 Oxford University Press published his *Tears of the Tree: The Story of Rubber—A Modern Marvel*. The man is an encyclopedia of rubber-related information, and he helped me several times as I wrote this book. I'm indebted to him.

I'm also indebted to Anthony Campbell, his mother Sallie Campbell, and their relatives Hubert Mitchell and Peter Lendrum. All are related to Wickham through a separate branch of the family tree. I'd been unable to find a photo of Henry and Violet Wickham while they were young, and without this I was having a hard time envisioning my main characters. Anthony and his mother were gracious enough to lend me their collection for this book. They and the others had stumbled upon the Santarém group photo and Wickham's sketch of five graves, among other treasures, while rummaging through some old family files. In addition, Anthony Campbell traced the lineage of his illustrious ancestor and was nice enough to provide it, too. Jenepher Allen and David Allen Harris appeared late in the game with the unpublished autobiography of great uncle Arthur Watts Allen, a distant relative of Wickham's. Allen's "The Occupational Adventures of an Observant Nomad" helped piece together a few mysteries of Wickham's years on the Conflict Islands. All helped me understand my subject immeasurably better.

Finally, I must thank those who keep me going and deal with me day by day. Wendy Wolf and Ellen Garrison, my editors at Viking Penguin, believed in and nurtured this little globe-trotting project, even as its scope expanded. Noah Lukeman, my literary agent, helps keep the bill collector at bay and has always been a valued sounding board and friend. And, as always, my love and gratitude go out to my wife Kathy and son Nick, who endure my transitory moods and do the happy dance with me whenever the latest project is safely put to bed.

PHOTO CREDITS

INDEX

A World on Fire
A Heretic, an Aristocrat, and the Race to Discover Oxygen

The discovery of oxygen began with nothing more auspicious than the survival of a mouse in a bell jar, yet it changed human thought and history as radically as the work of Newton or Darwin. This enthralling book tells the story of that breakthrough and the two men who brought it about in the late 1700s. In re-creating the lives of the English freethinker Joseph Priestley and the French aristocrat Antoine Lavoisier—the former exiled, the latter executed on the guillotine—*A World on Fire* gives us a riveting object lesson in the perilous place of science in an age of unreason.

ISBN 978-0-14-303883-2